WARSHIP 1993

WARSHIP 1993

Edited by Robert Gardiner

Frontispiece caption:
The Japanese Akizuki *class destroyer* Yoizuki *in Kure harbour, 16 October 1945. Some light anti-aircraft guns have been removed, but the ship still has her full radar fit: type 13 on main and fore topmast, type 22 (with its horn antennas) above the main director. The design of these impressive ships is described in detail in this issue.* (US National Archives)

© Conway Maritime Press Ltd 1993

First published in Great Britain by
Conway Maritime Press Ltd
101 Fleet Street
London EC4Y 1DE

All rights reserved. No part of this publication may be reproduced, stored in a retrieval system, or transmitted in any form or by any means, electronic, mechanical, photocopying, recording or otherwise, without the prior permission of the publisher.

British Library Cataloguing in Publication Data
Warship . .
  1993 –
  1. Warships
  623.825

  ISBN 0–85177–624–8

Design, typesetting and page make-up by
The Word Shop, Bury, Lancashire
Printed and bound in Great Britain by
Butler & Tanner Ltd, Frome

CONTENTS

Editorial	7
FEATURE ARTICLES	
Dalhousie: The Last Flagship of the Indian Navy and the Question of Imperial Defence *by Andrew Lambert*	9
USS New Ironsides: America's First Broadside Ironclad *by William C Emerson*	19
Flights of Fancy: Unusual Shipboard Aircraft Launch and Recovery Systems *by R D Layman and Stephen McLaughlin*	33
Super-Dreadnoughts: the Orion Battleship Family *by Keith McBride*	46
The First Hunter-Killers: British 'R' Class Submarines of 1917 *by David Miller*	65
Revolution Manqué: Technical Change in the Royal Navy at the end of the First World War *by D K Brown*	77
The French Flotilla Programme of 1922 *by John Jordan*	89
Seetakt *by P F Wright*	105
Japanese Midget Submarines: Kōhyōteki Types A to C *by Jiro Itani, Hans Lengerer and Tomoko Rehm-Takahara*	113
The Admiral Scheer at War *by Pierre Hervieux*	130
The Akizuki Class *by Jiro Itani, Hans Lengerer and Tomoko Rehm-Takahara*	143
USS Triton: The Ultimate Submersible *by Robert P Largess and Harvey S Horwitz*	167
REVIEW SECTION	
Warship Notes	188
Naval Books of the Year	206
The Naval Year in Review *by Ian Sturton*	219
Index	250

EDITORIAL

The editorial in *Warship 1992* aired the notion that the annual would be improved by being constructed around a single theme, although it also set out a number of possible drawbacks to such an approach. The one element missing from the discussion was the view of the readership, but in the intervening year this has become a lot clearer. We received barely a single letter in support of such a change, and indeed most expressed a positive preference for a mixed bag of material, in terms of time spread, navies covered, and approach. Of course, it is by no means certain that the majority view is represented by the small proportion who feel strongly enough to pick up a pen – or, more often than not, switch on the wordprocessor. Nevertheless, with virtually no evidence of a desire for change, we felt it wise to continue a 'steady as she goes' policy, at least for the time being.

As a result readers will find the current volume a familiar variety of material, covering technical aspects of warship history from the middle of the last century to almost the present day. Because of its origin as a journal, *Warship* has a circle of regular contributors whose particular hobbyhorses will be instantly recognisable, and in some cases articles follow on quite closely from work published in earlier issues. Keith McBride's studies of British dreadnoughts, for example, have moved forward a few years from the 12in ships covered in 1992 to the 13.5in battleships; Hans Lengerer's work on Japances special attack weapons now encompasses the first midget submarines, although he makes it clear that their origin was to be found in the Japanese concept of a decisive fleet engagement, rather than any *kamikaze* strategy. With twelve months separating one publication from the text, every article has to be inherently interesting, significant and above all self-sufficient, but if these conditions are met then the ongoing research provides a sense of continuity.

Other contributors' specialities produce novel variations on a theme. Dick Layman and Stephen McLauchlin are well known for their exploration of the less charted waters of naval aviation, and for this issue have discovered some wild and wonderful schemes for getting aircraft into the air from ships – and recovering them again. The significance of such ideas lies beyond the appeal of eccentricity, in that they help to place in context the relative merits or ingenuity of the devices which were adopted. So often, the technical historian presents the successful option as the only logical development, and navies and their administrators are castigated for not seeing what appears obvious with 20/20 hindsight. In the experimental stage of any technology, there are at least ten impractical ideas for any one good one, but unless all eleven are considered together it is impossible to make a judgement on the quality of the decision-making.

One might perceive a related theme in David Brown's outline of a technical revolution that might have been. At the end of the First World War the Royal Navy was working on a huge variety of novel ideas and technologies. When the money dried up, as it always does in peacetime, it was necessary to concentrate on essentials. Moreover, the riskier ventures seemed a poorer investment, and as a result some promising but, from the viewpoint of the time, far fetched schemes went to the wall.

Many of the pressures that shape technological development are not obvious in the documentary records from which so much history is written. Arguments and discussions may be framed in terms of logic and reason, but often decisions are as much influenced by a hidden agenda of ulterior motives, personal ambition or factional power struggles (the 'Not Invented Here' syndrome was a well understood prejudice during the Second World War). In these circumstances, the personal view of participants can give an entirely new slant, and Robert Largess and Harvey Horowitz have become adept practitioners of this approach. Having tackled the revolutionary *Albacore* in 1991, this year they are attempting to solve a number of mysteries surrounding the design background to the giant *Triton*; here again, the authors suggest, the answers may be obscured by personalities.

Besides the regular repertory company of contributors, we are always keen to introduce new names, and this year we have two in David Miller and Peter Wright. Although they have not been published in *Warship* before, neither is a novice in the field. David Miller has written a number of naval books, but makes a specialisation of submarines. For his study of the Great War 'R' class ASW boats, he went back to primary sources and discovered that many of the traditional views of their shortcomings were totally unfounded; furthermore, he reveals that their influence stretched much closer to the present than anyone had suspected.

Peter Wright usually writes about aviation, but was drawn to the story of the *Graf Spee* and her Seetakt radar by Bainbridge-Bell's epic air journey to South America to investigate it. The main outlines of the River Plate battle and its aftermath will be familiar to *Warship* readers, but it is well worth recording how the incident fitted into the overall history of radar. This was another event which cast a long shadow: the editor well remembers a conversation with M K Purvis shortly after the *Sheffield* was sunk in 1982, in which he expressed outrage at press suggestions that the Type 42s, for which he had been project leader, had aluminium superstructure, which would melt in a fire. He was the young Constructor sent to investigate the *Graf Spee*, and having seen the effects of fire after the scuttling, he vowed never to use aluminium for large structures in any warship he was responsible for. There are logical arguments for and against aluminium, but here again the apparent British preference for steel may have been settled by personal conviction rather than a purely rational policy.

Robert Gardiner

DALHOUSIE
The Last Flagship of the Indian Navy and the Question of Imperial Defence

Andrew Lambert considers the fate of the last major warship laid down for the Bombay Marine, the navy of the East India Company. The story of this ship also helps to explain the rationale for the abolition of that service.

Existing studies of the last days of the Indian Navy devote little space to the major issues at stake in the late 1850s. Commander Low dealt in great detail with the treatment accorded to the officers. He also followed up the careers of the surviving ships, down to the publication of his work in 1877. However, one unnamed 50-gun ship remained incomplete at the time the service was abolished, and Low provided no further comment on this vessel.[1] This was curious as she was by far the largest and most powerful unit ever ordered by the Bombay Marine. Her career illustrated many of the problems that faced the service in its last decade.

The history of the Bombay dockyard, where this vessel was laid down, is equally reticent about her, although in this case the silence is understandable, as this article will reveal. Instead there is a photograph of the ceremonial silver nail that would have been driven into the keel of the ship, had she been launched. This item, in perfect condition, had clearly never been used. It was marked 'Dalhousie 50-gun screw frigate', although neither caption nor text provide any further explanation.[2] This is the sum total of published information on this powerful vessel.

In the first half of the nineteenth century the Bombay Marine filled an important role in the defence of the East India Company, and by extension, British interests in the western Indian Ocean and the Persian Gulf. Essentially this was maritime police work, as there were no potentially hostile navies in the region, while effective command of the Indian Ocean theatre was guaranteed by the Royal Navy. Up to 1830 the Marine used small frigates, corvettes and other diminutive warships, many of which combined military with communication work. After 1830 the service rapidly turned over to steam, reflecting the increasing importance of communications from the Red Sea – the so-called 'overland' route to India via Egypt – and the difficulty of employing sailing ships in an area where the winds were adverse for much of the year. The lack of naval opposition allowed the Marine to dispense with heavily armed warships. During the following three decades the requirement for a specifically Indian force declined. Piracy, real or imagined, was largely suppressed by the Marine and the Royal Navy, while the increased presence of the Royal Navy in Indian and Chinese waters rendered the Marine an expensive and unnecessary item on the strained budget of the company. These trends were to reach a denouement in the years immediately after the Indian Mutiny (1857).

Naval Shipbuilding in India

The Bombay Dockyard had benefited from the pressures of the Revolutionary War (1794–1802), when the Admiralty forcibly curtailed the construction of large merchant ships for the East India Company in England. These vessels, termed 'East Indiamen', were the largest merchant ships of the day, competing for scarce timber resources with the Royal Navy. To keep up their fleet the Company enlarged their existing construction facility at Bombay Dockyard, which had two vital advantages – a large pool of semi-skilled labour and a supply of Malabar teak, soon recognised as the finest shipbuilding timber in the world. To recoup the outlay on new drydocks and infrastructure the Company offered to construct warships for the Navy, at a time when the Admiralty was at loggerheads with the private builders at home. Earl St Vincent, First Lord of the Admiralty, encouraged the idea

The ceremonial silver nail intended to be driven into the keel of Dalhousie.

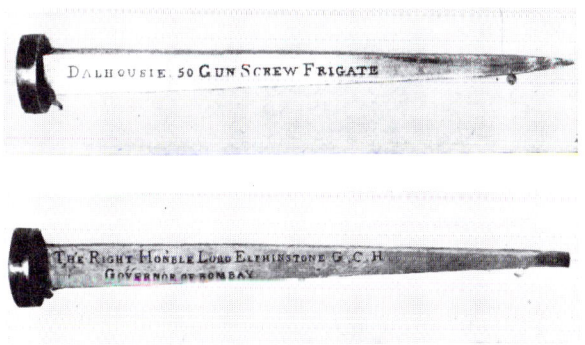

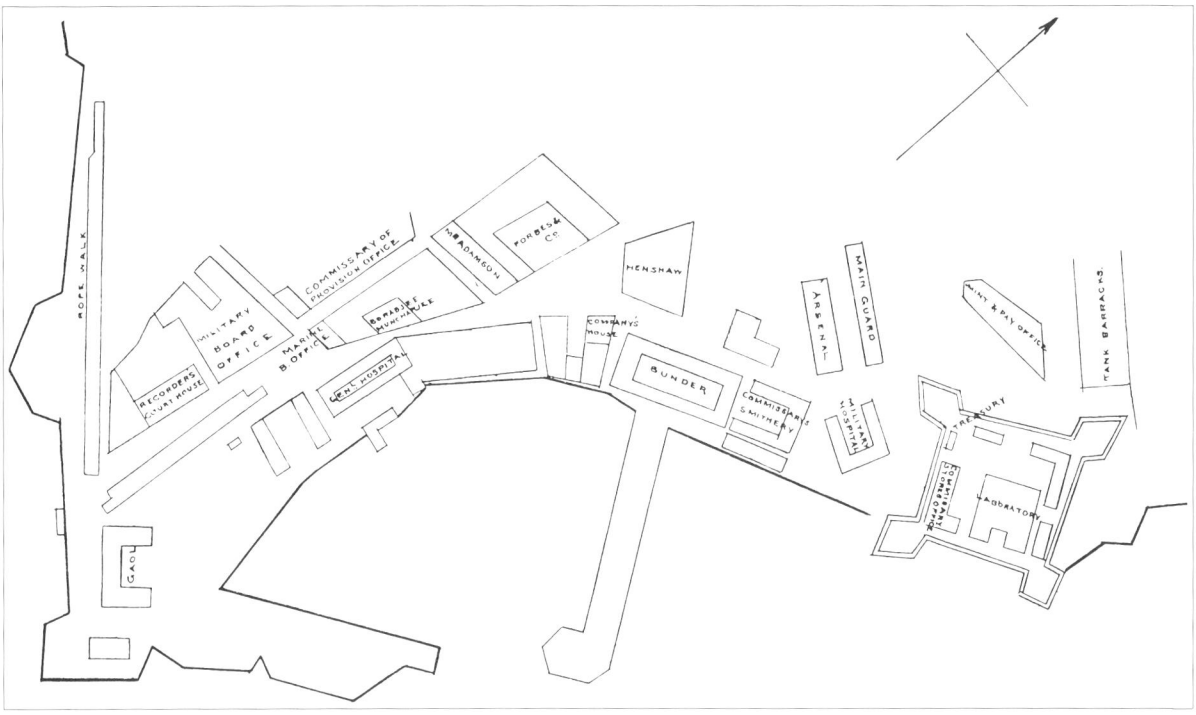

Bombay dockyard and related facilities in 1803.

The launch of the line of battleship Meanee *from the Bombay yard, from the* Illustrated London News *of 6 January 1849. The reputation of the yard was damaged by the discovery of an infestation of white ants in this ship.*

that one frigate and one battleship should be constructed annually at Bombay.[3] This policy continued until 1822, when problems in the British shipbuilding industry made it advisable to bring the Indian timber home to be worked into ships. The shortage of work in the merchant yards was threatening to throw large numbers of shipwrights out of employment, and possibly out of the trade. These men were not only a vital war resource, but were also highly politicised.[4]

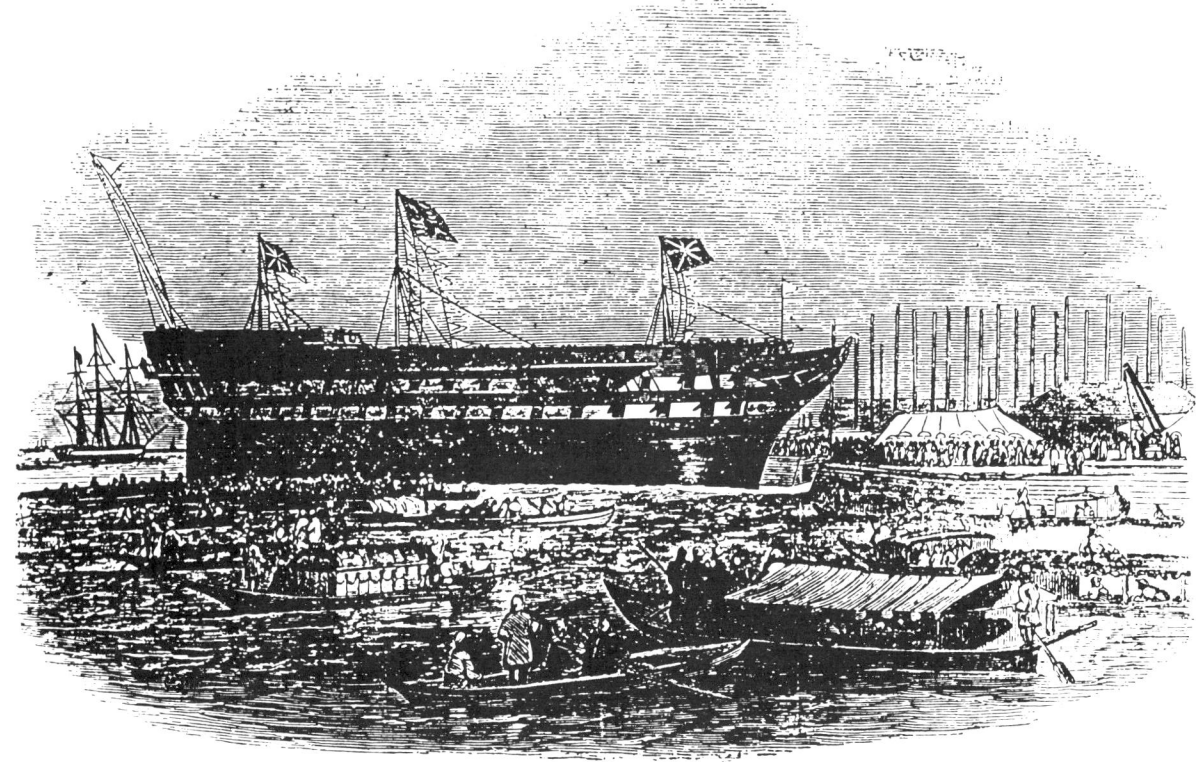

The Admiralty reopened construction in 1826, but stopped work again in 1829. However, the quality of teak justified the cost of transport. When the naval scare of the late 1830s encouraged a sudden increase in new construction Bombay was once again reopened, although this had more to do with teak than any intrinsic merit of building in India. However, the ten-year hiatus in orders had not been filled by other work, although ships were built for the Marine, the Imaum of Muscat and the East India Company. During the 1840s the Admiralty found costs running out of control and once more ended construction in 1847 after one battleship and two small frigates had been constructed. Their view of Bombay was not improved when the *Meeanee* became infested with white ants.[5] The problems of the yard were highlighted for the Company by the construction of the buoy vessel *Grappler* for Bengal in 1851. She cost 30 per cent more than had been estimated, because of what the Marine Board concluded were 'inexcusable oversights' by the chief builder amounting to, 'gross mismanagement'.[6] Against this background the order was placed for the largest wooden steam vessel yet begun in India.

Design Origins of the New Ship

These were complex, reflecting various elements of the Marine's role. In the era of smooth bore muzzle-loading cannon the defences of Bombay harbour were incomplete without a floating blockship. For many years that function, combined with service as a drill ship, was discharged by the veteran small frigate *Hastings*. By 1852 she was beyond economic repair, and became waterlogged after a gale in 1854. Sir Henry Leeke, the Commander of the Marine suggested replacing her with HMS *Hercules*, an old 74-gun battleship then doing duty carrying convicts to Australia. The Board of Control were not impressed. With common logic they argued that a ship built in 1815 would be a constant source of expense. In fact all oak-built ships suffered in the Indian climate, encouraging the Admiralty to employ Bombay ships in the Indian Ocean and Far East. To meet the need the Board called for an estimate to construct a new blockship.[7]

In this they were flying in the face of a logic at least as powerful as that they applied to turn down the *Hercules*. Blockships, only fitted for harbour service, were of very limited use given the dominance of the Indian Ocean by the Royal Navy. Building a new coast defence ship was anathema to the Royal Navy. Therefore, after a suitable period for reflection, the specification was altered to a screw propeller steam frigate of unspecified size, both to replace the *Hastings* and to be available for sea service, including the conveyance of troops.[8] Unlike the Marine's recent paddle frigates, a screw steamer had the great advantage of clear decks for troops and particularly horses. The attempt to move the 10th Hussars to Suez for service in the Crimea in mid 1854 had been delayed by the inability of the Indian paddle frigates to carry the regimental mounts.[9]

In addition, the Indian Navy was not the only service

A general impression of how Dalhousie *would have appeared can be gained from this view of the slightly larger 51-gun* Immortalité, *laid down in 1849 but not launched until 1859.* (CMP)

interested in building large screw frigates in India. In January 1855 the First Lord of the Admiralty, Sir James Graham, proposed to the President of the Board of Control, Sir Charles Wood, that *Imperieuse* class 50-gun steamers should be built at Bombay.[10] At the time the British dockyards were crowded with work by the Russian War, and Graham was deeply concerned by the challenge of Imperial France, despite the wartime alliance. He believed Britain could not afford to relax her efforts to outbuild the French in large screw warships, although the construction of ships of this type was unnecessary in the context of the Russian War.[11] Within six weeks Wood was proposing the new ship. After their experience with the *Meeanee* the Admiralty decided not to proceed, a wise decision in the event.

The design initially selected was that of the *Euryalus*, a curious choice as she had been converted from an incomplete 60-gun sailing ship. However the Surveyor of the Navy, Captain Sir Baldwin Walker provided all the necessary information to enable construction to commence, the drawings being despatched from London in August 1855, followed shortly after by drawings of the Miller & Ravenhill engined *Chesapeake*. The limited structural strength of wooden steam warships ensured that the arrangement of engine room beams was governed by the machinery layout.[12] Shortly after Wood had decided to order the ship, in late February, he replaced Graham at the Admiralty. This gave him an opportunity to obtain the assistance of Walker, the outstanding naval administrator of the era, and of the Surveyor's Department. This was a great benefit, for Walker had an unrivalled experience of the problems of creating a steam navy.

Dimensions of a Euryalus Class 50-gun Steam Frigate

Length, gundeck:	212ft
Breadth, extreme:	50ft
Depth in hold:	16ft 9in
Draught (fwd/aft):	20ft 7in/22ft 11in
Tonnage, old measurement:	2317
Displacement:	3356
Armament:	
Main deck:	30–8in 65cwt shell guns
Upper deck:	20–32pder 95cwt solid shot gun, or 1–10in 85cwt shell gun

Building the Ship

Bombay had some experience of building large wooden steamships. The 1400-ton paddle frigates *Punjaub* and *Assaye* had been completed in 1854. They were strong, well built ships, but their machinery proved to be a constant source of trouble. They were so unreliable that both were told off to be re-engined in 1860, although in the event both were sold to merchant owners, having been rejected by the Admiralty, and operated successfully as merchant sailing ships. This experience of steam machinery, British-built but installed at the Bombay steam factory, led the Governor of Bombay to demand better engines for the new ship.[13] There was also a feeling that the Bombay factory was not the best place to install large engines. Elphinstone relayed a suggestion that the ship be

Endymion, one of the last wooden screw frigates, laid down as a sailing ship in 1860 but converted to a steamer in 1866. (CMP)

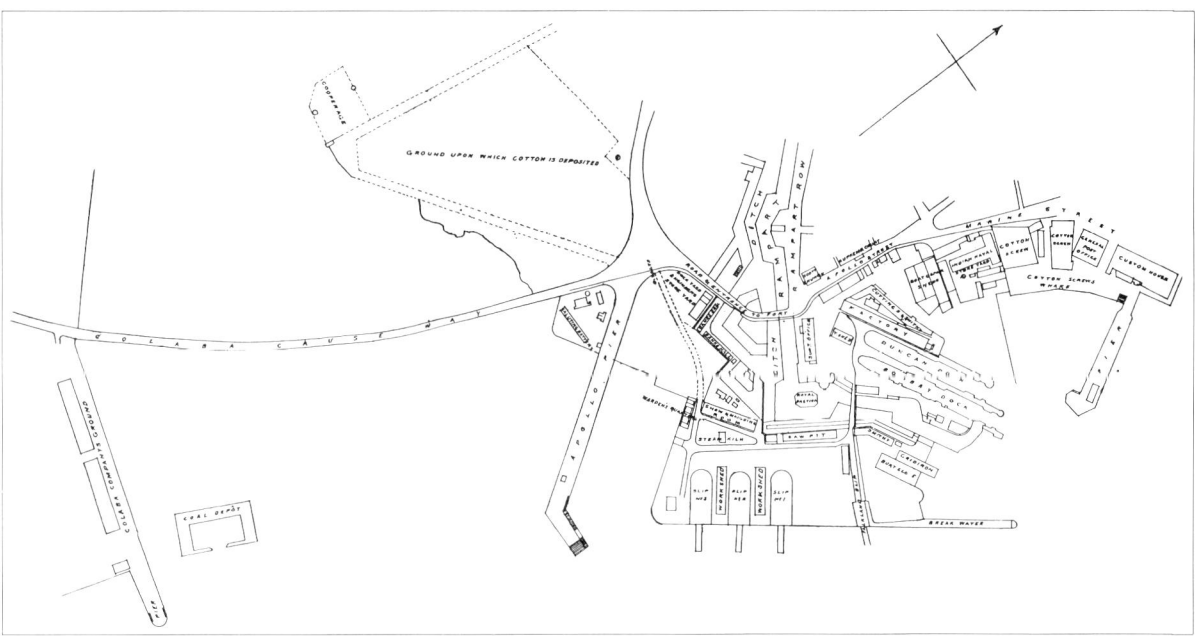

The Bombay dock and building yards in 1858.

sent to Britain to have her engines installed, and although the idea was rejected, it was only on the grounds of cost, time and risk. Little confidence was expressed in the factory.[14] At this stage the Board of Control had yet to open the contract to tender, although there was a good reason for hesitating. At the height of the Russian War, and the naval race with France, the engine builders favoured by the Admiralty, Maudslay, Sons & Field and John Penn & Sons, were unable to meet the demand at home, let alone take on a new customer. From 1850 the Admiralty had not allowed any other firm to tender for the machinery of any new ship of 50 guns or above, but by 1855 the pressure on the two Thames works forced a relaxation of this policy.[15]

The contract went to Robert Napier & Sons of Glasgow, among the best of the lesser firms, which had also provided machinery for some of the pioneer Indian steamships. The new engines were initially to be of 400 nominal horsepower, an old rating system based upon the cubic capacity of the single expansion twin-cylinder return connecting rod engine. By 1855 this was only one-third of the true horsepower. All large British wooden steam warships of the 1850s were only auxiliary steamships. The propeller was fitted with disconnecting gear, which allowed it to be detached from the drive shaft and hoisted up into the ship while proceeding under sail. This was considered vital for a service that had to cover great distances at a time when coal was scarce and engines inefficient. The engine drawings were vital to construction, as they determined the layout of the frames in and around the machinery spaces. Delays with the tender ensured these were only despatched in June 1857.[16] Napier's proceeded slowly with the engines; doubtless orders from the Admiralty and the French navy were executed with more haste. The engines and boilers were finally loaded aboard the merchant ship *Thomas Fielden* and departed Glasgow on 16 October 1857.[17]

Work in India proceeded even more slowly, giving rise to some concern. The 1856–57 expedition to Persia placed considerable demands on the Dockyard, converting ships to transport horses, building landing boats and maintaining the existing fleet ensured little work was done on the new frigate. The native Parsee Master Builder's estimate for the final cost of the hull and spars had been almost twelve lacs of rupees, but the cost of the timber alone was already threatening to run over this estimate. The estimate, the equivalent of £119,609 sterling, compared very unfavourably with the costs of *Imperieuse* and *Euryalus*, built at Deptford and Chatham respectively in 1852–53. They had cost £57,601 and £54,580 to bring to the same state. Part of the extra cost was traced to the timber, the British ships using oak and fir, while the *Dalhousie* was built entirely of Malabar teak. *Imperieuse* required 73,942 cubic feet of converted timber and *Euryalus* 68,828, whereas in India the estimate was for 198,177 cubic feet of unconverted timber. The East India Company's Surveyor did not consider the conversion would yield more than half that amount in timber ready for use. However, that did not explain the cost differential, despite 'the well known fact that building in Bombay with the best description of teak is much more expensive than building with oak and fir timber in England'. The Board of Control only agreed to continue work on the ship because so much timber had been provided. They demanded an explanation of the undue costs and agreed that the Dockyard required a thorough overhaul. Following the Governor's suggestion, an English civil officer would be appointed to run the yard, and a committee would report on the existing state of affairs.[18]

The ship was first attributed the name *Dalhousie*, in honour of the retiring Governor-General, in December 1857.[19] By that time the problems of the subcontinent had redirected the resources of the Dockyard to small steamers for river service, repairing troopships and gun carriages. No work was done on the *Dalhousie* during the period 1858 to 1861. Long after the Mutiny ended the backlog of work at Bombay precluded putting any men on the frigate. The Company that ordered her was about to be abolished, although it was allowed to run on for a few years. During this period of grace work on the new flagship of the Indian Marine was never effectively restarted. Furthermore Bombay's attempts to explain the extra costs were 'not by any means conclusive or satisfactory'. Finally London was having second thoughts about the ship. Her deep draught would, it was now believed, restrict her usefulness in Indian waters. Therefore serious consideration was given to selling her, or effecting an exchange with the Admiralty for smaller vessels. Despite these reservations she was to be completed.[20]

At Bombay Governor Elphinstone had a more direct appreciation of the problems. Reporting the ship was far from complete he enquired if the Admiralty would take over construction. Initially the idea was taken up, but the Board of Control quickly withdrew it after calculating the costs. The First Lord of Admiralty, the Duke of Somerset, was rather dismissive, although he did agree to buy the ship at a 'fair valuation'. He was aware that she had cost more than her true value to date.[21] Walker reported that it would be futile to attempt to complete the ship on the Admiralty's account. He returned to the idea of purchasing the complete ship.[22] To this end the Admiralty directed that a financial provision be made to purchase her in the estimates for 1861–62. Walker objected that the ship might not be ready in that year, and requested information on the exact state of the ship and the expected date of completion.[23]

The End of the Bombay Marine

The decision to sell the ship, in whatever condition, reflected the new realities of the Indian Administration. The Mutiny had added 70 per cent to the public debt, while the military charges were almost double the pre-1857 level of £12.5 million. This made a return to financial stability the first prerequisite for the continuation of British rule. As Secretary of State between 1859 and 1865 Wood was determined to prevent India becoming a financial dependency, taking full control of budgetary savings.[24] Before Wood came to office the Viceroy, Lord Canning, had suggested abolishing the Marine as an economy measure. With more local experience Elphinstone argued in favour of developing the service as a branch of the Royal Navy, under the control of the Indian Government. He pointed out that the Marine did not cost much, and that it would be more useful with small screw steamers in place of the large paddle frigates in service. His successor, Sir George Clerk, favoured reducing the Marine to the role of transport service. This would involve handing over slavery patrol duties on the East Coast of Africa and Arabia, which had cost little short of £1,000,000 in 1859, to the Royal Navy.[25] The conjunction of the Viceroy's idea and the Secretary of State's determination to reduce expenditure made the end of the Marine inevitable, although it was not completed until 1863.

The events of 1857–58 emphasised that the greatest threat to British rule in India was internal. The only other danger, however visionary, came from Russia and to meet this it was considered necessary to have some defence for Bombay harbour. During the Russian War reports of a Russian squadron at Singapore led Elphinstone to reinforce the harbour defences. The French war scare of 1859 indicated that the need remained. In 1860 Wood initially sanctioned improved fixed batteries, but the short range of contemporary artillery made it essential that they be placed out in deep water. This made the cost higher than had been anticipated, leading him to consult the Admiralty on the cost of more than one floating battery, possibly armour plated, steam powered and carrying at least sixteen heavy guns. The Admiralty was then considering such vessels for home defence and kept the Indian request on file for some time. However, coast defence remained a

Indian naval defence was eventually entrusted to the coast defence turret ships Abyssinia *and* Magdala. *This is* Magdala's *sister* Cerberus, *which served even further afield – in Australia.* (CMP)

defective and wasteful strategy for Britain, both at home and in India. Walker sent estimates based on the floating batteries built for the Russian War, but the Admiralty did not approve any coast defence designs in the early 1860s. The file was finally closed in June 1862.[26] In view of the economic pressures on the Indian administration in this period it is hardly surprising nothing was done. Two small coast defence turret ships, based on Admiralty designs, were finally provided in 1867, the British-built *Magdala* and *Abyssinia*.

Construction Transferred to Britain

The Surveyor's request for additional information on the state of the *Dalhousie* was answered in 1861. The Master Builder, Jehangir Nowrojee, reported that the frame, except the stern frames affected by the double sternpost, had been set up, the whole chain bolted up to the orlop deck, part of the lower planking was in place and five diagonal iron riders had been fitted. These last were vital features of the wooden steamship, allowing the construction of vesels over 200ft long. In addition almost seventy beams had been converted for the orlop, gun and upper decks. Commodore Wellesley, Commander-in-Chief of the Marine, revealed that he had not had a man to spare for work on the ship since January 1858, and did not anticipate having any in the foreseeable future. Therefore the ship would not be completed for many years, while the cost of completion had risen to an unacceptable degree. The old estimate of £120,000 for the hull and spars had been far too high, and since then the workforce had been given a 30 per cent pay rise. Wellesley estimated that another £40,000–50,000 would have to be added to the original figure, along with interest charges during the period required to complete the ship. This would expose the Indian Government to unacceptable losses, as the Admiralty could not pay more for the *Dalhousie* than the cost of a comparable British-built ship, at most £80,000 when the final price of the Bombay product would probably reach £200,000. Wellesley proposed dismantling the ship, on which only £20,000 had been spent to date, principally on the timber, and sending it to be purchased at a fair valuation. This would limit the losses to no more than £10,000 and the cost of freighting the engines. If the Admiralty rejected this plan the only alternative was to dismantle the ship and sell the materials locally.[27]

The new Controller of the Navy (until recently termed the Surveyor), Captain Sir Robert Spencer Robinson, reported in favour of Wellesley's solution. Instructions were then sent to the India Office to have the ship dismantled and freighted to Woolwich Dockyard, at the expense and risk of the Indian Government.[28] India no longer required large steamships, having turned over the deep water role to the Royal Navy, and *Dalhousie* symbolised the end of the Marine. However, she was not destined to join the Royal Navy as a 50-gun screw frigate. By the time her frame and engines reached Woolwich the large wooden steamship had become obsolete. The materials at Bombay were loaded aboard the merchant ship *Castilian*, after some cutting and dismantling of the host ship. She departed Bombay on 28 January 1862 with a full inventory. Arriving at Woolwich in June, her cargo had been landed by the end of the month. The timber matched up to the inventory, but it was decided to leave the machinery in the packing cases, to save the labour of unpacking and the space they would occupy when assembled.[30]

The materials of the *Dalhousie*, both the large quantity of converted Malabar teak, and a set of well made, if old-fashioned, engines, was too useful to be frittered away in petty repairs. For the next four years nothing was done, aside from appropriating the boilers to another ship. In June 1865 John Willis, an East India Merchant, offered to buy the timber still lying at Woolwich for £1500. Admiralty enquiries revealed that only three pieces of sided timber, equal to five loads, had been used from the original supply, at a cost of £5; the remainder was still in good condition. In 1863 the Admiralty had paid the Indian Department:
Timber £6748. 3.0.
Ironwork £2110.14.8.
Total £8859.17.8.

In consequence the offer was rejected.[31] However, this enquiry served to remind the Controller's Department that the materials of the *Dalhousie* still existed.

A New Incarnation

In January 1866 the Board requested designs for a new paddle wheel corvette, similar to the old *Gladiator*, but capable of carrying 500 troops. After lengthy discussions on the merits of paddle and screw vessels Spencer Robinson translated this enquiry into an order for a screw corvette capable of carrying 300 troops on a troop deck. One of the reasons for the design was that 'there are engines of 400hp that can be adapted to the screw ship'; these were the *Dalhousie* engines.

The name ship of the class, designed by the Chief Constructor Sir Edward Reed, was the *Juno*. This role had hitherto been filled by the old paddle frigates, but they were then at the end of their economic lives, and in designing these replacement ships the major issue was whether to fit them with screws or paddle wheels. The availability of one set of screw engines, fitted for the disconnecting and hoisting propeller, encouraged the adoption of the former.[33] The second ship of the class, *Thalia*, was ordered on 22 September 1866. The estimated costs were a modest £63,800.[32] For her role *Thalia*'s teak hull, suitably coppered, would be far less susceptible to

DIMENSIONS OF THE CORVETTE-TROOPSHIP THALIA

Length between perpendiculars:	200ft 0in
Length on keel:	171ft 5⅝in
Extreme breadth:	40ft 4in
Depth in hold:	22ft 0in
Tonnage, old measurement:	1459³⁴/₉₄
Tonnage, new measurement:	2240

The unusual troop-carrying corvette Juno. *The ship's role is revealed by the long line of small ports on the main, or berth, deck.* (CMP)

marine growths than iron, an important consideration in her role of carrying troops or relief crews out to the distant stations. Her armament consisted of two 7in muzzle-loading rifles, four 64pdr smooth bores and eight 9pdr field guns for any troops that might be aboard.

In October 1866 Woolwich Yard was short of work, being run down prior to closure. To fill out the remaining time of the artificers, and use up the available materials the Controller recommended that the half-built 90-gun battleship *Repulse* should be converted into an ironclad, and that the *Thalia* be commenced immediately. *Thalia* was laid down on Number 4 slip at Woolwich on 17 October 1866 and launched on 14 July 1869. She proved to be a poor sailing ship, while by 1874 her engines were reported to be of 'an inferior description'. They were to become a constant maintenance chore.[34] These two defects restricted her usefulness, condemning her to a brief active career in the Far East. By 1891 she had been turned over to serve as a powder depot, and in February 1915 was commissioned under her own name as a storage hulk in the Cromarty Firth. Throughout the First World War she flew the flag of Rear-Admiral Edward Pears and was only paid off in July 1919. The hulk was sold to the Rose Street Foundry Company on 16 September 1920.[35]

Conclusion

The career of the *Dalhousie* provided an illustration of the many problems that beset the Indian Navy in the middle of the nineteenth century. Changes in maritime technology rendered the old fleet of sailing cruisers useless, but new steamships were both expensive, and beyond the capacity of Indian resources, forcing the Company to spend heavily in Britain. In the attempt to reduce expenditure after the Mutiny the British Government agreed that the Royal Navy should take over the remaining deep water duties of the Marine, allowing the service to be abolished. At the Treasury the parsimonious Gladstone demanded to know how much the new service was costing. The Duke of Somerset informed him that the Indian Ocean station had been reinforced by one 50-gun frigate, one large corvette and three sloops, and that these ships, with their reliefs, would cost on average £99,599 per annum. Almost one half of this figure was accounted for by the frigate. However, Somerset observed that the greater part of this work was, as it had always been, Imperial rather than Indian in character. Beyond Indian coastal waters he considered that only the Persian Gulf could be treated as a purely Indian interest; the Red Sea, East Africa and the route to China were all issues for the Imperial Government. The Indian Government did possess ships in the later nineteenth century, notably the *Abyssinia* and *Magdala*, but they never left Bombay harbour.[36] No iron warships of this type were built in India.

The rising cost of construction at Bombay during the period 1830–1860 reflected several factors. Primarily they can be traced to the lack of effective control, both from London and within the yard. Between 1800 and 1820 the

Juno's *sister, the* Thalia. (PRO)

Chief Native Builder, Jamsetjee Jehuboy, in concert with the British Master Shipwright at the Royal Dockyard (a small maintenance facility), had maintained a firmer grip on the working methods and costs of the yard. The British officer was responsible for ensuring that the ships built by the Company met the specification, and instructed the workforce in the new methods and designs then being adopted in Britain. The only reason for the added cost of Indian-built warships lay in the unequal exchange rate adopted, and the Bombay Government's policy of charging dock hire during periods when no work was in progress. With the closure of the Royal Dockyard and the death of Jamsetjee costs rose unchecked. Labour was hired using an inefficient system and timber prices were driven up by a combination among the purveyors allied to the failure of the Company to establish and maintain an effective policy for forest management until 1850.[37]

The failure of the Dockyard was recognised by the Company, almost at the end of its existence. In the sailing ship era it had been a major resource for the Company, and the British Government, but changing perceptions of Indian and British defence requirements, allied to technological change left it vulnerable to competition from more efficient yards in England. As late as 1860 the Governor of Bombay was arguing that the yard could be made efficient by improved management, but with the end of the Marine, and the changeover to iron the effort was not justified.[38] As the *Dalhousie* demonstrated, India could not justify the yard on any ground when it could not match British costs. Iron shipbuilding required a massive investment, and with the end of the Indian Navy and the poor track record of the yard there was never any question of putting scarce resources into such a project.

One issue that requires some elaboration is why the Board of Control selected a 50-gun screw frigate of all the available types. The 50 gun Fourth Rate was the product of confused tactical thinking after the War in 1812. The Americans and French adopted the type as an ideal commerce raider, an intermediate type between the battleship and the old Fifth Rate frigates of the French wars. The British copied the type, but saw it primarily as an addition to the battlefleet, or as a flagship for detached service. In this last role *Dalhousie* might have found some employment, but only if one of the major navies had sent raiders into the Indian Ocean, and the station was not reinforced from Britain. After 1815 this had become less likely. The French no longer possessed the facilities at Mauritius, and no other power had a secure harbour in the area.[39] The real work of the Marine – policing and trade protection – called for smaller craft, something that was widely acknowledged after the Mutiny. Had the Board of Control selected a more suitable design in 1855, rather than giving way to delusions of grandeur, it is possible the service might have been better equipped to meet the needs of the mutiny, and perhaps survived. In the event *Dalhousie* was the wrong ship, built at the wrong time. A vessel more akin to *Thalia* would have been more closely attuned to the existing requirements.

Notes

1. C R Low, *History of the Indian Navy*, 2 vols (London 1877), particularly in Vol II at p256.
2. R A Wadia, *The Bombay Dockyard and the Wadia Master Shipbuilders* (Bombay 1955), p272.
3. Earl St Vincent to The Chairman and Deputy Chairman of the East India Company, 31 March 1802 and 17 April 1802: *The Letters of Lord St Vincent 1801–1804*, D B Smith (ed), 2 vols (London 1921 and 1926), Vol II, p238 & 241–3.
4. The history of Royal Navy shipbuilding at Bombay 1810–1832 is covered in A Lambert, *The Last Sailing Battlefleet: Maintaining Naval Mastery 1815–1850* (London 1991), pp178–188. For Shipwrights and radical politics see I Prothero, *Artisans and Politics in Early Nineteenth Century London* (Manchester 1979). For the decision to abandon shipbuilding in India, Admiral Sir Thomas Byam Martin (Controller of the Navy) to Lord Melville (First Lord of the Admiralty), 5 July 1821; and Byam Martin to the Duke of Clarence (Lord High Admiral), 19 April 1828: Byam Martin MSS, British Library Additional Manuscripts 41,395 fE 84–6 and 41,397 f121.
5. Marine Department to the Governor of Bombay No 21, 4 April 1848: India Record Office E4/1086.
6. Marine Board to Governor of Bombay No 3, 8 January 1851: *loc cit* E4/1092.
7. Marine Board to the Governor of Bombay No 39, 24 August 1852: *loc cit* E4/1099.
8. Marine Department to Governor of Bombay No 16, 21 February 1855: *loc cit* E4/1102.
9. Elphinstone (Governor of Bombay) to Sir Charles Wood (President of the Board of Control), 19 July 1854: Wood MSS, India Record Office F78/28a f163.
10. Sir James Graham (First Lord of the Admiralty) to Wood, 8 January 1855: Graham MSS Microfilm Ms.
11. A D Lambert, *Battleships in Transition: The Creation of the Steam Battlefleet, 1815–1860* (London 1984), especially pp25–52 for a more detailed treatment of the problems of integrating long term construction policy with wartime exigencies.
12. Board of Control to Governor of Bombay 8 August 1855: India Record Office E4/1103.
13. The problems of the machinery of these two ships was so embarrassing that no indication is left in the Marine Department files of the names of their constructor! M Elphinstone to Wood, 2 April 1855: *loc cit* F78/28a f294.
14. Marine Board to Governor of Bombay No 14, 20 February, 1856: *loc cit* E4/1104.
15. Lambert (1984), pp55–9.
16. Marine Board to Governor of Bombay No 42, 4 June 1856: India Record Office E4/1105.
17. Marine Board to Governor of Bombay No 84, 18 November 1857: *loc cit* E4/1109.
18. Marine Board to Governor of Bombay No 45, 8 June 1857: *loc cit* E4/1109.
19. Governor of Bombay to Board of Control No 72, 4 December 1857: *loc cit* L/MAR/C504.
20. Board of Control to Governor of Bombay No 263, 14 April 1859: *loc cit* L/MAR/C501.
21. The Duke of Somerset (First Lord of the Admiralty) to Wood (Secretary of State for India) 20 August 1860: India Record Office, Wood MSS F78/80 f251.
22. Surveyor of the Navy to the Board of Admiralty, 9 November 1860: Public Record Office (PRO) Admiralty (Adm) 92/21, p453, ref S5605.
23. Surveyor to Admiralty, 5 December 1860: PRO Adm 92/21, p472.
24. R J Moore *Sir Charles Wood's Indian Policy: 1853–1866* (Manchester 1966), pp227–30 for a brief resume of the financial problems facing the Government of India after the Mutiny.
25. M Maclagan, *'Clemency' Canning* (London 1962), p260 referring to a letter from Canning to Lord Stanley, then Secretary of State, of 8 February 1859 and Elphinstone to Canning 18 March 1859: Canning MSS, University of Leeds Archives Dept 17/75 and Sir George Clerk to Canning, 14 June 1860: *loc cit* 18/5 and Clerk to Canning, 8 April 1861: *loc cit* 18/46.
26. Elphinstone to Wood, 10 May 1854: Wood MSS F78/28a f127–141 and Elphinstone to Canning, 20 June 1860: Canning MSS 17/79 and India Office to Admiralty, 29 September 1860, with several endorsements through to 1862: PRO Adm 87/77, S6094.
27. India Office to Admiralty, 1 April 1861; enclosing Commodore Wellesley to the Secretary of State 22 February 1861: *loc cit* Adm 1/5770.
28. Secretary of State to Governor of Bombay, 16 April 1861: India Record Office L/MAR/502A.
29. India Office to Admiralty, 11 March 1862: PRO Adm 1/5802.
30. India Office to Admiralty, 26 August 1862: PRO Adm 1/5803.
31. Willis to Admiralty, 20 June 1865; Controller to Storekeeper 20 July 1865; Correspondence with Woolwich Dockyard, June–July 1865: PRO Adm 1/5942.
32. Controller-Admiralty Correspondence, January–February 1866: PRO Adm 1/5980. Programme of Works, Controller of Navy Department, August 1866: PRO Adm 1/5981.
33. Programme of Works, Controller of Navy Department, 1866: PRO Adm 1/5982.
34. Admiralty Work Book; *Thalia*: PRO Adm 135/472.
35. Controller's Submission, 3 October 1866 and minute of Admiral Dacres, First Sea Lord: PRO Adm 1/5982. J J Colledge, *Ships of the Royal Navy: an Historical Index* (2nd Ed London 1987), p346. The Navy List, various.
36. Gladstone to Somerset 2.1.1863: Somerset MSS Buckingham Record Office MS14/42. Somerset to Gladstone 4 and 10.1.1863: Add. MSS 44,304 f137–141 with Gladstone marginalia of 8.1.1863. For the careers of *Magdala* and *Abyssinia*, see O Parkes, *British Battleships* (London 1956).
37. E P Stebbings, *The Forests of India* (London 1922), Vol I, especially pp63–90.
38. Elphinstone to Canning, 18 March 1859: Canning MSS 17/75.
39. G S Graham *The Politics of Naval Supremacy* (Cambridge 1965), particularly pp31–43 and 53–61.

USS NEW IRONSIDES
America's First Broadside Ironclad

William C Emerson discusses the design, construction and career of this underestimated warship. Although slow and somewhat ungainly, she nevertheless participated in more battles, fired more shots, and was hit by large calibre shot more often than any warship of her time.

When the American Civil War broke out in April 1861, neither side was equipped with ironclad vessels. Only two countries had such vessels – France with *La Gloire* and Britain with HMS *Warrior*. Hoping to develop a knockout punch for his greatly inferior navy, Confederate naval secretary Stephen R Mallory authorised ironclad construction, starting with the conversion of the burned-out hulk USS *Merrimack* into an ironclad steamer – the CSS *Virginia*.

News of the Confederate actions soon reached Washington. Congress authorised the appointment of a special board to examine and report on the subject of ironclads, and appropriated $1.5 million to be used for the construction of one or more armoured ships. The board considered having ships constructed at more experienced shipyards in England or France, but concluded . . . 'we are of the opinion that every people or nation who can maintain a navy should be capable of constructing it themselves'.

In August, the Navy Department issued an advertise-

One of the few full views extant of the New Ironsides *showing details of her barque rig. At one point in her career, her tall smoke stack was cut down to improve visibility from her pilot house, located just aft the main mast. However, smoke so choked those inside that the stack was rebuilt.* (Smithsonian Institution)

ment requesting proposals for ironclad steam vessels of war. Because the inland waterways in the South were shallow, all proposed designs were required to have a draught of no greater than 16ft.

The text of the 7 August 1861 advertisement was as follows:

IRON-CLAD STEAM VESSELS

The Navy Department will receive offers from parties who are able to execute work of this kind, and who are engaged in it, of which they will furnish evidence with their offer, for the construction of one or more iron-clad steam vessels of war, either of iron or of wood and iron combined, for sea or river service, to be not less than ten nor over sixteen feet draught of water, to carry an armament of from eighty to one hundred and twenty tons weight, with provisions and stores for from one hundred and sixty-five to three hundred persons, according to armament, for sixty days, with coal for eight days. The smaller draught of water compatible, with other requisites, will be preferred. The vessel to be rigged with two masts, with wire rope standing rigging, to navigate at sea.

A general description and drawings of the vessel, armour and machinery, such as the work can be executed from, will be required.

The offer must state the cost and the time for completing the whole, exclusive of armament and stores of all kinds, the rate of speed proposed, and must be accompanied by a guarantee for the proper execution of the contract, if awarded.

Persons who intend to offer are requested to inform the Department of their intention before the 15 August, instant, and to have their proposition presented within twenty-five days from this date.

The board, made up of prominent naval officers appointed by the Secretary of the Navy, reviewed a total of seventeen proposals of which three were accepted.

The first design accepted was the most controversial. Submitted by John Ericsson of New York, and with a contract price of $275,000, this ship became the famous *Monitor*. It was built mostly of iron and featured a very low deck upon which rested a revolving turret carrying two heavy guns. Protection of the turret and hull was of 1in laminated iron plates for a total thickness of up to 11in. Length was 172ft, breadth 41ft, displacement 1255 tons, and draught was 11.5ft. While Ericsson's design failed to meet some requirements of the advertisement, it proved to be a great success and some forty monitors were built or were building by the end of the fighting.

Also accepted was a design by Bushnell & Co of New Haven, Connecticut, at a contract price of $235,250,

At her best, the New Ironsides *was an ungainly looking ship. This drawing of her under sail appeared in the biography of William A Cramp.*

which became the *Galena*. She had a very rounded shape, was barque rigged, constructed of wood, and carried a broadside battery. Protected with 4in thick iron bars, she had a length of 180ft, a breadth 32ft, and draught of 10ft. She was not considered a success, her armour being penetrated numerous times at Drewry's Bluff in May 1862.

A third design accepted, by Merrick & Sons, became the *New Ironsides*. Merrick, a well known builder of steam engines in Philadelphia, having no shipways, sub-contracted the hull construction to William Cramp and Sons, also of Philadelphia. Merrick offered to complete the vessel at a cost of $780,000, in nine months from the signing of the contract. The contract was signed on 15 October 1861.

Table 1: *DESIGN PARTICULARS*

Dimensions:	249ft 6in loa, 57ft 6in breadth, 15ft draught (mean)
Displacement:	4120 tons
Machinery:	4 Martin boilers, 1800hp
Speed:	6kts (steam power only), 7kts (steam and sail)
Coal capacity:	350 tons
Armament:	14–11in Dahlgren smooth bores, 2–8in Parrott rifled muzzle-loaders, 2–5.1in Dahlgren rifled muzzle-loaders, 1–12pdr smooth bore
Complement:	449 officers and men
Contract signed:	15 October 1861
Launched:	10 May 1862
Commissioned:	21 August 1862
Fate:	Burned 16 December 1866

General Characteristics

The *New Ironsides* was a frigate-built vessel with a hull of 243ft (this length is often misrepresented as 232ft or less), displacement of 4120 tons, and an extreme breadth of 57½ft. To satisfy the shallow draught requirement, she drew only 14ft, exclusive of her keel (15ft total). Gunport sills were 7ft above the waterline, allowing use in a moderate seaway. Her hull was massively built of oak, much of which was only cut after the contract was signed, allowing little time for proper seasoning. Contemporary accounts remark on her lack of fine lines or beauty, partly the result of the shallow draught, which required a broad ship with a flat bottom. She had a noticeable degree of tumblehome, with her sides tapered inward at an angle of about 17 degrees from a point about three feet above the waterline. She was built with three decks; the berth deck ran forward and aft of the engine room, but did not cover it. Above this, and covering the engine room, was the gun deck, which ran the full length of the ship. The uppermost deck, the spar deck, was above this. She was rather roomy, having six feet of clearance between decks.

She was powered both by steam and by sail. Her engines produced about 1800 horsepower, and her best logged speed without sails was 6kts, well under the contract speed of 9½kts. While fully rigged when travelling between ports or at sea, her rigging was considered a liability in the close quarters of the inland waterways of the South. Consequently, once the *New Ironsides* was in an appropriate theatre of operations, her masts and spars were removed and signal poles substituted.

She was armoured with solid iron plates up to 4½in thick, the maximum manufacturable in the US at that time. She had a small armoured pilot house mounted on the spar deck which could comfortably hold but three persons. Steering problems, severe enough to delay her speed trials, were never completely resolved. In shallow water her flat bottom, single propeller, and poor underwater lines resulted in a very disturbing lack of control. Despite the size and complexity of the project, *New Ironsides* was launched on 10 May 1862, seven months from the signing of the contract.

Table 2: *HULL DIMENSIONS OF USS NEW IRONSIDES*

Length overall	249ft 6in
Length on load water line	242ft 2in
Length between perpendiculars	230ft 0in
Beam, moulded	55ft 9in
Beam over planking	56ft 7in
Beam over plating (4½in)	57ft 4in
Beam over plating at spar deck	46ft 2in
Depth to planksheer above base line	26ft 7in
Depth to top rail above base line	29ft 3in
Depth to upper port sill above base line	24ft 2in
Load line above base line	14ft 5in
Load line above bottom of keel	15ft 0in
Timber frames at throats, moulded	1ft 6in
Timber frames at turn of bilge, moulded	1ft 1in
Timber frames at vertical sides, moulded	0ft 9in
Timber frames at spar deck, moulded	0ft 7in

The hull was entirely designed by William Cramp, and was 230ft in length between perpendiculars, giving her a waterline length, exclusive of her ram, of 243ft. Extreme depth of hold was 24ft 9in, with a tonnage of about 350 tons. With the exception of the deck, which was yellow pine, the hull was constructed entirely of white oak in the conventional manner of the day. *New Ironsides* was built very heavily, with outside planking below the lower edge of armour being 12in thick, and tapering off at the lower turn of the bilge to 5in.

Frames were very heavy and were without first futtocks. The spaces between the tightly fitted frames were filled in solid from the throat of the floor timbers to the planksheer and caulked before planking. Frame timbers ranged in thickness from 7in to 18in. Frame scarphs were each bolted with three 1¼in iron bolts. The heels of cant frames were bolted through the deadwood with two 1in copper bolts. Timber for the curved futtocks was principally made of tree roots.

U.S.S. New Ironsides

Inboard profile and rigging plan, reconstructed from National Archives documents. Shown are the below deck arrangements and her long propeller shaft. (Drawing copyright David J Meagher)

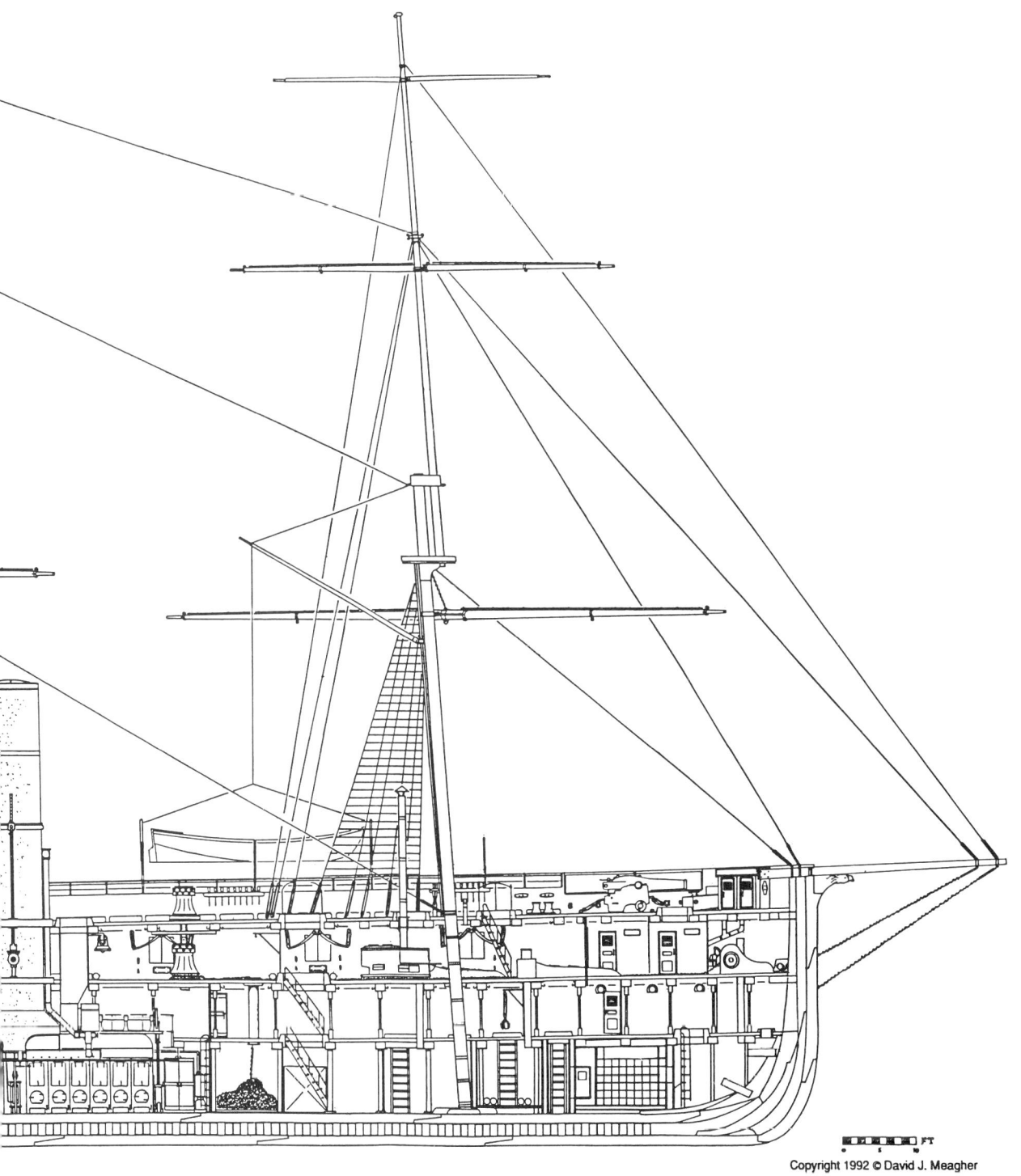

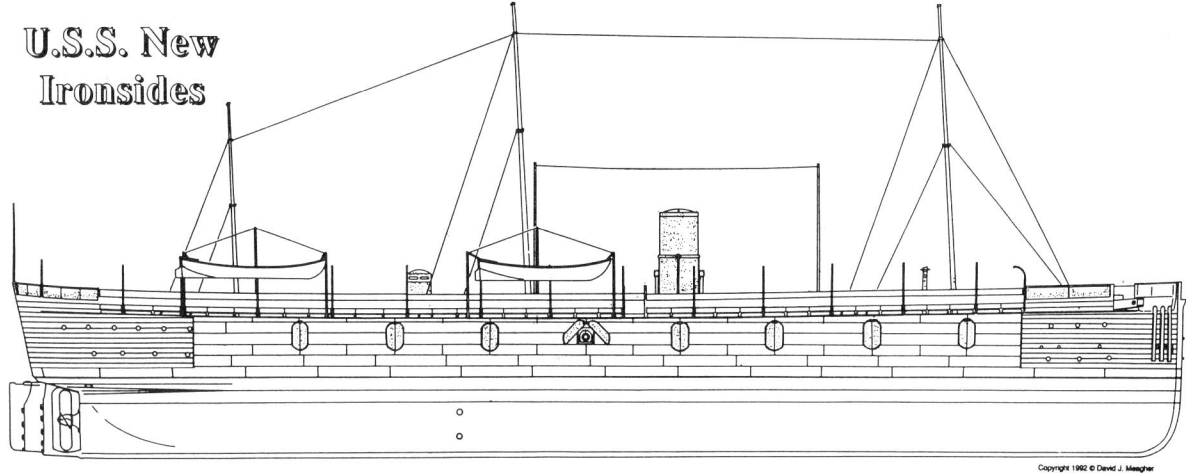

Outboard profile of New Ironsides *as she was rigged when going into battle. Her masts and spars were for transit only: considered a liability in combat, they were replaced with signal poles.* (Drawing copyright David J Meagher)

The hull had an extensive arrangement of iron bracing. Diagonal braces 4½in by ¾in were worked inside the frames from near the plank sheer to the turn of the bilge, one brace every three frames. Another set was run horizontally. Where braces crossed over frames they were bolted with iron 11/16in bolts, and where they crossed other braces, were hot-riveted together. Three separate iron braces 5in wide and 1in thick were worked in from the forward square frames around the hull aft. These braces ran on the outside of the frames and were bolted to each frame. They were then covered by the outside planking.

Understandably, much very large timber was required for the construction of the *New Ironsides*. Floor timbers, for instance, were two to a frame and ran from bilge to bilge. Timbers for these required a tree large enough to be 22in in diameter at a height of 45ft from the ground. Some

Sheer and half-breadth plan redrawn from originals in US National Archives. (Drawing copyright David J Meagher)

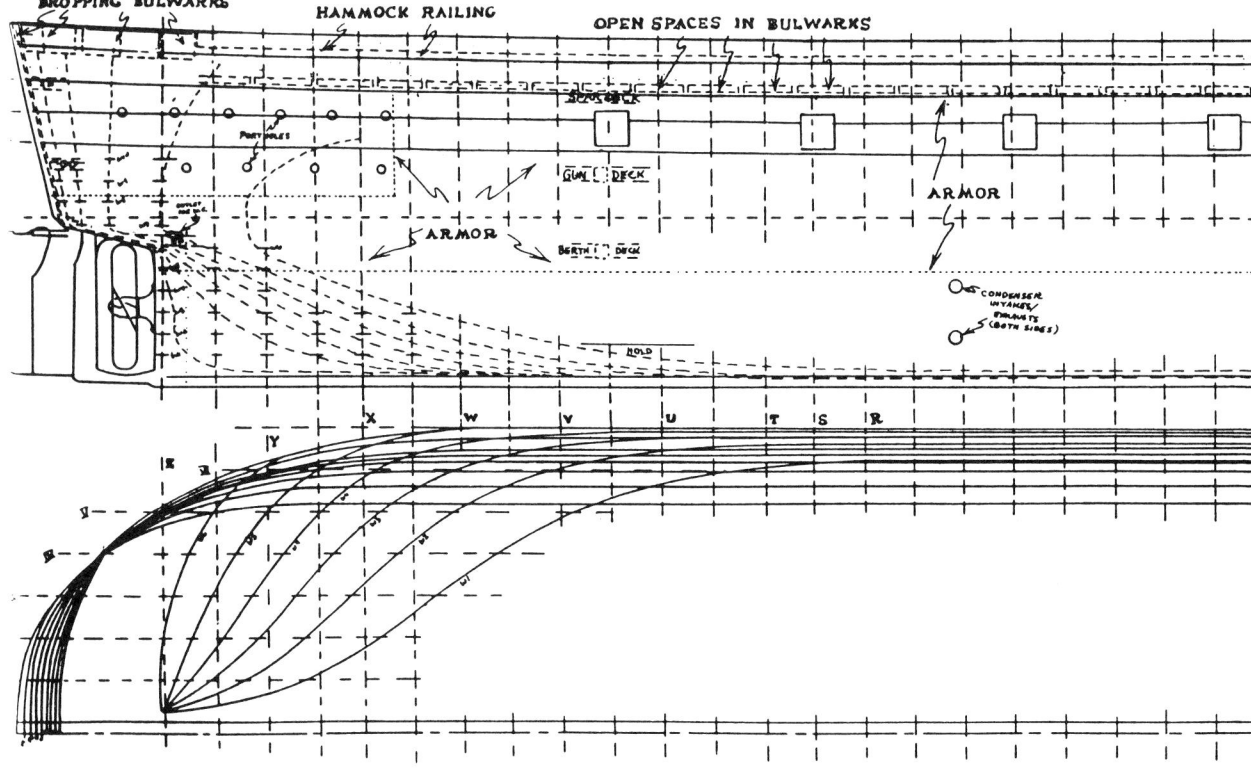

of the timbers used were 60ft long, and even in 1861 trees of this size were difficult to find. Cramp advertised extensively through the local counties and paid a premium for such timber. He would later claim the timber used in construction was growing in the forests of Pennsylvania when the contract was signed; however, much came from a US government stockpile at the local Navy Yard.

As originally designed, the *New Ironsides* had a superabundance of buoyancy. This was fortunate, as much additional weight was added to the ship above the original specification. Additions included armoured bulkheads, iron gunport shields, an armoured pilot house, and changes to the armament – a total of about 550 tons. In later years, Cramp would claim that he had used great foresight in providing an extra foot of draught for these unexpected additions.

The bottom of the hull was payed with turpentine and tallow and then coppered with 30oz and 32oz copper sheets. Coppering went from the keel up to the 10ft line above the keel, where the iron armour started.

Armour

The *New Ironsides* was protected in an early 'armoured citadel' arrangement, similarly to HMS *Warrior* (1860). Rather than armouring the entire hull, as had been done with *La Gloire* (1859), this arrangement protected only guns and engines in an armoured box in the central area of the ship. Beyond this, the ends of the ship were unarmoured and exposed to enemy shot and shell. Even with the ends damaged, the fighting strength of the vessel would be, in theory, little damaged.

Side armour in the *New Ironsides* was 4½in thick, and ran for 170ft amidships, up to spar deck level, protecting the gun deck. In order to protect against raking fire, the ends of the citadel must also be armoured, as was *Warrior*. While the original contract called for no such armour, increasing concerns about vulnerability caused the Navy Department to alter the design, adding armoured bulkheads at each end of the citadel, between the gun and spar decks. These bulkheads had 2½in iron covered with 12in of oak, and access was provided by heavy sliding doors. The distance between the bulkheads was 163½ft, somewhat shorter than the outside armour.

A belt of waterline armour extended entirely around the vessel, from 4ft below to 3ft above the 15ft waterline. This armour was 4½in thick down to the second plate below the waterline, where the thickness was reduced to 3in. The 4½in thick side armour extended up seamlessly from the waterline armour. At the bow the belt terminated in a large ram. The gunports were protected by heavy iron shutters, fitted in halves, and worked from inside by means of lever and tackle. The port openings were 38in high by 42in wide – the lack of height limited the elevation of the guns, reportedly, to only 4–4½ degrees, preventing the guns being used at their maximum range.

One inch of armour was fitted on the spar deck beams, and over this 3in of yellow pine decking was laid. This light protection from plunging shot was a concern in battle with forts, which were generally at a higher elevation. Various schemes were tried to add protection to the decks, including untreated (and very malodorous) animal skins, and sand bags (as many as 6000 were used). Happily, no shots ever penetrated the spar deck.

The specifications called for solid armour plates 4½in

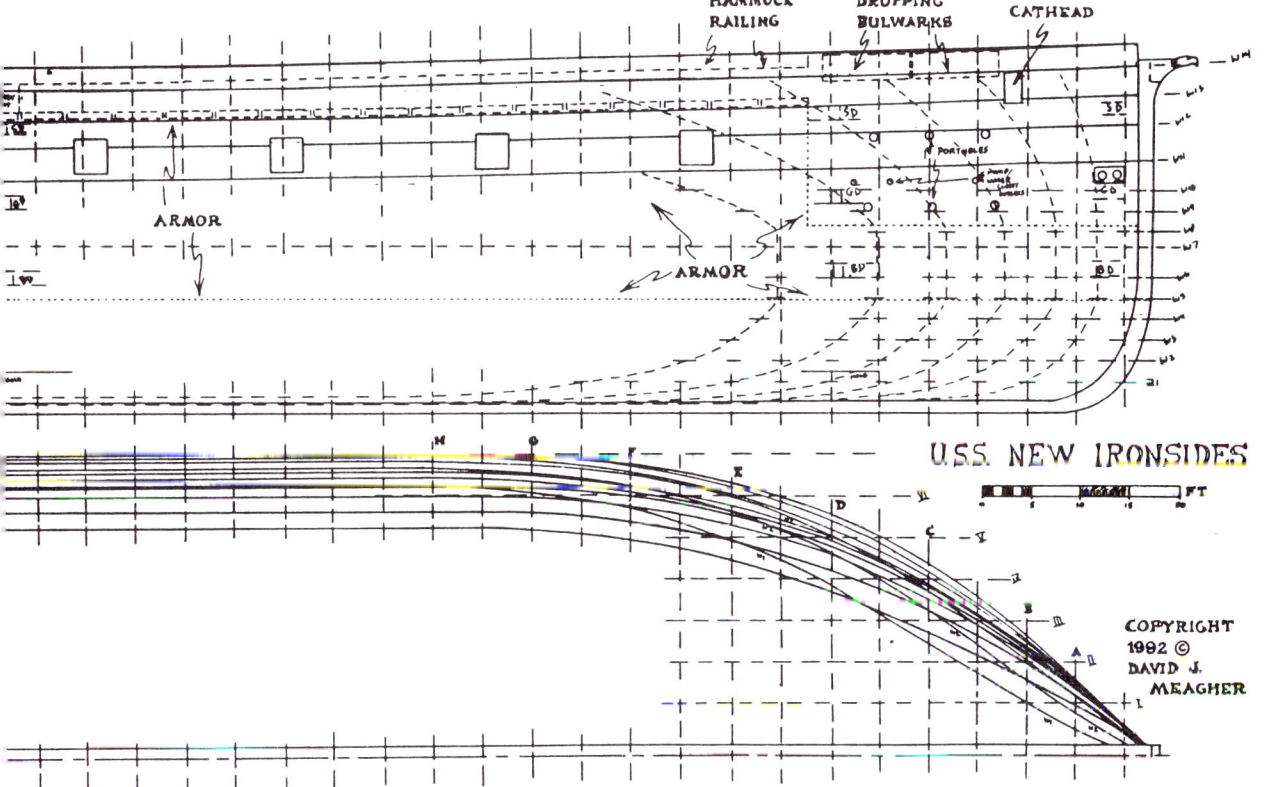

Carte de Visite photograph by P F Cooper of Philadelphia, showing the ship with a full set of masts and yards; may have been taken in 1862. (US Naval Historical Center)

thick, 28in wide and 15ft long. Testing had previously shown that armour plates less than this thickness did not provide adequate protection from naval ordnance of the day. Each plate was planed at its edges, and had a groove 1in deep and 1½in wide recessed all around, producing a tongue-and-groove joint. This arrangement was intended to give added strength, with adjacent plates taking part of the shock of a strike, but in practice, the joints just made replacement of a single armoured plate more difficult. In an attempt to improve the deflection of hostile fire, the side armour of the *New Ironsides* was at one point coated with a heavy layer of grease. Happily, for officers and men using her Jacob's ladder to come aboard, this practice was quickly abandoned.

A pilot house, 6ft in diameter and 8ft tall was added on the spar deck in the summer of 1862. Protected with 4in of cast iron armour, it housed only three comfortably. When the *New Ironsides* was flagship of the combat fleet off Charleston, this limitation proved difficult – there was insufficient room for the ship's captain, who was forced to station himself on the gun deck, where there was little visibility!

The armour plates were hammered of charcoal scrap iron, and were held in place by large wood screws made of galvanised iron. Because these screws did not penetrate into the ship, there was no opportunity for fragments to be knocked around the compartments when struck on the side by a heavy shot, as was the case with a number of monitors whose armour was through-bolted.

Body plan of the New Ironsides *from original Navy drawings, showing her wide beam and flat bottom. Not shown is her armour, which protruded some 4½in.* (US National Archives)

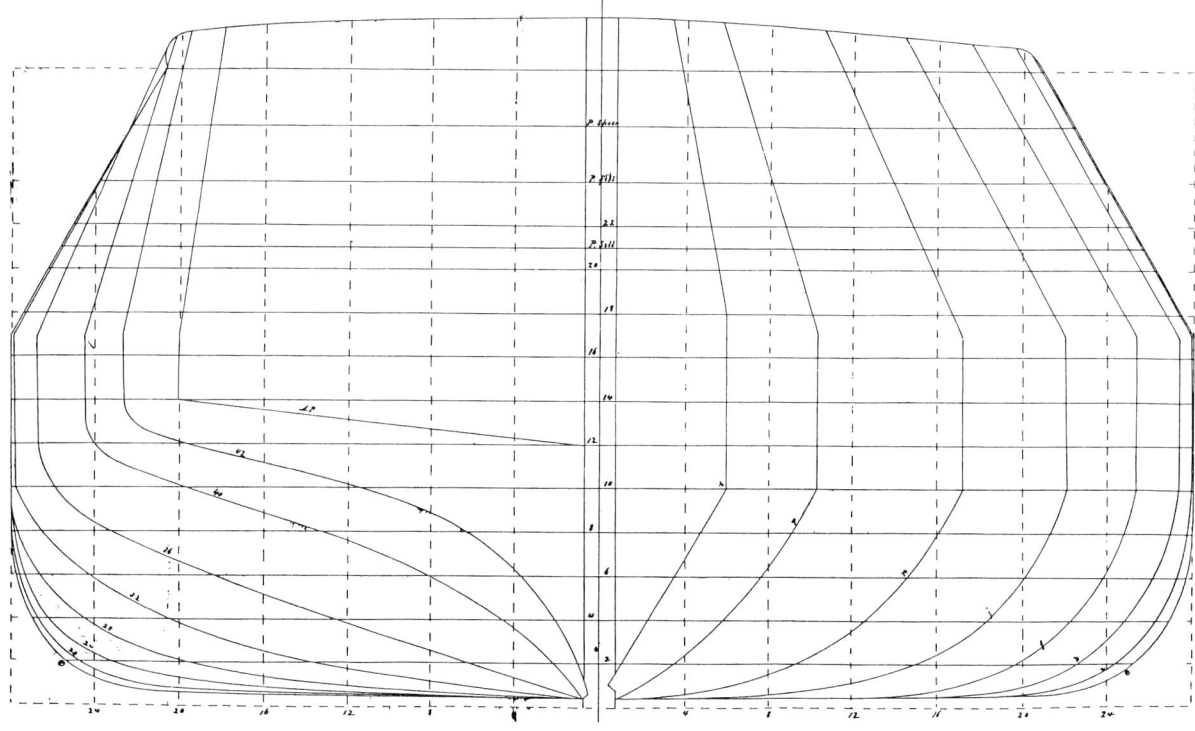

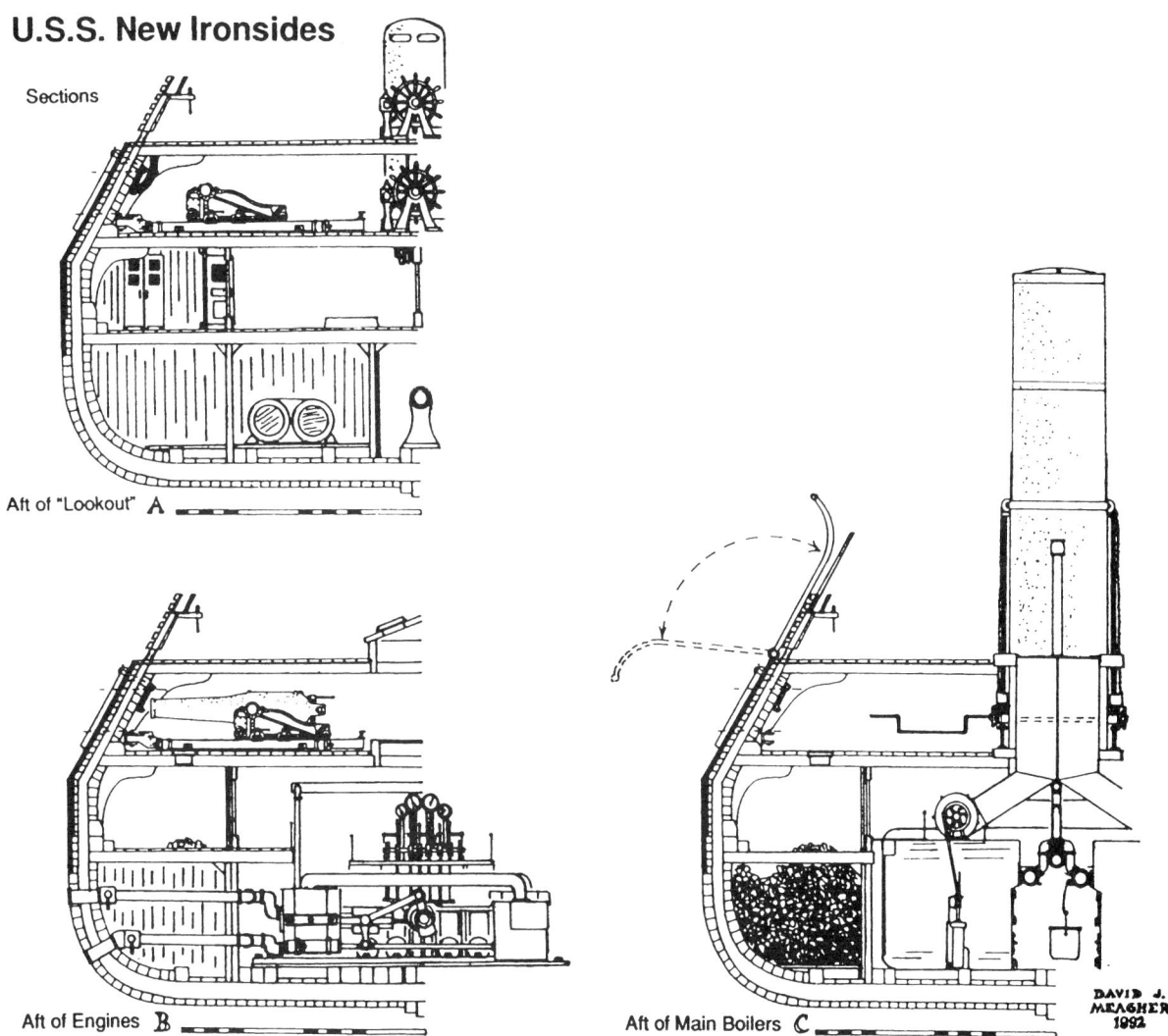

Three cross-sectional views of the ship. Top view is aft of pilot house; bottom left shows engines; right view is aft of main boilers. Note the extreme tumblehome and the mounting of her side armour. (Drawing copyright David J Meagher)

Armament

Originally, the ship was to carry sixteen 9in Dahlgren smooth bore cannon, eight to a side. These were very heavy naval guns of the day, but during building the navy began to consider the even heavier 11in Dahlgren guns used aboard *Monitor*. Eventually they settled on fourteen 11in Dahlgren smooth bores, plus two 150pdr (8in) Parrott rifles. The exact reasoning behind this mixed arrangement is unknown, but the mix persisted throughout the career of the *New Ironsides*. What is known is that the 150pdr Parrotts were carried in the No 2 gun positions, port and starboard.

Difficulty in obtaining seasoned timber during the war led to the substitution of iron for wood in the construction of gun carriages and the first trial of iron was made in the carriages of the *New Ironsides*. After many problems, especially in controlling recoil, these carriages were made to work, and eventually were considered better than those of wood. A crew of thirty-five manned each gun in the main battery: 25 for the gun itself, and 10 stationed at the tackle for the ports. These latter crew also relieved the side-tacklemen in serving the guns in continuous and rapid fire.

Two 5.4in (50pdr) Dahlgren rifles were carried on the spar deck, one each at the bow and stern, but these were seldom used. At least one brass howitzer was carried on the spar deck as well.

Considering the difficulties associated with handling heavy shot and manoeuvring guns weighing 16,000lbs, the rate of fire was impressive. In one 50-minute period of combat, 230 shots were fired from the *New Ironsides*. Though sighting and accuracy were somewhat compromised, such a rate was far superior to that of the monitors, whose rate of fire was about twenty rounds for an equivalent period.

Wash drawing by Clary Ray, c1900. Although smaller and slower than her European contemporaries, her heavier armament would have given her an advantage in combat. (US Naval Historical Center)

Machinery

Steam was supplied from four horizontal fire-tube boilers, placed facing each other, with the fireroom between and smoke connections in the centre. Each boiler had six furnaces, along a 17ft front. This provided a total grate area of 350sq ft, and a heating surface total of 8776sq ft. The tubes were wrought iron (not copper or brass), lap welded, 3¼in in diameter with ⅛in walls. The four boilers were so arranged as to connect into one 40ft smoke pipe. This was made in two parts with the upper part able to telescope up when in use, and down when under sail alone.

John Dahlgren standing before one of his rifled 4.4in or 5.1in guns. Similar guns were mounted at the bow and stern on the spar deck of the New Ironsides. *These guns were seldom used, being fully exposed to enemy fire during battle. The trunnions of these guns were attached with a wide metal band or capsquare.*

Drawing of the 11in Dahlgren guns and carriges, redrawn from Navy plans. The gun carriages were of iron, the first used on any US ship, and problems with control of recoil caused a number of delays in getting the New Ironsides *into battle.* (Drawing copyright David J Meagher)

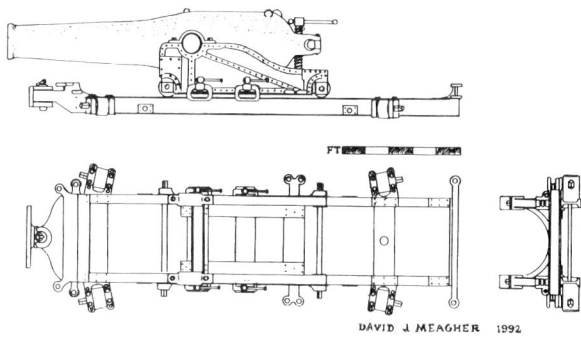

Motive power came from two horizontal direct-acting engines, mounted in line. In these early days of steam engineering, the concept of using waste steam to operate ever larger pistons at lower pressure was not yet a reality, and consequently the *New Ironsides* was provided with single expansion engines. Each engine had a single piston at 50in diameter with a 30in stroke. The cylinders were placed athwartships and each had two piston rods. A single surface condenser was positioned between the engines. It consisted of some 2400 brass tubes tinned securely with each exposing 7ft of tube for cooling.

The drive shaft was of wrought iron forged in one piece, with double cranks at right angles, and with arms to counterbalance the weight of the cranks. The screw was of brass, 13ft in diameter and was four-bladed, with a pitch of 20ft. While the size of the propeller was small by contemporary standards, it was the maximum diameter that could be mounted in such a shallow draught ship. Her maximum speed was 6kts using steam alone, and 7kts under both steam and sail.

Coal capacity was 350 tons, which was said to provide for nine days at full steaming. Coal was carried in hanging bunkers over the boilers as well as below the berth deck. Her steering consisted of a compound rudder arrangement which was not fully successful, her steering being especially poor in shallow water. While a more conventional rudder was designed, the replacement was apparently never carried out.

Boats

The *New Ironsides* apparently carried six boats, two each at bow, waist and after davits. Contemporary reports indicate that the boats included a launch, 2nd, 3rd and 4th cutters, and a dinghy. It is likely that she also carried a 1st cutter and a 2nd launch. Based on contemporary standards, the launches were likely 30ft in length, the 1st and 2nd cutters 27ft, the 3rd and 4th cutters 24ft, and the dinghy 18ft.

Battle experience

The *New Ironsides* took part in a large number of battles against forts, but never against other armoured vessels. Before going into action her deck was covered with sandbags, and the iron bulkheads at her ends were reinforced with more of the same. Her draught was found to limit her approach in operations against forts and she seldom got closer than a mile from her target. Still, her

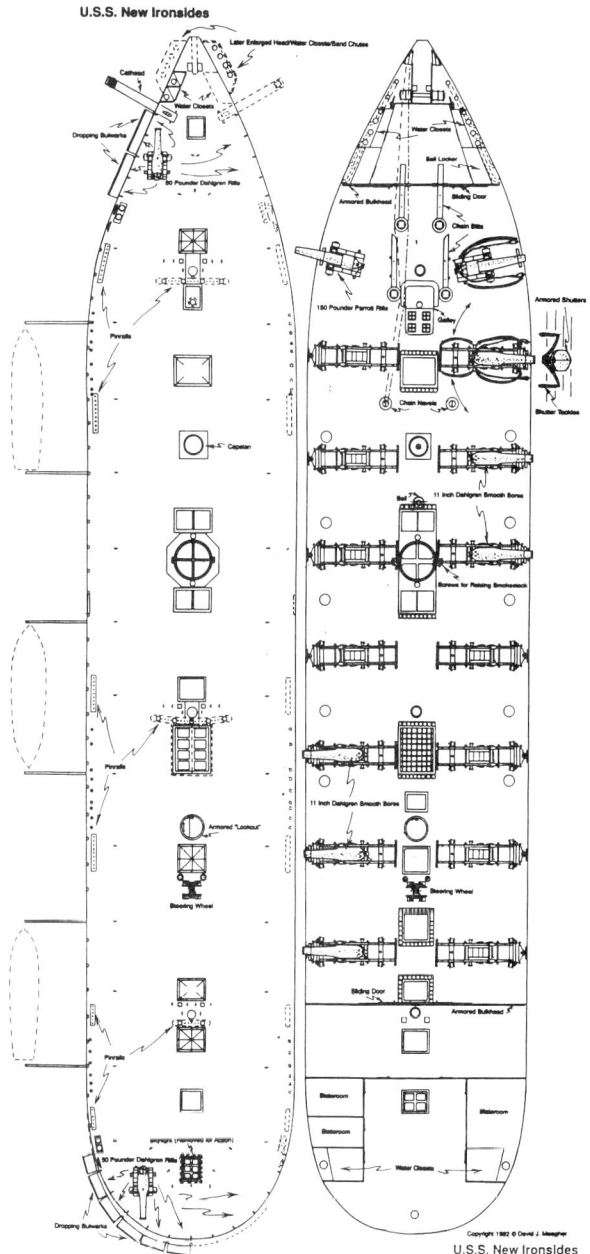

Spar and gun deck plans redrawn from originals in US National Archives. (Drawing copyright David J Meagher)

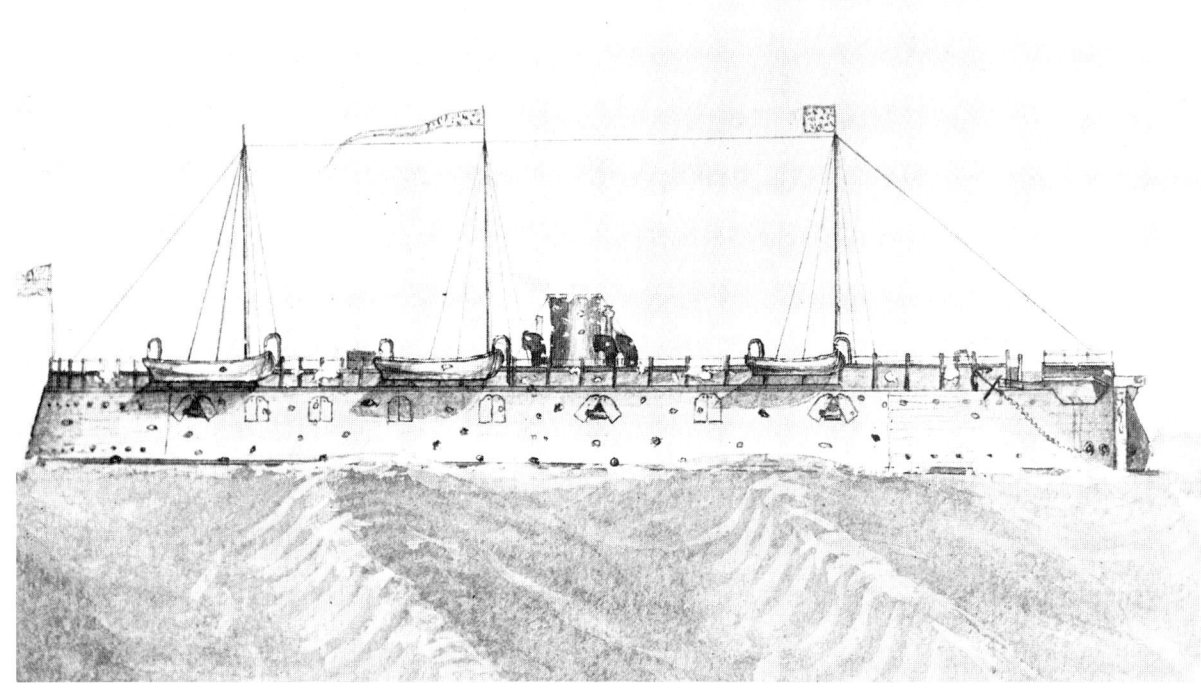

Drawing of the New Ironsides *as she appeared off Charleston, South Carolina during the war. Once in the theatre of battle, her top hamper was removed and replaced with signal poles. While her armour was never penetrated, she received substantial damage to her unarmoured ends, railings, and stack, as depicted here.* (US Naval Historical Center)

support was invaluable, as the monitors' slow fire was far less effective. Her rapid firing and large ammunition supply were credited with considerable damage to fortifications. Despite the range, she was often hit in return by heavy ordnance.

The naval assault of well defended forts was a risky business. Fire from the Confederate forts was aided by range markers, and was very effective. When *New Ironsides* and a number of monitors attacked the forts at Charleston on 7 April 1863, the Southern gunners fired 2209 shot, shell and rifle bolts, and even with the *New Ironsides* in the fray, only 144 rounds were returned! In that battle, the ship was hit more than fifty times, and lay, for nearly an hour, directly over a 'torpedo' (mine) containing over a ton of powder. While the Confederates frantically tried to explode the device, their efforts were frustrated by a broken wire, accidentally run over by a wagon!

Despite numerous battles and repeated strikes by heavy ordnance, damage to the *New Ironsides* was minor. Her armour was never penetrated, though gunport covers were occasionally knocked off. While her unprotected wooden ends were riddled, the damage proved easy to repair. Only a few minor injuries took place in these battles. While her side armour was badly dented, these dents were never greater than half the armour thickness. Perhaps her most severe test came on the night of 5 October 1863, when a near-successful attempt was made by the Confederate torpedo boat CSS *David* to destroy her. Under cover of darkness, this semi-submerged vessel exploded a spar torpedo against her hull, throwing up a huge column of water and swamping the *David*. The explosion was under the curve of her hull, near No 6 gunport on the starboard side, and took place at the level of the berth deck, where the outside planking was 12in

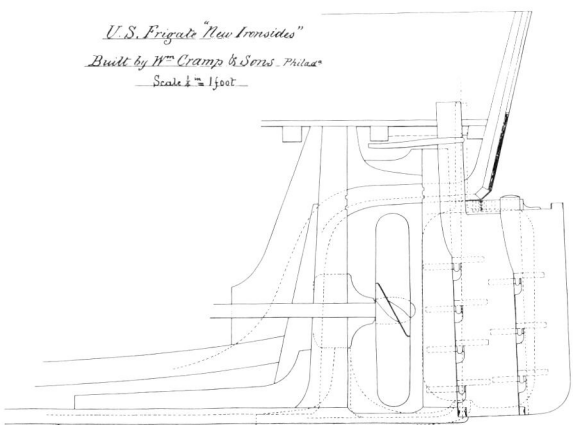

Drawing of the articulated rudder. The aft portion of the rudder was attached with gearing to the forward portion to produce a two-step effect when turning. Never very successful, a standard rudder was designed (note dotted lines) but apparently was never fitted. (US National Archives)

USS NEW IRONSIDES

▲
The New Ironsides *in company of the double-turreted monitor* Monadnock, *from an 1866 issue of* Harper's Magazine. (US Naval Historical Center)

Drawing of the CSS David, *the vessel which nearly sank the* New Ironsides. (Drawing copyright David J Meagher)
▼

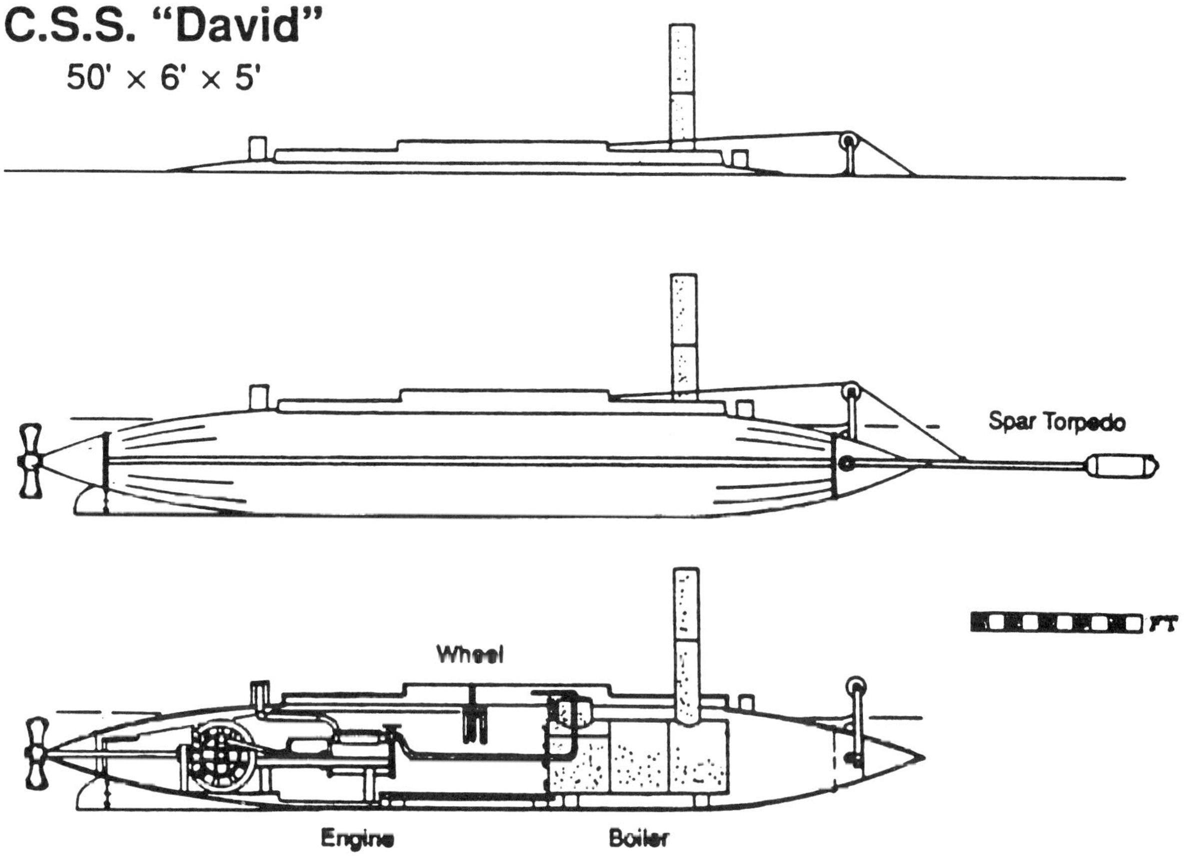

C.S.S. "David"
50' × 6' × 5'

Scale model of the ship as she appeared immediately after commissioning in 1862. (Author)

thick, and where a 16in beam backed the frames. The force drove in her side 6in and 'broomed' the end of the beam, but the damage was minor and she was able to stay on station for some eight months before returning to Philadelphia for repairs.

Conclusions

It was a notable achievement that, despite the new technology required to build her, the *New Ironsides* was bombarding Fort Sumter only 11 months and 2 days after the signing of the builders' contract. And she was durable, judging from her ability to withstand long periods of blockade duty, bombardments from coastal batteries, and the explosion of a large torpedo against her hull. In 1865, Admiral David Porter praised her, saying 'I have never yet seen a vessel that came up to my ideas of what is required for offensive operations as much as the *Ironsides*.'

In many respects, the *New Ironsides* was the most successful of the Civil War ironclads. With the ability to go to sea, to fight her guns in moderately severe weather, and with sufficient firepower to face her European rivals, she, more than the monitors, was the forerunner of the modern United States Navy. Yet, the Union failed to capitalise on her design, with no more of her type commissioned during or after the war. While smaller and slower than her European rivals, she was more heavily armed and in a fight could have easily held her own. Yet she was far too slow to be able to force action, and her enemies could have easily avoided a fight.

As strong as she was, soon after the war she fell to a most dreaded enemy of wooden ships. On a windy evening in December 1866, she was destroyed by fire in Philadelphia.

Sources
Conway's All the World's Fighting Ships, 1860–1905, Conway Maritime Press (London 1979).
Dictionary of American Naval Fighting Ships, Naval History Division, Navy Department (Washington 1959).
Library of Congress, Washington DC 20540 (photos).
National Archives, Record Group 19, Cartographic & Architectural Branch, Washington DC 20408 (plans and *New Ironsides* original Specification).
Naval Historical Center, Washington Navy Yard, Washington DC 20374 (photos).
Still Pictures Branch (NNSP), National Archives, Washington DC 20408 (photos).
Augustus C Buell, *The Memoirs of Charles H Cramp*, J B Lippincott Co (Philadelphia and London 1906).
George E Belknap, 'Reminiscences of the *New Ironsides* off Charleston', *The United Service; a Quarterly Review of Military and Naval Affairs* 1 (1879).
William H Roberts, 'The Neglected Ironclad: A Design and Constructional Analysis of the USS *New Ironsides*', *Warship International* xxvi (2, 1989).
'The USS Armoured Frigate *New Ironsides*', Editorial, *Journal of the Franklin Institute* (February 1867).

FLIGHTS OF FANCY

Unusual Shipboard Aircraft Launch and Recovery Systems

Well before and even well after the development of the flight-deck carrier and the shipboard catapult, many ingenious, sometimes bizarre, schemes were advanced to permit the operation of aeroplanes from warships. Nearly all have vanished into limbo. R D Layman and Stephen McLaughlin describe some of these often weird ideas.

Given that flying aircraft from ships was a worthwhile matter – and many naval officers accepted this as self-evident from the earliest days of aviation – the practical business of getting an aeroplane aloft from a vessel, and perhaps of recovering it also, presented officers and aviators with unprecedented problems. First and foremost was figuring out what sort of appliance would do the job with a minimum of interference with the normal business of running a ship. Early flights were made from wooden platforms or trackways supported by stanchions over the upper deck, and with few exceptions these cumbersome additions were mounted forward,[1] where they could not help hindering the use of a warship's guns. This situation was also unsatisfactory from the aviator's point of view, as the superstructure limited the length of the take-off run.

Eugene Ely's take-off from the US cruiser Birmingham, *14 November 1910. The event has rightly been hailed as the birth of shipboard aviation, but it also demonstrated how minimal was the gravity thrust imparted by an inclined platform. Moments after this photograph was taken the aircraft struck the water, splintering the leading edges of its propeller, and Ely was barely able to regain height and reach a nearby beach. The experiment was to have taken place while the ship was under way, but Ely took off when a squall approached while the cruiser was weighing anchor. (National Archives)*

A Wright brothers' aeroplane on the monorail of their gravity-powered catapult, date and venue unknown. The structure housing the thrust-imparting weight is to the left. (R D Layman collection)

This second problem was exacerbated by the fact that many early flights were made from warships hove to or anchored, in the mistaken belief that this conferred a greater degree of safety. But aviators already knew that take-off was best (and most safely) achieved against the wind. It was soon apparent to all concerned that a ship under way would create a useful artificial 'wind', which, coupled with even a light breeze, would greatly assist an aeroplane's take-off. This was demonstrated in 1912 when Royal Navy Lieutenant Charles Samson made the first underway take-off from the battleship *Hibernia*.

This sort of wind-assisted take-off reached its apogee during the First World War, when it was found that relatively high powered, light aircraft could leap into the air after astonishingly short runs from platforms trained into the wind aboard moving ships. By the end of the war nearly every British capital ship and several cruisers were equipped with these 'flying-off platforms', which, mounted on turrets or revolving turntables, in no way interfered with gunnery.[2]

But it was clear long before the war that the wind would not always be enough; pioneer naval aviators had to get aloft in aircraft that seem ridiculously underpowered today, and of course larger aircraft never could make use of the short flying-off platforms. Something more than the wind was needed.

Gravity, Pulleys and Drogues

The first aid enlisted was the very force aircraft sought to overcome – gravity. Launching platforms were sloped downwards at various angles in the hope that a 'downhill' run would add some impetus to an aeroplane. In reality, the acceleration imparted was minimal, as American aviator Eugene Ely discovered after nearly coming a cropper in the world's first shipboard take-off from the cruiser USS *Birmingham* in 1910.

Earlier, brothers Orville and Wilbur Wright had created a successful aircraft catapult, harnessing gravity in the form of a falling weight. Their device was a round-about way of solving a problem of their own making – their refusal to abandon skid carriages, which produced take-off retarding friction. The Wrights therefore perched their aeroplane on a small two-wheel truck, or dolly, that ran along a wooden, iron-faced monorail 70ft to 125ft (21m–38m) long. At the base of the rail was a pyramidal 'tower' formed by four 25ft (7.6m) timbers. At its apex was suspended a 1400lb (635kg) iron weight, connected to the truck by a line running through pulleys. With the aircraft engine revved up, the weight was dropped and its 450lbs (204kg) of thrust was transmitted to the truck, which sped the aeroplane down the rail for 50ft or 60ft. At that point the line automatically disengaged and propeller thrust took over to complete the lift-off.

This system worked quite well, and it made a considerable impression upon the American naval attaché in France, Commander F L Chaplin, who saw it demonstrated in September 1909 during the first major European aeronautical exhibition and competition, the *Grande Semaine d'Aviation de la Champagne*, near Rheims. He mentioned it favourably in his official report, suggesting that a Wright catapult be installed on a *Connecticut* class battleship for experimental purposes.

Chaplin was probably correct in his estimate that the Wright device was feasible for shipboard use. The low rail would not have interfered with gunnery, and the 'tower' could be erected and disassembled fairly quickly. But his recommendations were ignored.

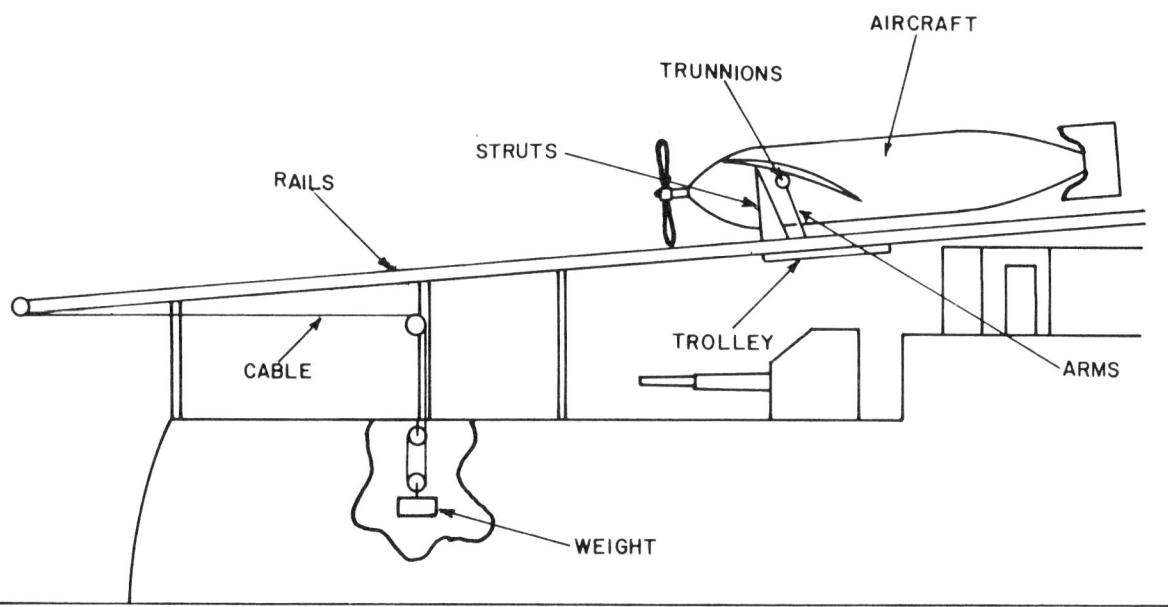

The Sueter/Boothby/Paterson gravity catapult. (By Stephen McLaughlin based on a sketch in Dick Cronin's Royal Navy Shipboard Aircraft Developments 1912–1931)

In February 1911 a patent for a somewhat similar gravity-powered starting device was issued to three pioneers of British naval aeronautics, Captain (later Rear-Admiral Sir) Murray F Sueter (soon to become director of the Admiralty's new-found Air Department), Captain F L M Boothby and H G Paterson. In this scheme, the aircraft was placed on a trolley mounted on two parallel rails angled downward over the quarterdeck. A weight suspended under the rails was connected to the trolley by a line wrapped around a pulley at the stern. Take-off force was generated by dropping the weight into a deck well. Alternatively, the weight was to be suspended from the mast. In its basics, this system was identical to the Wrights' and probably would have worked equally well.

The possibility of using a trolley running on rails, but propelled by mechanical means, rather than gravity, was raised by Winston Churchill, an early advocate of naval aviation. In a minute of 26 October 1914 to Sueter, Churchill, then First Lord of the Admiralty, proposed putting aircraft on towed barges and launching them from 'a trolley on rails actuated by an accelerating windlass'.[3] Although the minute was couched in imperative terms, nothing came of the idea.

Another fascinating variation on this basic idea was suggested by RNAS Wing Captain Oliver Schwann,[4] commanding officer of the seaplane carrier *Campania*. In an 8 August 1915 report to Sueter describing the first take-off of a seaplane on a dolly from the ship's flight platform, Schwann noted that the take-off was successful but risky, because the platform allowed a maximum run of only 152ft (46m). Schwann concluded that 'either outside power must be given the seaplane or it must be given an inclined way to run down.'[5]

'The easiest method of obtaining the pull from outside power,' he continued, '. . . is to have a drogue or other appliance which is being towed overboard by a rope led thro' a block at the bow and secured to the seaplane, and which is released at the same time as the seaplane is released'

A Sopwith Schneider on a dolly taking off from HMS Campania, *probably in August or November 1915. The short length of the flying-off deck led* Campania's *captain to a proposal for applying external power to aircraft. (Courtesy of John B Hollingworth)*

Schwann's proposal for creating external power was never explored, but the flying-off deck was, as he suggested, given an 'inclined way' sloping about 4 degrees from the horizontal when *Campania* was remodelled to extend the length of the deck.

The gravity catapult returned for one last appearance in the early 1920s during British tests of a self-propelled, wireless-controlled flying bomb. Development of such a drone, designated for reasons of secrecy as Aircraft Target, had started in 1916 but had lapsed by the end of the First World War. Revived interest in 1920 resulted in production by the Royal Aircraft Establishment of a craft called the RAE 1921 Target, designed to carry a 200lb (91kg) warhead at a speed of 103mph. After unsuccessful tests of it aboard the aircraft carrier *Argus*, experiments were continued aboard the destroyers *Stronghold* and *Thanet*. They were fitted with trestle-like structures, angled sharply upward, mounted at the extreme bow. The drones were launched from these by the thrust of a large bag of water dropped into the sea.

Eight launches, with varying degrees of success, were made from *Stronghold* between 13 September 1923 and 26 February 1925. It is unclear how many launches, if any, were made from *Thanet*. The gravity devices were replaced by cordite-powered catapults on both destroyers for tests of an improved version of the drone, the RAE Larynx, during 1926–27, but the project was abandoned soon after.

Fly-by-Wire

In the search for alternatives to platforms or launching mechanisms, one idea would recur repeatedly in varied form for many years: the concept of launching and recovering aircraft along a line or lines slung from masts or stretched outboard of the hull.

What may have been the first such suggestion came from USN Captain (later Rear-Admiral) Charles F Pond, commanding officer of the armoured cruiser *Pennsylvania* when in January 1911 Eugene Ely made the first shipboard landing on that vessel. Pond became an instant convert to aviation, but in his official report on the experiment expressed hope that in the future

> There will be no necessity for a special platform. The flight away from the ship may be made either from a monorail or from a stay, and either from forward or aft, but preferably forward, while the return landing may be made on the water alongside . . .

Taking this idea to heart, Lieutenant Theodore G Ellyson, the USN's first aviator, devised a scheme whereby a seaplane would take off from a line stretched from a ship's foremast to its bow. Parallel lines on each side would steady the aircraft until aileron control could take over. Ellyson estimated that the whole procedure – rigging, take-off and unrigging – could be done in 15 minutes.

Eight months later, with the concurrence of Captain Washington I Chambers, *de facto* chief of infant USN aviation, Ellyson conducted a test of this system at Lake Keuka, New York, enlisting the help of pioneer aircraft designer and pilot Glenn Curtiss, the man who had taught him to fly. They erected the three parallel lines on wooden supports extending from a small platform on the beach to a piling in the lake. The Curtiss pusher 'hydroaeroplane' provided for the experiment had a metal-lined groove cut

HMS Stronghold *with her unusual catapult installed for experiments in launching flying bombs*. (National Maritime Museum)

into its single float for a more secure grip on the central line. As a safety precaution, steadying lines were attached to each wingtip, to be handled by men running alongside the seaplane. These proved unnecessary when on 7 September 1911 Ellyson took the aircraft off the line after a 150ft (45m) run.

Chambers publicly praised the success of the experiment, but privately realised that the system was workable only under ideal conditions that would rarely be experienced by a ship at sea. The Lake Keuka test was the first and last, and Chambers turned his attention to development of a compressed-air catapult.

Lieutenant Theodore G Ellyson at the controls of the Curtiss seaplane shortly before taking off on the wire system he devised, 7 September 1911. At left, holding a wingtip steadying wire, is Lieutenant John Towers, US Naval Aviator No 2. (National Archives)

James M Thorp's wire retrieval/launching system. (By Stephen McLaughlin based on a sketch in US Patent Office *Official Gazette*).

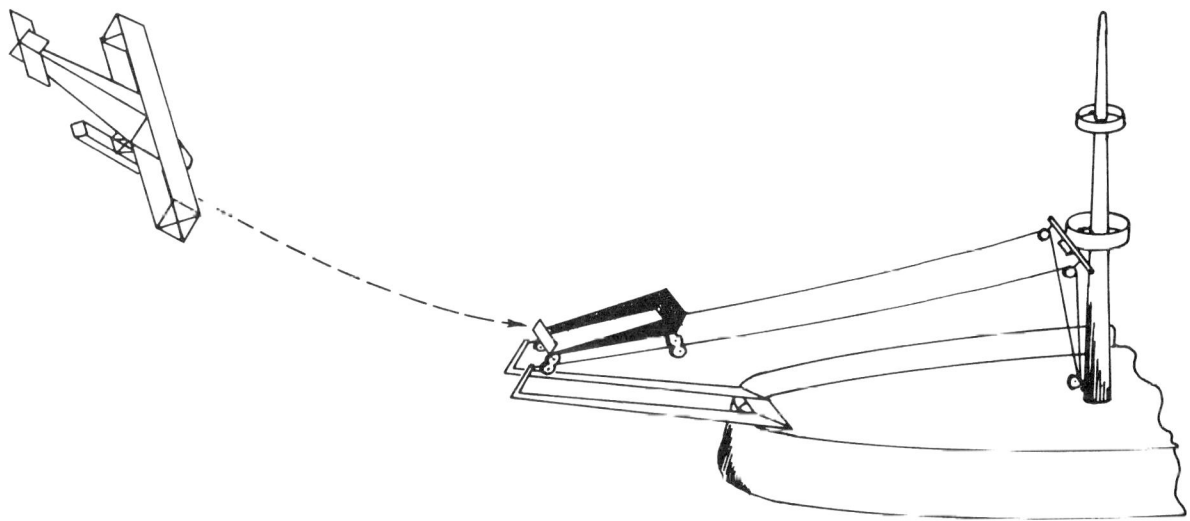

Célestin Adolphe Pégoud in the cockpit of a Blériot monoplane suspended from the Blériot wire system that the aviator tested successfully in 1913. (R D Layman collection)

Sopwith Pup B.5940 after crashing at the Isle of Grain during an unsuccessful attempt to connect with the wire system seen suspended from the pole in the background. (J M Bruce/G S Leslie collection)

Two years later one James M Thorp of Alameda, California, came out with a scheme similar to Ellyson's but added a landing system. Thorp's patent application, filed 1 October 1913, describes a pivoting, hinged support extended from the stern of a ship. On this is mounted a wheeled platform (called a car under the definition still prevalent in 1913) that runs along two parallel lines suspended from a yardarm on an aft mast.

After the support was swung to whatever horizontal and vertical angles desired, the aircraft would alight upon the car; the impetus of this would drive the car up the lines until braked by gravity. For launching, the aircraft would be reversed and hoisted on the car up the mast for a combined power/gravity take-off.

With regard to launching, Thorp's scheme seems at least as reasonable as Ellyson's, although more cumbersome, but his apparent belief that an aircraft could come down safely on the 'car' was nothing short of ludicrous.

The line systems described so far relied partially on gravity, but in 1913 French aviator Louis Blériot devised one that utilised only aircraft engine power. Intended as a means of taking off from short fields, it consisted of a steel cable, slung horizontally between two tall poles, clasped by shearlike hooks mounted on top of an aircraft's fuselage. Revving the engine to full power sped the aircraft along the line until flight velocity was reached, at which point the aviator opened the hooks by pulling a line extending from the cockpit. After a short dip, the craft became airborne. For return, the aircraft was manoeuvred under the line until the hooks touched it, whereupon a spring mechanism responding automatically to pressure opened the hooks to reclasp it.

The system was tested in 1913 by early aerobatic flier Célestin Adolphe Pégoud at a land field near Buc. Flying a Blériot monoplane, he made several successful take-offs and returns on an 80m (262ft) line, a length found to be more than adequate.

This scheme was one of several considered by the French navy as it explored ways of operating aircraft from ships. For shipboard use, the line would be suspended outboard of the hull by booms. Like torpedo nets, it could be run in and out, and even when extended would not interfere with gunnery or normal ship operations. However, it would have required the ship to steam into the wind for aircraft launching and recovery, and only a highly skilled aviator could have made use of it if the ship were rolling or pitching.

Despite these drawbacks, French naval authorities took the concept seriously enough to propose tests of its feasibility. Measurements were taken of the distances between masts of the old battleships *Charlemagne*, *Saint Louis* and *Jauréguiberry*, and the last of these was chosen for an experimental installation. The space between her military masts would have allowed a line 52m (170ft) long, and presumably it would have been rigged high enough to assure a drop of an aircraft well above sea level.

An installation aboard *Jauréguiberry* was authorised on 30 December 1913, but for unknown reasons was never carried out.

French interest in the idea revived briefly after the First World War, and the aircraft firm Société Anonyme Pour l'Aviation et ses Derives designed a tiny biplane, the SPAD 38, equipped with a line-clasping device nearly identical to Blériot's. However, the concept was soon outmoded by the rapid development of the catapult.

British tests of a line system were undertaken by the Royal Air Force at the Isle of Grain air station between March 1920 and February 1921 as a spin-off of experiments in launching and retrieving aeroplanes by rigid airships. In December 1918 it was suggested that the 'possibility of aeroplane returning to ship and securing to a wire' be investigated.

Little is known of these experiments, for no adequate documentation of them has been found. The scheme was very similar to the French plan, using a line rigged outboard of a ship's hull, but differed in that the aeroplane would alight by snagging a loop suspended from it. The tests were carried out by at least one Sopwith Pup, and perhaps by other aircraft, equipped with what was termed the 'Overhead Landing Gear, Grain Type Mk I'. At least one Pup crashed while trying to snag a line stretched between tall poles, and the idea went into limbo.

The major appeal of line systems was that, assuming they could be rigged and unrigged quickly and easily, they would have little or no impact on a ship in terms of weight, space or interference with guns or deck gear. On the aeronautical side, a repeated argument for them, advanced as early as 1918,[6] was that they could eliminate the need for aircraft undercarriages, whether wheel or float, with the resulting loss of weight and drag improving aerodynamic efficiency. This was demonstrated in the 1930s when routine removal of undercarriages from the Curtiss F9C-2 Sparrowhawks operated by the US Navy's airship *Macon* lightened them by 250lbs (113kg), increased their speed by 24mph and permitted them to mount auxiliary fuel tanks, lengthening their range by 50 per cent.[7]

Despite such advantages, it was not until well into the Second World War that a line system was actually installed on a ship and used both experimentally and operationally. Devised by a US Army artillery officer, First Lieutenant James H Brodie, it was originally conceived as a means of operating anti-submarine aircraft from ships without the need for flight decks or catapults.

Conceptually, Brodie's system was identical to the French navy's plan for the Blériot system – a line slung outboard of the hull – although there is no reason to suppose that Brodie was influenced by this, or even aware of it. Brodie's scheme was more mechanically sophisticated than Blériot's; a hook atop the fuselage was a main component, but instead of clasping the line directly it latched on to a sling dangling from a small line-mounted trolley. As in the Blériot system, the aircraft raced along the line until take-off speed was achieved, whereupon the pilot pulled a release lanyard and became airborne. For recovery, the aeroplane was flown below the line and the hook engaged a sling suspended from a second trolley. An arresting brake slowed it to a halt, for whilst the Blériot system relied on friction to stop the aircraft, even the light aeroplanes using the Brodie system were much heavier and more powerful than the Blériot monoplane of 1913.

Receiving permission to test his device, Brodie had an installation rigged at New Orleans in September 1943 – a 600ft (182m) line running between 40ft (12m) booms mounted atop kingposts. Aircraft used in the tests were Taylor L-2s, slightly modified versions of the well-known civilian Cub light aeroplanes, adopted by the US Army for the liaison/observation/spotter role. After several flights to and from this rig were made, it was transferred to the mercantile MV *City of Dalhart*, from which ten flights were made by the heavier Taylor L-4. Despite these successful demonstrations, the US Navy rejected the system as being operable only in fair weather and feasible only for aircraft too small to be useful against submarines.

The US Army, however, became interested in the idea as a means of operating light observation/spotter aircraft during landing operations – an interest heightened by such use of light aeroplanes from a temporary take-off deck on *LST-525* in the Mediterranean. After further tests ashore, a formal Army request for the Brodie system was made in April 1944 and Navy authorisation was given on 8 June for installation of a Brodie rig on *LST-776*, then under construction at New Orleans.

When commissioned on 20 July 1944 this vessel sported tall masts fore and aft between which the line was suspended to port 40ft above the waterline. The masts were braced by a series of diagonal supports placed to starboard. During sea trials in August two Army aircraft flew to the line and subsequently made several successful take offs and recoveries.

In late September *LST-776* joined the Amphibious Training Command at San Diego, where another land rig was set up and training of pilots in its use began. At the same time, a special catapult designed for landing ships was installed on *LST-776* and comparative tests of it and the line were made. The fixed catapult, installed midships and angling diagonally to starboard, was faster at launching but the Brodie system was preferred because it occupied no deck space and could recover aircraft. The catapult was soon removed.

Training ended 30 November and on 3 December

LST-776, *equipped with the Brodie system and the athwartship catapult, under way during experiments in the Pacific in late 1944. One of the light spotting aeroplanes is suspended from the line far forward, another is stowed on the forward deck and a third is perched on the catapult.* (United States Naval Institute)

LST-776 sailed for Pearl Harbor, arriving twelve days later. In preparation for the attack on Iwo Jima, training of Marine Corps pilots on the Brodie system was undertaken on land and at sea. Two aircraft were lost while attempting to alight at sea. A problem arose when it was found that the hooks adequate for L-4s were not strong enough to support the Convair OY-1, the Marine Corps' light observation/spotter. New ones were improvised, but were faulty, two aircraft being lost from them, the first at Saipan.

For the Iwo Jima operation, *LST-776* embarked seven OY-1s and eleven Marine pilots. Trying to beach on the island, she was damaged by fouling wreckage and then colliding with another LST. An OY-1 was flown to shore on 27 February 1945, but another fell into the sea when its hook failed. The others were flown ashore on 1 March.

Later in March *LST-776* proceeded to Leyte Gulf, where Army pilots were trained on the line. On 19 March she sailed to support the Okinawa operation, carrying fifteen Army pilots, nine L-4s (three of them with hooks) and two more in crates. Her 'hooked' aircraft flew observation missions on 26–30 March during the preliminary landing on the Kerama Retto island group and then spotted for Army artillery bombarding Okinawa from an island eight miles distant.

All the aircraft were disembarked or flown off during 2–12 April. *LST-776* ended her aviation career later in 1945 in Manila Bay, where she was training Army pilots for the expected invasion of Japan.

Although Vice-Admiral Richmond Kelly Turner reportedly condemned the Brodie system as 'another hare-brained scheme', it must be considered an outstanding success statistically and mechanically. Hundreds of flights were made to and from it – more than 500 at the San Diego land installation alone[8] – with apparently the loss of only five aircraft (three caused by pilot error, two by faulty hooks) and no loss of life.

But although it was the acme and seeming vindication of the 'fly-from-a-line' principle, it was never used for maritime purposes but solely in the service of ground forces. More damningly, it came too late; by the last years of the Second World War the helicopter had been perfected, and it put paid permanently to line systems as it did to the catapult.

Masts and Booms

An alternative to launching by means of a line stretching from a mast was simply to fly from the mast itself. Perhaps the earliest such proposal was made by *Enseigne de Vaisseau* Lafon, one of the first generation of French naval aviators. In March 1911 the Ministry of Marine asked Lafon and four other aviators to submit ideas for the conversion of the torpedo boat carrier *Foudre* into an aviation vessel. Lafon's idea was to use a yardarm as a launching and recovery point, and, apparently on his own initiative, he conducted tests of the scheme at Châlons. Little information is available on these tests beyond the fact that they took place. Lafon was advocating the scheme, whatever it may have been, again in early 1914, but again no further information has come to light.[9]

Perhaps influenced by Lafon, the French navy did install and experiment with a mast launching system on the battleship *Lorraine* after the First World War. In early 1924 a launching device consisting of three rails was mounted on the starboard side of the tripod mast, the rails angling forward. The idea was that a light aircraft – in this case, an Henriot H29, a modified version of the standard H27 – would fly off the rails, and then achieve flight speed before it hit the water. The rails were about 8.5m (28ft) long, and were 24m (79ft) above the water.[10]

If the trials were successful, the French navy intended to install two such launching devices on each of the tripod-masted battleships, one to starboard, as on *Lorraine*, the second angled to port, positioned slightly below the first. A new, undercarriageless aircraft with a watertight fuselage, allowing it to alight at sea, would also be developed. It was also planned to install shelters and hoisting mechanisms for the aircraft.

The first – and only – trial of this system came on 27 February 1924 in the bay of Saint Raphael on the Côte d'Azur. The pilot, *Lieutenant de Vaisseau* Teste,[11] revved his engine to full power, sped down the rails and into space; now in freefall, the plane had to develop enough speed to stay airborne. It did not. Teste found himself dropping too far; his wheels touched the water, and the plane flipped over. Fortunately, he was quickly rescued.

A report on this trial was submitted in April; it noted, among the causes of failure, that the wind had been light (just over 2mph), and recommended that some force supplemental to that of the aircraft's engine be developed. But the idea was abandoned, and French battleships were eventually equipped with catapults and seaplanes.

One of the most outlandish schemes for shipboard operation of aircraft was put forward by one James B Harriss, a resident of Newark, New Jersey. In June 1917 Harriss applied for a US patent for a method of 'launching and landing of aeroplanes'. Although not purely a mast system, a mast was a key element of it.

In part, Harriss' system was quite reasonable. A pair of rails, hinged at the base, would when stowed be affixed vertically to the mast of a ship. For launching, the rails would be pivoted to a horizontal position; Harriss' sketch shows them resting on stanchions on the forward turret of a warship. In this position, the rails would provide the take-off run for the aircraft.

So far, so good; Harriss' launching method would have been quite reasonable for light aircraft of 1917. Where he got into trouble was in the recovery method. The rails would be raised until they were vertical; in this position, they would extend above the mast that served as their brace, with a pivoting net at the top. Along would come the aircraft, equipped with a sort of bumper on its nose; it would fly straight into the net, whereupon the aircraft would drive the rails down to the horizontal position, its momentum being absorbed as it did so.[12]

The Harriss landing technique has the virtue of novelty, but has no other points in its favour. Hitting an object, even a net, at flying speed would have been a considerable shock to both plane and pilot, and the sort of precise alignment of bumper and net required would have called for remarkable piloting skill, especially if the ship were rolling – a small net a hundred or so feet in the air

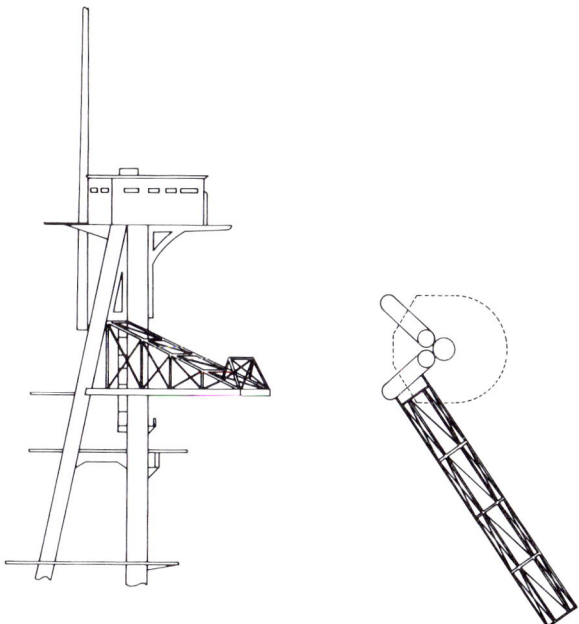

The mast launching system experimentally installed on the French battleship Lorraine *in 1924. (By Stephen McLaughlin)*

Harriss' idea for an aircraft fitted with a forward buffer that would permit it to engage the rails when they were elevated. Wheels on the floats would gradually engage the rails as they were lowered to the horizontal. (By Stephen McLaughlin, based on a sketch in US Patent Office Official Gazette)

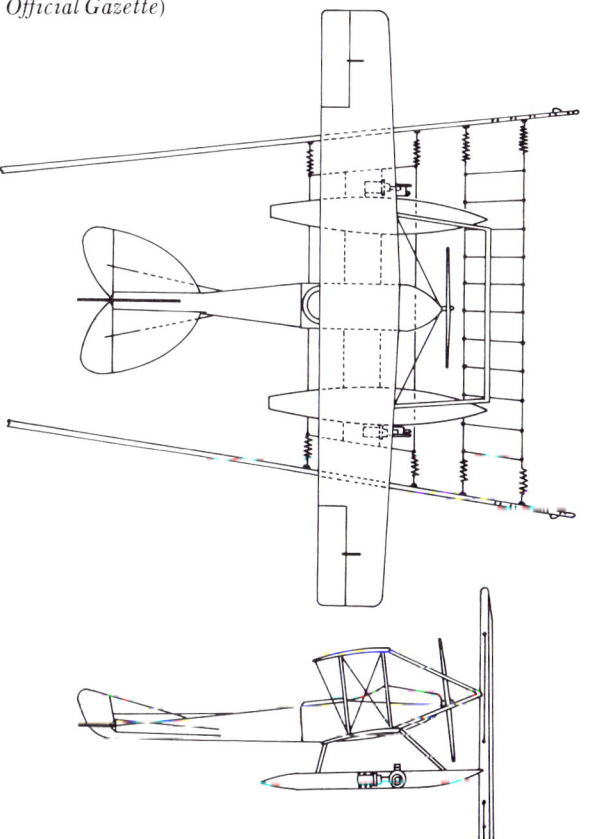

would cover a considerable arc, even if the ship were rolling only slightly. Not surprisingly, nothing ever came of Harriss' scheme.

Arresting/Retarding Systems

Gravity figured in early schemes for landing aeroplanes aboard ships as well as for launching them – just as a 'downhill' run could help impart impetus, an 'uphill' run could help check it.

Stern-mounted inclined platforms were among the ideas advanced in France during the conversion of *Foudre*. *Enseigne* Lafon took this a step further by proposing inclined platforms both fore *and* aft, which would have given the vessel a profile resembling a shallow inverted V broken by a midships superstructure.[13]

In 1912, an Italian aviation pioneer, *Capitano del Genio Navale* Alessandro Guidoni, in drawing up plans for conversion of the old cruiser *Piemonte* into an aviation vessel, included a sharply inclined platform. Forty metres (131ft) long and raised high above the upper deck over the after portion of the ship, it would have served for both launching and landing.[14]

made by early British and American carriers of fore-and-aft lines on the deck. These had a dual purpose: they steadied the aircraft during landing and prevented it from swerving to either side, and, by friction, helped to brake its momentum.

This idea may have been foreshadowed in a US patent application filed 9 March 1911 by Byron C Riblet of Spokane, Washington, for a 'Landing and Launching Platform for Aeroplanes'. It was described as a raised 'skeleton platform . . . comprising a series of closely arranged spaced and separated longitudinal flexible connections extending in the same . . . direction . . .'[15] However, it is unclear whether aircraft were to function from this structure longitudinally in the direction of the lines or laterally across them.

In an added touch to his 1911 proposal for a dual-platformed *Foudre*, the aerially indefatigable *Enseigne* Lafon suggested a braking device he termed a 'rolling carpet' (*tapis roulant*) that would act in the manner of a treadmill to exert opposing force to the velocity of an aircraft alighting upon it.[16] This idea surfaced again six years later when the American marine engineer and naval architect John L Bogert included 'accelerating and retarding belts' in his 1917 design of what was in effect a well

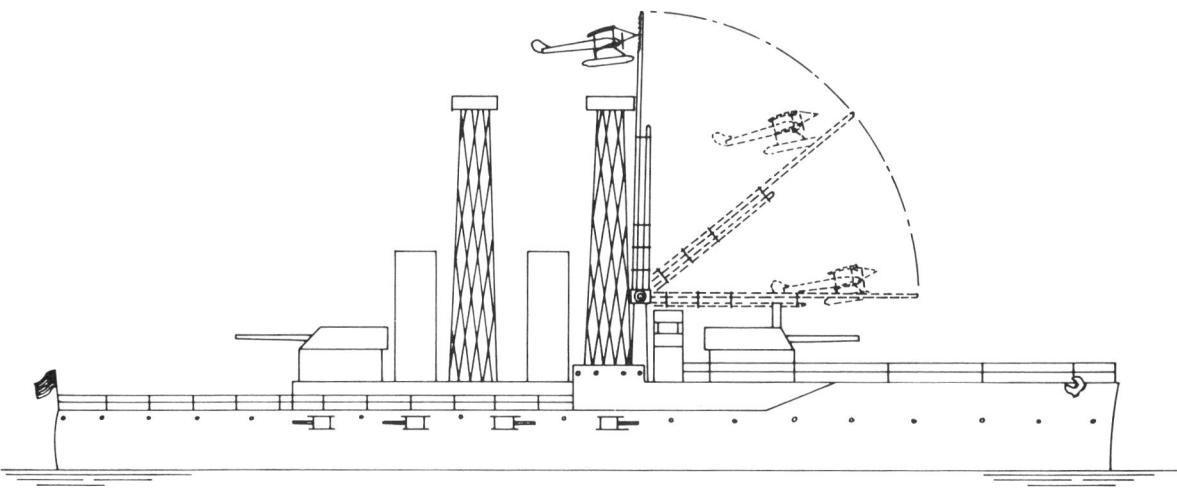

James B Harriss' scheme for an elevating launching/retrieval system. (By Stephen McLaughlin, copied with modifications from a sketch in US Patent Office *Official Gazette*)

Early on, however, it was realised that the minimal braking effects of gravity would have to be supplemented or entirely supplanted by more dynamic means. The concept of snagging an aircraft on a series of weighted lines placed at a right angle to its landing path arose independently and almost simultaneously on both sides of the Atlantic. This method was used for Eugene Ely's landing aboard *Pennsylvania* and with equal success on land by Claude Grahame-White in England and *Enseigne de Vaisseau* Jean Conneau in France. In all of these, the rope lines were suspended a short distance above the landing surface and weighted at each end by sandbags. From these crude improvisations evolved the athwartship arresting concept now long since standard on carriers.

Before this system was perfected, considerable use was

thought-out flight-deck carrier. Bogert also devised an arresting system embodying a grapnel suspended from an aircraft to snag a line, connected to a braking drum, rigged athwart the stern of a ship. He continued to tout both these schemes until as late as 1945.[17]

In March 1921, as the US Navy was converting the collier *Jupiter* into its first carrier, *Langley*, the highly innovative Rear-Admiral Bradley A Fiske (who had patented an aircraft torpedo-dropping mechanism in 1912) opined that a flight deck should be 'a nice soft cushion'. He did not specify of what material the 'cushion' should be made, but some anonymous soul later interpreted it as loose earth – which would have created a very literal 'floating airfield'.[18]

Outlandish as that sounds, it was one of several ideas

mooted at Britain's Royal Aircraft Establishment in 1945 during discussions of alternatives to the metal or wooden flight deck in the still continuing quest for the undercarriageless aeroplane. The stimulus for this was the advent of the jet aircraft. Shipboard aeroplanes had continually grown larger, heavier and faster, but the jet threatened to accelerate this trend enormously. To accommodate these new aircraft, flight decks would have to become longer and stronger, requiring increasingly larger and more expensive ships.

Ultimately, more powerful arresting gear and the steam catapult helped alleviate the problem – at least in part. But these solutions were far from evident in early 1945. Thus the RAE discussions, from which emerged the idea of a flexible or elastic deck on which jets, thanks to their lack of propellers, could land *sans* wheels. The rationale, as ever, was that the elimination of wheels would decrease aircraft weight and improve aerodynamic performance.

Other priorities at the RAE delayed tests of the concept until December 1947, when a simulated flight deck made of a rubberised material was erected at Farnborough. The first successful landings were accomplished by Lieutenant-Commander Eric M Brown, RNVR, in a de Havilland F20 Sea Vampire jet.[19]

ings. It was estimated that their complete elimination could reduce aircraft weight by 9–11 per cent, resulting in a 45-minute increase in flight endurance and an increase in speed of up to 23mph.[20]

These minor gains were far outweighed by the enormous cost that would have been involved in refitting carrier decks and land fields to accept wheel-less aircraft, as well as the cost of developing such aircraft. And so the 'rubber deck' concept represents a technological success doomed by the impossibility of its practical application.

Beyond the Fringe

Amongst the most bizarre and unrealistic early ideas for launching systems were two envisioning the use of centrifugal force. One, by an inventor whose name apparently has been lost to history, was presented to the US Navy Department in 1911. It was a rotating mast that would swing the aircraft around at increasing velocity until flight speed was achieved, when it would be released. Return would be made by reverse, slowing rotation.[21] The proposal apparently did not touch upon the dizzying effects of such a launch on the pilot.

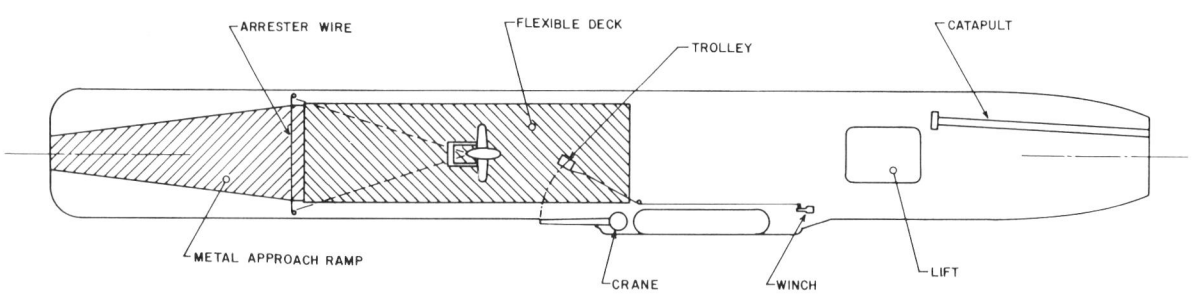

Layout of the flexible deck as installed on HMS *Warrior. (By Stephen McLaughlin, based on a drawing in Norman Polmar's* Aircraft Carriers*)*

Carrier tests were carried out the next year aboard HMS *Warrior*, recently returned from the Royal Canadian Navy. A rubberised platform 190ft long, raised 2½ft above the normal flight deck, was installed just aft of the island superstructure. Aft of the flexible structure was a 150ft long metal approach ramp with a single arresting line at its forward end. An aircraft snagging this line would be halted after a short run along the flexible surface. Then a crane aft of the island would place it aboard a trolley connected by a line to a forward winch, and aircraft and trolley would be hauled off the flexible deck, clearing it for the next landing.

The first landing aboard *Warrior* using this technique was made by Brown on 3 November 1948 in a Sea Vampire. In tests continuing until May 1949 more than 200 landings were made. Mishaps were few and minor – in fact, it was found that an aircraft missing the arresting line would sometimes bounce off the flexible deck and regain flight.

Aircraft in both the land and carrier tests retained their normal undercarriages, which were retracted for land-

A similar scheme was given to the French navy about the same time by a civilian identified only as M Maux. He proposed placing a 15m (49ft) mast in the middle of a 40m (131ft) platform on the stern of *Foudre*. The aeroplane, attached to the mast by a hook under its centre of gravity, would start under its own power, swinging around the mast at increasing velocity and height until flight speed was reached. Unlike the American scheme, the aircraft was not expected to return on the mast.[22]

Even more fantastic was a scheme reportedly attributed to Glenn Curtiss as thoughts about the catapult began to percolate. If Curtiss was indeed the originator, this ordinarily rational man must have taken temporary leave of his senses, for the scheme was a throwback to antiquity's trebuchet or onager: '. . . the plane was to be mounted on a thirty-foot cantilever arm which would be forced to rise quickly [apparently by a spring mechanism] and throw the aircraft in a tangential path from the top of the arc'.[23]

Still another idea from never-never land is ascribed to an unidentified officer of the Russian Black Sea Fleet,

A de Havilland Sea Vampire about to alight on the flexible deck of HMS Warrior *in February 1949. The landing gear is retracted and the aircraft has snagged the single arrester wire.* (Fleet Air Arm Museum)

who conceived of attaching an aircraft to an electromagnet at the end of a boom extending from shipboard. When a speed of 30kts was achieved by the vessel steaming full ahead into the wind, power to the magnet would be cut and the aircraft would zip away.[24] One is left wondering how the electrically sensitive components of an aircraft, especially its all-important engine magneto, could have been insulated against the effects of its attaching device.

In Retrospect

The fact that so many schemes for shipboard aviation were proposed before and during the First World War indicates how widespread was the belief that aircraft were or soon would be useful adjuncts to warships. That several were conceived by naval professionals is a further refutation of the myth that navies were oblivious to the value of aviation.

That being said, there can be no denying the fact that many inventors had more enthusiasm than experience, and most of their ideas were founded on naive assumptions regarding the possible performance of aircraft or their pilots. But before deriding such infeasible ideas, it is well to remember that there were no models, no precedents for shipboard aviation; everything – including ideas like the catapult and flight deck – was an experiment.

There are several conclusions we can draw from all these inventive schemes. One has to do with the fundamental difference between theory and practice – methods that seemed quite reasonable in theory, or were even achieved under controlled conditions, were impractical under operational conditions.

Some of these ideas died because they had only limited applications. The Brodie system is a clear example of this phenomenon. The need to launch and recover light spotting aircraft was simply not that great, so an otherwise successful system vanished into obscurity.

Other ideas, like the flexible deck, were technologically valid, but were impractical from the standpoint of cost-effectiveness. Re-equipping carriers and air bases with flexible landing surfaces, and developing a new breed of aircraft that would take advantage of the reductions in weight brought about by the elimination of landing gear, was simply too expensive to be worthwhile, especially in a postwar era of tight budgets.

Yet the potential benefits of eliminating the landing gear and its mechanism continue to appeal, as shown by British Aerospace's SkyHook proposal – an indirect descendant, perhaps, of the fly-from-a-wire schemes of the teens and 1920s. SkyHook calls for the operation of suitably modified Harriers from ships by means of an articulated crane with a docking mechanism that would both launch and recover the VTOL aircraft.[25] The undercarriage would be completely eliminated, allowing improved aircraft performance.

The counter-arguments remain as always that such aircraft would be unable to alight safely anywhere on land or sea, and that the costs of developing the system and the aircraft required would be expensive. To date, although SkyHook has been successfully tested on land, no navy has adopted it. Yet the idea has a strong appeal, and it may yet enter the operational arsenal of naval aviation.

Notes

[1] Aft platforms were installed in November 1910 on the Hamburg–America liners *Kaiserin Augusta Victoria* and *Pennsylvania* for scheduled take-offs that did not occur and on the US armoured cruiser *Pennsylvania* in January 1911 for Eugene Ely's experimental landing.

[2] The ease with which aeroplanes could fly from these platforms retarded British development of the catapult, with the result that it did not become standard on RN ships until years after it was in common service in the US Navy.

[3] From the papers of Royal Air Force Group Captain Hugh A Williamson, quoted in S W Roskill (ed), *Documents Relating to the Naval Air Service*, Navy Records Society (London 1969).

[4] Schwann is credited with developing the first successful British seaplane. In 1917 he anglicised his name to Swann.

[5] Public Record Office, Adm 1/8430, quoted in Roskill, *op cit*.

[6] See, for example, Charles C Turner, *Aircraft of To-day: A Popular Account of the Conquest of the Air*, Seeley, Service & Co (London 1918).

[7] Richard K Smith, *The Airships Akron and Macon: Flying Aircraft Carriers of the United States Navy*, US Naval Institute (Annapolis 1965).

[8] Vice-Admiral George van Deurs, 'The Navy and the Brodie', *United States Naval Institute Proceedings* 102, No 10 (October 1976).

[9] Francis Dousset, *Les porte-avions français: des origines (1911) à nos jours*, Éditions de la Cité (Brest 1978).

[10] Robert Dumas and Jean Guiglini, *Les cuirassés français de 23,500 tonnes*, Editions des 4 Seigneurs (Grenoble 1980).

[11] A seaplane carrier, *Commandant Teste*, was later named in honour of this pioneer of French shipboard aviation.

[12] United States Patent 1,428,163, granted 5 September 1922.

[13] Dousset, *op cit*. It is interesting to note that the Japanese carrier *Akagi* featured a slightly inclined landing deck upon completion in 1927.

[14] Erminio Bagnasco, *La portaerei nella marina italiana: idee, progetti e realizzazioni dalle origini ad oggi*, Revista Marittima (Rome 1989).

[15] *Official Gazette*, 19 November 1912, US Patent Office (Washington DC).

[16] Dousset, *op cit*.

[17] For details of these schemes, see R D Layman and Stephen McLaughlin, 'The Father of the Flat-top?', *Warship 1991* Conway Maritime Press (London 1991).

[18] In the mid-1960s an elderly employee of the Public Record Office told the senior author of this article of an even odder idea purportedly studied by the Royal Navy in the 1920s: coating the deck of a carrier with glue as a retarding agent. He declined further discussion, with the explanation that the scheme was still under the Official Secrets Act. This may well have been a leg-pull, but the statement was made in all seriousness.

[19] Brown was an ubiquitous Fleet Air Arm test pilot who has flown probably more different types of aircraft than any other person in aviation history.

[20] Norman Polmar, *Aircraft Carriers: A Graphic History of Carrier Aviation and Its Influence on World Events*, Doubleday & Co (Garden City, NY 1969).

[21] Lt Harold Blaine Miller, *Navy Wings*, Dodd, Mead & Co (New York 1937).

[22] Dousset, *op cit*.

[23] Miller, *op cit*.

[24] Norman Polmar, 'The Soviet Aircraft Carrier', *United States Naval Institute Proceedings* 100, 855 (May 1974).

[25] Some of these are discussed in R D Layman and Stephen McLaughlin, *The Hybrid Warship*, Conway Maritime Press, (London 1991).

SUPER-DREADNOUGHTS

The Orion Battleship Family

Keith McBride unravels the convoluted design history of the three classes of British battleships with 13.5in guns – ships which seemed so much more powerful than their predecessors that they were dubbed 'super-dreadnoughts'.

From 1905 to 1908, the British dreadnoughts authorised all carried 12in guns, in a disposition with wing or, later, echelon turrets and, outwardly, did not differ greatly from their progenitor. Behind the scenes, there was concern that the 12in gun might be outclassed, and experience with the 50-calibre Mark XI suggested that any further improvement would require an increase in calibre. British gunnery thinkers felt that high explosive shells were the best sinkers of ships, being able to open up light plating, start fires and damage the target's structure. As early as 1906, designs had been sketched out to carry eight 13.5in/50-calibre guns on 19,500 tons, with fore and aft and echelon turrets in the *Invincible* disposition. The 13.5in was probably chosen because the calibre had been used in the *Royal Sovereign*s of 1889, and much design data existed which could be extrapolated. At the time, this gun was ruled out as having a range beyond that of practical fire control, and the 12in Mark XI was adopted instead.

The 13.5in reappears in 1908, with 'M', a 21,500-ton

The Orion *leads* Monarch, Conqueror *and* Thunderer *on a North Sea sweep some time in 1918.* (IWM)

super-*Neptune*. This project had one foot less freeboard and one foot more draught than her predecessor, and, later, thicker armour. In parallel with her, 'L' of 22,500 tons had her turrets on the centreline, while at some point the guns had been shortened from 50 calibres to 45, possibly as a result of inaccuracy and rapid wear experienced with the 50-calibre 12in. This must have resulted from tests, as none of the 50-calibre 12in ships were yet in service. Like its predecessor of 1889, the new 13.5in fired 1250lb shell, permitting much bigger bursters if desired. The change to a centreline turret disposition avoided the strain and excessive blast damage experienced with echelon turrets; end-on fire was neither sought nor obtained, since the old-type sighting hoods were retained and stops were fitted to the superfiring turrets, precisely to prevent superfiring.

In the meantime the international scene had worsened. Because of the 1906 economic slump and Britain's apparently overwhelming lead in dreadnoughts, only one and a battlecruiser were authorised in 1908–09, when the Germans authorised four, the Austrians four and the Italians one. The 1909–10 programme was set at four, of which one battleship and the battlecruiser were to carry the '12in (A)' – the new 13.5in 45-calibre.

Then a naval 'panic' began. The Government convinced themselves that the Germans could have seventeen dreadnoughts by the spring of 1913, instead of the officially announced thirteen; the Conservative opposition said there could be twenty-one, and public meetings, speeches and petitions abounded, all reflecting the steady build-up of tension between Britain and Germany.

Eventually, the Government decided on four ships, plus four 'contingency' ones, for 1909–10, five for 1910–11 and five for 1911–12, apart from any ships that the 'Colonies' might provide. The 'contingency' ships were duly authorised a few months later; the distinction between them and the original four helped Asquith, the Prime Minister, to manoeuvre between the rival wings of his party. As it happened, the Germans *did* have the thirteen ships which they had announced, and no more, when spring 1913 arrived.[1]

The Orion *Class*

The 1909–10 ships initially comprised two modified *Neptune*s, one 'L' battleship and one 'C–V' 'Cruiser' in parallel with her. 'L', which was designed by Ernest Mooney and A W Worthington, had her five turrets disposed: two 'superfiring' forward, one amidships, between the boiler and engine rooms, and two 'superfiring' aft. Like most big British ships of the day, she was a 'long forecastle' design, of 22,500 tons 'Legend' or 'Navy List' displacement, which included all stores, equipment, ammunition and 900 tons of coal. As with previous dreadnoughts, the 4in secondary armament was in the superstructure – eight forward and eight aft. Perhaps because the Germans had gone from 11in to 12in guns in 1908, armour was increased to 12in/9in/8in on the sides, with 11in turret faces. Apart from the normal angled armoured bulkheads, there was an additional one right forward. One hopes this was not to protect the wardroom,

Thunderer *fitting out at Thames Iron Works: 'A' and 'B' turrets and the top of the conning tower are visible.* (CMP)

Thunderer *moving away from Thames Iron Works to be fitted out downstream at Dagenham, cheered by the workforce that built her and their families.* (CMP)

Thunderer *leaving Dagenham for Sheerness for compass adjustment.* (CMP)

which in the 1909–10 ships was forward. Eighteen boilers provided steam for 27,000shp, which was expected to give the usual 21kts with ease. On 12 May, 1909, when the 21kt design was laid before the Board in preliminary form, it was accompanied by a 23kt one of 24,250 tons and 7 per cent more cost. This was rejected by the Board of Admiralty over the strong objections of Sir John Fisher, who had his protest that the Germans were going for higher speeds and that the extra cost would be well worth while, recorded in the Board minutes.[2]

The idea of a fast, fully-armoured battleship had appeared in the 'X4' project of 1905, and was to become reality with the *Queen Elizabeth*s of 1912–13. The final 'L' design was laid before the Board on 30 July 1909. In his covering minute, Philip Watts the DNC wrote:

> If the *Colossus* (which is 510ft × 85ft × 27ft × 20,000 tons and estimated to cost £1,800,000) be made to carry 12in (A) guns instead of ordinary 12in guns, everything else being kept the same, her dimensions would become 525ft × 86ft × 27ft × 21,000 tons, cost £1,860,000.
>
> If, then, the armour and protection be increased to that provided in design 'L', her dimensions will become 530ft × 87ft × 27ft × 21,500 tons, cost £1,910,000.
>
> If, further, the ship be modified and the turrets all placed on the middle line as proposed, we obtain design 'L', her dimensions being 545ft × 88ft × 27ft × 22,500 tons and estimated cost £2,000,000.

As usual, he added a list of docks which could accommodate 'L'; there were seventeen of these, plus a floating dock and six more docks building or projected.

One bad feature shared by all the eight 1909–10 capital ships was the placing of the tripod mast aft of the fore funnel. The reason was apparently the desire to use the tripod legs as supports for the 50ft boat derricks. Jellicoe, who was Controller from 1908 to 1910 was apparently responsible for the revival of this bad feature, which had been tried in the *Dreadnought* and dropped in the *Invincible*s and the 1906 to 1908 ships because of bad smoke interference with fire control. The *Lion*s, with their great boiler power, were rectified at the huge contemporary cost of £60,000 each, but the battleships had to suffer; at Jutland, the *Conqueror*'s captain had to rely entirely on information passed down from the control top. The larger boats were stowed in a well between the funnels, partly overhung by blast screens. Fitting in the new 21in torpedoes and their tubes also led to a lot of problems, particularly as the torpedo designers had not yet finished their work. These weapons were regarded as of great value; as of 1909, accurate gunfire was not expected at extreme ranges, and it seemed possible to strike an initial blow before the gun battle was fully joined.[3] The 13.5in turrets were not much larger than those for the 50-calibre 12in, and were much better disposed, while the guns themselves were expected to be superior in everything except muzzle velocity. However, they weighed 76.1 tons each against 66.5 for the 12in. All but one of the class had the new swashplate hydraulic engines, which trained their turrets more smoothly than previous systems. However, the *Conqueror*'s gun mountings and hydraulic system, and those of the later *Ajax* and *Benbow*, were Mark III, III* and III**, made by the Coventry Ordnance Works, a new firm set up by the big steelmakers, which the Admiralty was 'feeding', to bring pressure on the Armstrong/Vickers ordnance duopoly. To outflank those firms' patents, the Coventry designers had to use six-cylinder hydraulic engines and other less-than-ideal devices. Outwardly, the Mark III series mountings could be distinguished from the usual Mark II series by their curved turret sides.

Armour was thicker and more extensive: 12in on the lower side tapered to 8in 4ft below the waterline, with a 9in upper belt and an 8in strake from the fore to the after barbette, which previous ships lacked. As in the *Colossus*, there were three engine rooms, one containing the machinery for the two inner shafts, and three boiler rooms. Up to 3300 tons of coal and 850 tons of oil could be carried. Since the number of guns remained the same, the complement was assessed at 717 against 708 – which proved a vast underestimate for both ships. Metacentric height varied from 5.1ft at extreme light draught to 6.7ft at extreme deep load.

The designers were fairly happy about the watertight integrity of the class; if all compartments above the lower protective deck and before the fore armoured bulkhead were open to the sea, they would trim about 18in by the head; if the same happened aft, about 3.5ft by the stern, the effect being practically negligible.

Of the four ships, the *Orion* – the 'lead ship' – was ordered from Portsmouth Dockyard, the *Monarch* from Elswick – the 'old firm' – the *Conqueror* from Beardmore's and, after much thought and the visit of a delegation including local MPs to Reginald McKenna on 12 August 1909, the *Thunderer* was ordered from the Thames Iron Works, the last of the Thames-side yards.

There had been troubles over wages and rates, and grave doubts whether this Blackwall yard was capable of building a dreadnought – when they launched the *Duncan* class *Cornwallis*, she careered across the basin and hit the wall. The Iron Works had been ordered to strengthen their slipway before being given the *Black Prince* contract in 1902, but had only done the minimum necessary. Nevertheless, they were given the *Thunderer* and completed her without mishap. George Macrow, the firm's great naval architect,[4] boasted that they had defied all forebodings, but it proved a pyrrhic victory; the firm went bankrupt shortly after. The special fitting-out wharf prepared for the *Thunderer* at Dagenham proved very useful to Ford's when they moved there in the twenties and it is still called Thunderer Jetty.

In general, construction went smoothly, and the ships were completed in 1911–12. Hulls averaged 300 tons light – no-one quite knew how – and, apart from an initial tendency to roll, cured by enlarging the bilge keels, and the smoke interference problem, they were well-liked. They steamed well and were good sea-boats. With her Pollen fire control outfit[5] and her very accurate 13.5in guns, the *Orion* speedily became the best shooting ship in the fleet, but this only lasted nine months. Then the *Thunderer*, with the inferior Admiralty equipment, but with only the second director afloat, arrived. In a test

A good portrait of Centurion, *showing the revised layout of the* KGV *class; probably taken just prewar.* (CMP)

against targets towed at 12kts on a steady course at 9000yds range, the latter scored six hits to the *Orion*'s one. This was taken as conclusive proof of the superiority of the director and the adequacy of the Admiralty equipment. The test evidently represented the Admiralty idea of how battles would be fought. The new ships were so different from the 12in ones that they got the semi-official designation of 'super-dreadnoughts'.

The King George V *Class*

For 1910–11, four 'L1' battleships and one 'cruiser' (not yet termed a 'battlecruiser') were agreed. At first, the battleships were virtually repeats of the *Orion*s, the only significant change being that those 4in secondary guns in the fore superstructure were to be protected by 3in armour; a parallel change was made in the cruiser *Queen Mary*. This added some topweight, and the ships were lengthened by 10ft to carry this; beam was 1ft more; draught was left unchanged to avoid docking troubles.

A social change was that the officers' accommodation was moved back to the traditional position aft while the sailors were moved forward; the change made in the *Dreadnought* and succeeding ships had brought little advantage. In practice, most of the officers were no nearer to their action stations, the *Dreadnought*'s thieves had made full use of their opportunities, and the sailors in

Audacious, *just before the outbreak of war.* (IWM)

general felt that the change had been made to spare the officers the vibration of the quadruple screws, while Jolly Jack could be expected to put up with anything! Tradition was now restored, and everyone was happy.

Initially, the same main armament had been planned as for the *Orions*, but tests of the 13.5in Mark V had shown it to be much stronger and harder-wearing than expected. In January 1911, the Director of Naval Ordnance advised Watts, the DNC, that it would stand 20 atmospheres pressure instead of the designed 18 and have a life of 450 rounds. This technical bounty could be used to fire heavier shell, fire existing shell at higher velocity or use MDT, a more advanced propellant. The gun breech chamber was too large for MDT and tests showed heavier shell to be the best proposition. A 1500lb shell proved inaccurate, but a 1400lb one gave a useful improvement in 'remaining velocity', especially at long ranges, and was adopted. The changes to loading arrangements, plus the extra weight of shell, came to about 13 tons per turret, and on 11 March authority was given for the Board Margin to be drawn on for the *King George V*s and *Queen Mary*. The performance figures for the new shell were given for ranges of 4000, 8000 and 12,000yds, 'short', 'medium' and 'long' range by 1911 ideas. Extreme range of the 13.5 was between 23,000 and 24,000yds.

Another reversion, which was rather more than cosmetic, was that of the positions of the fore mast and fore funnel. The original design had the *Orion* arrangement, but in late 1911 orders were given that the next class along, the *Iron Dukes*, should have their upperworks arranged as God intended. This undoubtedly resulted from the departure of Jellicoe from the post of Controller, after Winston Churchill and Sir Francis Bridgeman became First Lord and First Sea Lord, respectively. Apparently someone asked whether it was too late to change the *King George V*s. Ernest Mooney was consulted; he replied on 6 December that:

> The same change in the positions of funnels and masts &c to those proposed for 1911–12 could be made in the *King George V* class, but would probably cost not less than £20,000 per ship and might possibly result in some delay[6] in completion.
>
> If it is desired to make these alterations, action should be taken as soon as possible, as boilers are now being fitted into *King George V* & *Centurion*, and the cost of alteration will increase from day to day from now onwards. The alterations would involve carrying 45ft instead of 50ft steam pinnaces; 8 boats of the latter type are already on order for the 4 ships of the *King George V* class.

Philip Watts approved, as did the Board members; Bridgeman said 'it is fully concurred in and is most necessary'. He thought it possible to squeeze the 50ft boats in as they need not be in place in action. Churchill added on 20 December that: 'It is very satisfactory that these marked improvements can be introduced into the 1910–11 vessels. I approve the changes & the expenditure. Every effort not inconsistent with reasonable economy must be made to avoid delay in completing the ships.' With the growing importance of fire control, the reduction of smoke interference was vital. By 10 January 1912 it had been found possible to retain the 50ft boats.

The change involved considerable internal rearrangement, and changes to the conning tower and bridge

Audacious sinking: the sternwalk is still just above water and the crew are being transferred. The photographs, of which this is part of a sequence, were probably taken from the Olympic, *despite strenuous efforts to prevent photography.* (IWM)

The new Iron Duke *and* Marlborough *with three of the* 'KGV's, *probably at Spithead in July 1914.* (IWM)

layout. The former mast served as the base of a derrick post.

Fire Control – An internal change linked with this was the provision of Pollen fire control equipment. Since the turn of the century, scientific fire control equipment had been under development in Britain and elsewhere. Two parallel systems were being tried by the Royal Navy; the official RN system, and the 'Aim Control' (AC) or 'Argo' system invented by Mr A H Pollen, manager of the Linotype Corporation, who had a keen interest in the Navy, and, of course, a deep knowledge of precision manufacture. He had appreciated the complexity of the problem of firing from one moving ship at another, and had developed a 'Table' for obtaining data on the moving target, and a 'Clock' or computer, for converting this data into deflection, to take account of the lateral movement of the target, and 'Rate', to allow for changes in range. Both of these items changed at continually varying rates, and the Pollen apparatus was necessarily complex and expensive.

The first working Pollen installation was fitted into the *Orion*, which also had a 15ft base rangefinder instead of the usual 9ft one. It was decided to fit the Pollen equipment to all five capital ships of the 1910–11 Programme. A revolving hood was fitted on the top of the conning tower, and the class was first fitted with pole masts, as a big control top was not considered necessary. Later, first stiffening flanges and, finally, full tripods had to be fitted. Unfortunately, during 1912, for reasons which are not entirely clear, the Admiralty fell out with Mr Pollen, and the 1911–12 and later ships were fitted with the inferior Dreyer system, partly pirated from the Pollen one.

There were considerable troubles over gun mountings, especially the *Ajax*'s Mark III*, and an effort was made to standardize fittings and equipment as much as possible; there had been trouble over this with the *Orion*s. Special precautions were taken to ensure the correct alignment of the turret roller paths, for gunnery reasons. Gyro compasses were also introduced and were valuable for gunnery as well as navigation.

Apart from these points, the 1910–11 ships were much delayed by a fierce engineering strike. Trials went reasonably well, though the name ship ran hers in a gale, which revealed many leaks in the deck, and her galley boiler spat fire at its operators. Incidentally, the breadmaking plant was cannibalised from pre-dreadnoughts. The *Centurion* steamed well, but vibrated, and her 30-hour trial at 18,500shp ended at 0545 on 10 December 1912 when she collided with a merchant ship. Apart from the usual 8– and 30–hour trials, a special 4–hour one at 31,000shp was required for the class, and 21.5kts or more attained.

Names – There was a very old tradition that the first First Rate built in a new reign should bear the name of the new sovereign: *Royal Charles, Royal Anne, Royal George*, etc. The first battleship built in the preceding reign had been the *King Edward VII*, but many reactionaries complained that *Royal Edward* would have sounded so much better. Logically, the first of the 'LI's should have been the *Royal*

George, but memories of events at Spithead in 1782 were too strong, despite a later *Royal George* having had a successful career. The names of the other ships were those of old ironclads recently disposed of, the *Centurion* taking that of an 1889–90 second class ship. In service, they were not reckoned such good sea-boats as the *Orion*s: possibly their greater length and equal draught made them roll more.

The Iron Dukes

Throughout the period from 1904, there had been an undercurrent of opposition to the use of such small guns as 4in for the secondary battery, and this was reinforced by the growth in destroyer size and the range and speed of torpedoes. It was only about 1910 that the Royal Navy realised that daylight destroyer attacks were likely during a fleet action, and in that year, Rear-Admiral Mark Kerr, one of the naval thinkers of the day, wrote to Philip Watts about various design matters, saying, *inter alia*:

> . . . as you know, I am the originator of that theory of torpedo defence which does away with the secondary battery, substituting shrapnel in the primary armament. Our neighbours use their torpedo flotillas in fleet actions and it would be embarrassing to have to take one's guns off the target for this purpose. Many are in favour of keeping the 6in guns to 'hail' on the enemy if met at close range in poor visibility; they claim (and the Japanese applaud them) that the showers of spray from 'shorts' will hamper the telescope sights . . .

No doubt there were other suggestions, official and otherwise, on the same subject.

According to Mark Kerr, Watts then prepared alternative plans for the forthcoming 1911–12 battleships, so that he had something in hand when the First Lord, Reginald McKenna, asked him if he could put 6in batteries into the 1912–13 ships. The story according to the Ships' Covers (three of them for this class), is more complex; at the outset, 'Battleship 'LII' was simply 'LI' with all her 4in guns behind 4in armour and four broadside tubes instead of two and a stern tube, this resulting from a study by the newly-established Naval War College. These changes brought her up to 560ft × 90ft × 27.75ft; 23,750 tons Legend, costing £2,050,000.

This question of secondary armament protection, and indeed of the very existence of secondary armaments, was fiercely discussed at the time. On one occasion the Institution of Naval Architects discussed it, and a sick member had himself carried into the lecture hall on a stretcher to make his contribution. The 'protectors' argued that the guns might be disabled by long-range gunfire, or even small shell or machine-gun fire from the attacking torpedo craft; their opponents said that it would be impossible to protect the secondary guns against heavy fire while any thinner protection, even shields, would simply explode shells that might otherwise have gone their way harmlessly. In America, they thought seriously about retracting secondary guns into the ship during the

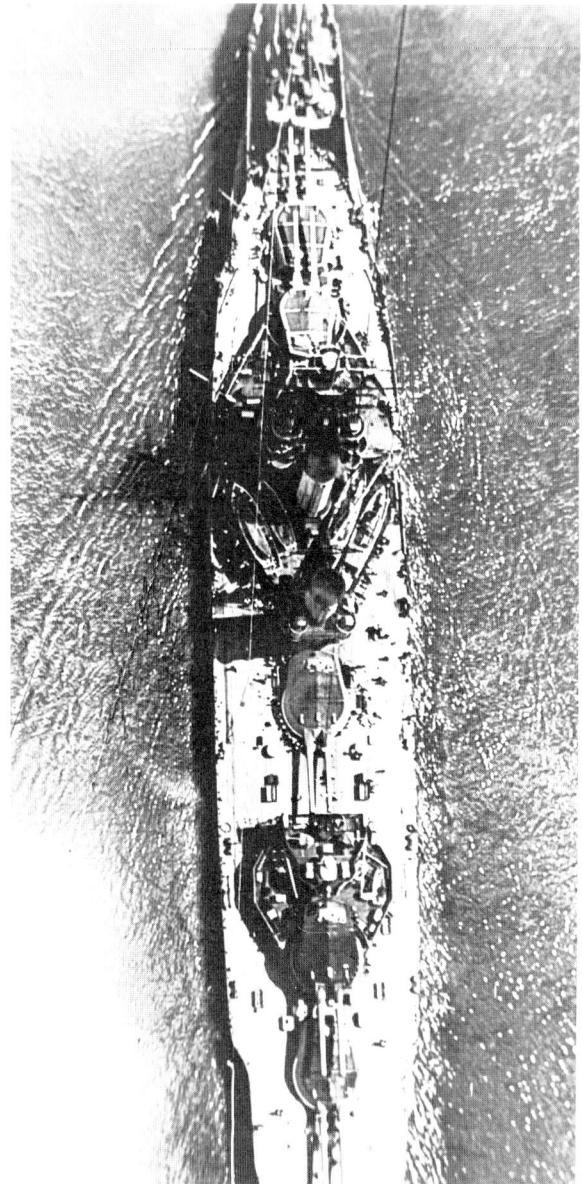

An unusual overhead view of an unidentified Iron Duke, *showing the layout of deck and superstructure.* (CMP)

course of heavy gun action, or, at the other extreme, abolishing secondary guns entirely. This is linked up with the 'all or nothing' argument about armour, and in the *Nevada* and her successors, the secondary guns got 'nothing'.[7]

Experience at Jutland was that secondary guns behind 6in armour suffered badly from heavy gunfire; it was the Germans' good luck and the British 4th Destroyer Flotilla's bad, that the German Battle Squadron I which they encountered had its secondary batteries and their controls intact. Improbably, the only people to try protecting their secondary guns adequately were the Italians, the apostles of light armour, who, many years later, used 11in face plates in their *Littorios*.

Table 1: *Super-Dreadnoughts, Projects and Comparative Foreign Ships*

	'L' (21kt)	'L' (23kt)	Orion	King George V ('LI')	'LII'	'LIII'	'MIII'	'MIV'	'MV'	Iron Duke	Erin	Canada	König	Texas	Bretagne
LBP (ft)	540	575	545	555	560	575	580	580	580	580	525	625	573 (LWL)	565	541
LOA (ft)	576	600	581							622.75	559.5	661	575.75	573	544.5
Breadth ext (ft)	88	88		89	90	90	90.5	90.5	90	90.5	91.6	92	96.75	95	88.6
NLD fore perp	27.5	28	27.5	27.5	27.75	28	28	28	28	?29.5	28.67	29 (Mean)	27.25	?28.5	28.54
NLD aft perp	27.5	28	27.5	27.5	27.75	28	28	28	28	?28.5	28.15				29.85
Disp NLD (tons)	22,500	24,250	22,500	23,000	23,750	24,500	25,000	25,000	24,750	25,000	22,940	28,600	25,390	27,000	23,940
Sinkage (tons per inch)	82	86	84	86	88	90	91	91	91	?91	83	101	91.44		
Freeboard NLD (ft)															
fore	26	26	26	26	26	26	26	26	26	?24			22		21.5?
min	16.5	17	16.5	16.5	16.5	16.5	16.5	16.5	16.5	?16.5			9.5		
aft	17.5	18	17.5	17.5	17.5	17.5	17.5	17.5	17.5	?17			13		13.8?
Deep load draught	31.5	31.5	31.25	31.25	31.5	31.5	31.5	31.5	31.5	33.5/32	31.5				
Depth amidships	44	45	44								43.5				
Axial gun heights															
forward	?	30.5/40.5	30/40	30.5/40.5	30.5/40.5	30.5/40.5	30.5/40.5	30.5/40.5	30.5/40.5				26/35		
amidships	32?	23.5	24	24	24	24	24	24	24				27		
aft	23	32/22.5	32.5/23	32.5/23	32.5/23	32.5/23	32.5/23	32.5/23	32.5/23				27/19		
SHP at NLD	27,000	35,000	27,000	27,000	28,500	30,000	30,000	30,000	30,000		26,500	37,000	31,000	28,100	29,000
Coal NLD (tons)	900	900	900	900	900	900	900	900	900		900	1150	850		900
Bunker max (tons)	3000	3100	3300	3000	3000	3000	3000	3000	3000		1900/2120	3300	3600	2892	2680
Oil max (tons)	850	900	850	850	850	850	850	850	850		800	520	700	400	300
Complement (private ship)	717	?	717	759	759	759	759	759	?759		1064 (flag)	1167	1033	952	1193
Armament	10–13.5in	10–13.5in	10–13.5in	10–13.5in	10–13.5in	10–13.5in	10–13.5in	10–13.5in	10–13.5in		10–13.5in	10–14in	10–12in	10–14in	10–13.4in
(rounds per gun)	(80)	(80)	(80)	(80)	(80)	(80)	(80)	(80)	(80)		(?)				
	16–4in	16–4in	16–4in	16–4in	16–4in	16–5in	16–6in	16–6in	12–6in		16–6in	16–6in	14–5.9in	21–5in	22–5.5in
	(150)	(150)	(150)	(150)	(150)	(150)	(150)	(150)	(150)		(150)				
Torpedo tubes	2–21in	2–21in	2–21in	2 beam, 1 stern	4 beam	4 beam	4 beam	4 beam	4 beam		4–21in	4–21	1 bow, 4 beam	4–21in	4–18in
(torpedoes)	(?)	(?)	(?)												
ARMOUR (ins thick)															
Side height (ft awl)	16.5	17	16.5	16.5	16.5	16.5	16.5	16.5	16.5				13		
Side depth (ft bwl)	4	4	4										5.6		
Height of protective deck (ft awl)	2	2	2										1.64		
Side forward	8	8	8								6		8	6	
Side amidships	12/9/8	12/9/8	12/9/8	12/9/8	12/9/8	12/9/8	12/9/7	12/9/8	12/9/7		12/9/8/5	9/4.5(?24)	10/13.8	12	9.84
Side aft	Nil	Nil	Nil	2.5	4/6	4/6	4/6	4/6	4/6		4		6	6	

SUPER-DREADNOUGHTS

Bulkheads fwd	6/5	6/5		4/6	4/6	4/6	4/6	4/6	?4.5/3	10/8	9/6	
Bulkheads aft	10/9	10/9		10/2.5	4/6	4/6	4/6	4/6	?4.5/3	10/8	9/6	
Barbettes	10/9/6/2	10/9/7/6/5/4	10/9/6/3/2				10/3		10/4	13.8	12	10.6
Turret shields	11	11	11				11		10	11.78	149.84 to 15.7	
CT forward	11	11	11				11		11/6	13.8	12	11.8
Director/CT aft	3	3	3				6		6	8		
Signal tower	11	11	11				11					
CT tube fwd	5/1	5	5/1				5/3					
CT tube aft	3	3	3				4					
Machinery	Nil	Nil	Nil				2		Nil			
Magazine/shell rooms	2/1	2/1	2/1				2/1		2/1.5	1.5T		
Uptakes/vents	1.5/1	1.5/1	1.5/1				1.5/1					
Upper deck	1.5	1	1.75/1				2/1.75/1.5		1.5 1 fore & aft			0.5
Main deck fwd and aft	1.5	Nil	1.5				1.5		1.5 aft only			
Middle deck	1/2.5/1	1	1				1		1			0.4
Lower deck fwd and aft	4/3	2.5/1	2.5/1/3/4				1/2.5		2 fwd, 4 aft, forecastle over battery	2.5/4.7		1.2
												2.75/1.5

WEIGHTS (tons)

Water (10 days)	64	64	64				1	70	70			
Provisions (4 weeks)	41	41	41				70	45	45			
Officers' stores	43	43	43				45	45	45			
Men and effects	92	92	92				45	100	100			
Masts and rigging	107	115	107				100	120	120			
Anchors and cables	128	140	128				120	145	145			
Boats	50	50	50				145	50	50			
WO stores (4 months)	95	100	95				50	100	100			
Torpedo net defence	50	55	50				100 75	75	75			
Armament	3950	3950	4000				4360/4380	4322	4450	3102	2582	4207
Machinery and stores	2460	3000	2420				2580	2580	2580	2703	2390	2323
NLD coal	900	900	900				900	900	900	850		900
Armour	6400	6900	6460				7750	7650	7750	4202	8965	7562
Hull	8000	8700	7950				28560	8410	8500	10,118	8063	8482
Margin	100	100	100				100	?100	100	7595 ?2102		
Total NLD displacement	22,480	24,250	22,500				25,000	24,750	?25,000	25,470	27,000	23,560
Cost £million	£2.0	£2.15	£1.96	£2.0	£2.05	£2.1	£2.15	£2.13	£2.15	£2.4	?£2.2	£2.6

Notes

NLD = Normal load draught, sometimes called Navy list draught or Legend; awl = above waterline; bwl = below waterline

WARSHIP 1993

Drawings of the 'MIV' design. (drawn by John Roberts from originals in the National Maritime Museum).

Profile (some time before 1911 when the position of the mast changed). Twelve 6in, two of which are at an upper level and probably unprotected, are shown, together with eight 12pdrs; the latter were removed at the end of 1912. There is no main mast as yet.

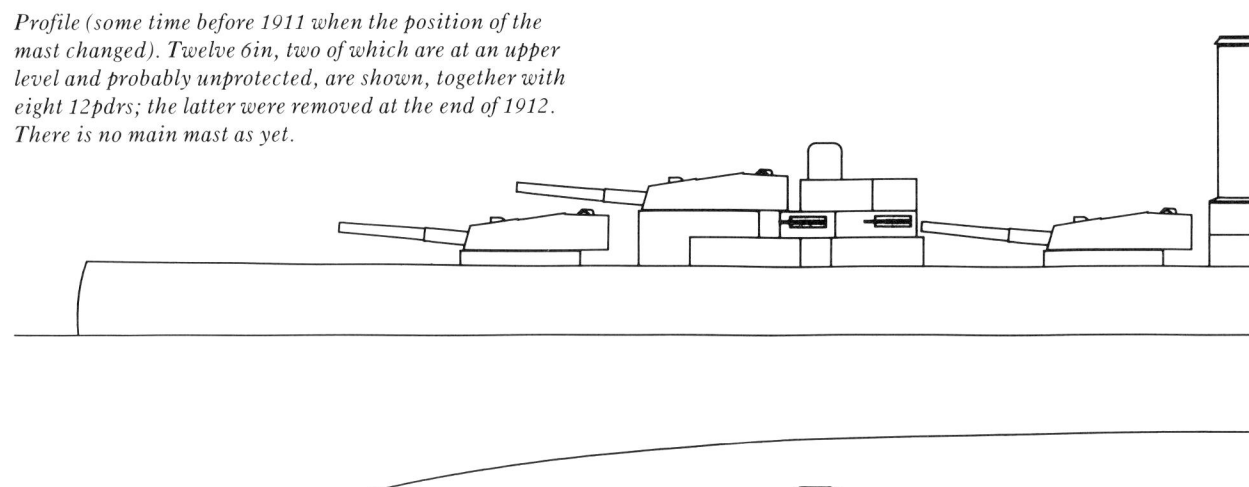

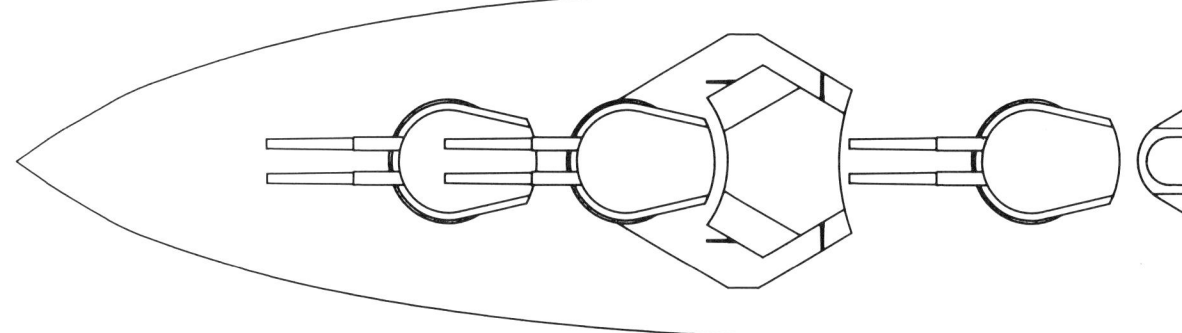

A forecastle deck plan.

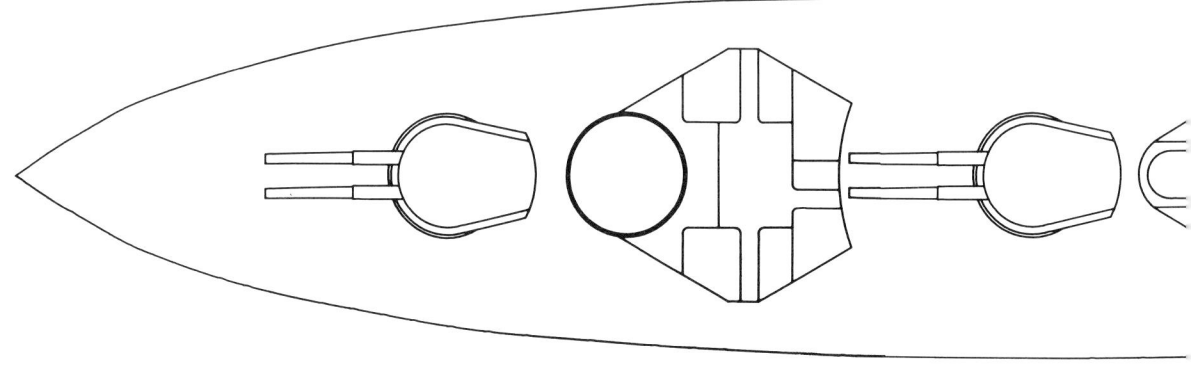

An upper deck plan.

SUPER-DREADNOUGHTS

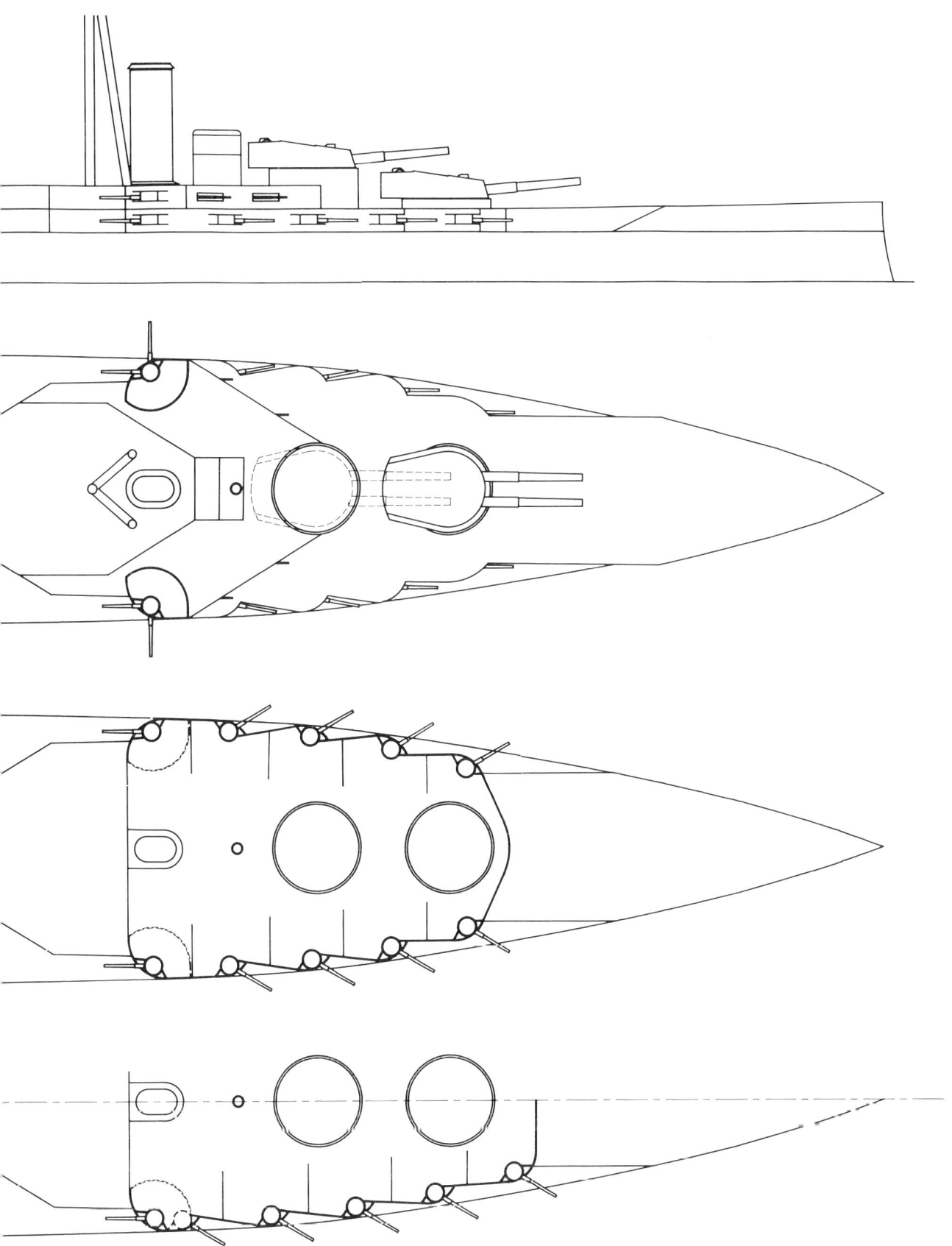

Jellicoe firmly believed 4in guns to be adequate against destroyers, but he had had a 5in gun designed about 1906–7 by the Coventry Ordnance Works on hearing that 'the American dreadnought' (the *Delaware*) was to have them; he was overruled at the time. Nevertheless, it was proposed for 'LIII', sixteen 5in sending dimensions up to 575ft × 90ft × 28ft; 24,500 tons, cost £2,100,000. This reflected another argument as to what was needed to stop a destroyer quickly and at night. A bigger gun hit harder, further off, but less often, meant more topweight, bigger gun crews and to some extent bigger ammunition spaces. The 5in was again rejected, and the new 45-calibre Mark XII 6in proposed. It weighed 6.75 tons and required 11 men against seven; moreover its greater weight meant a new design series, with the 6in at main deck level for stability reasons.

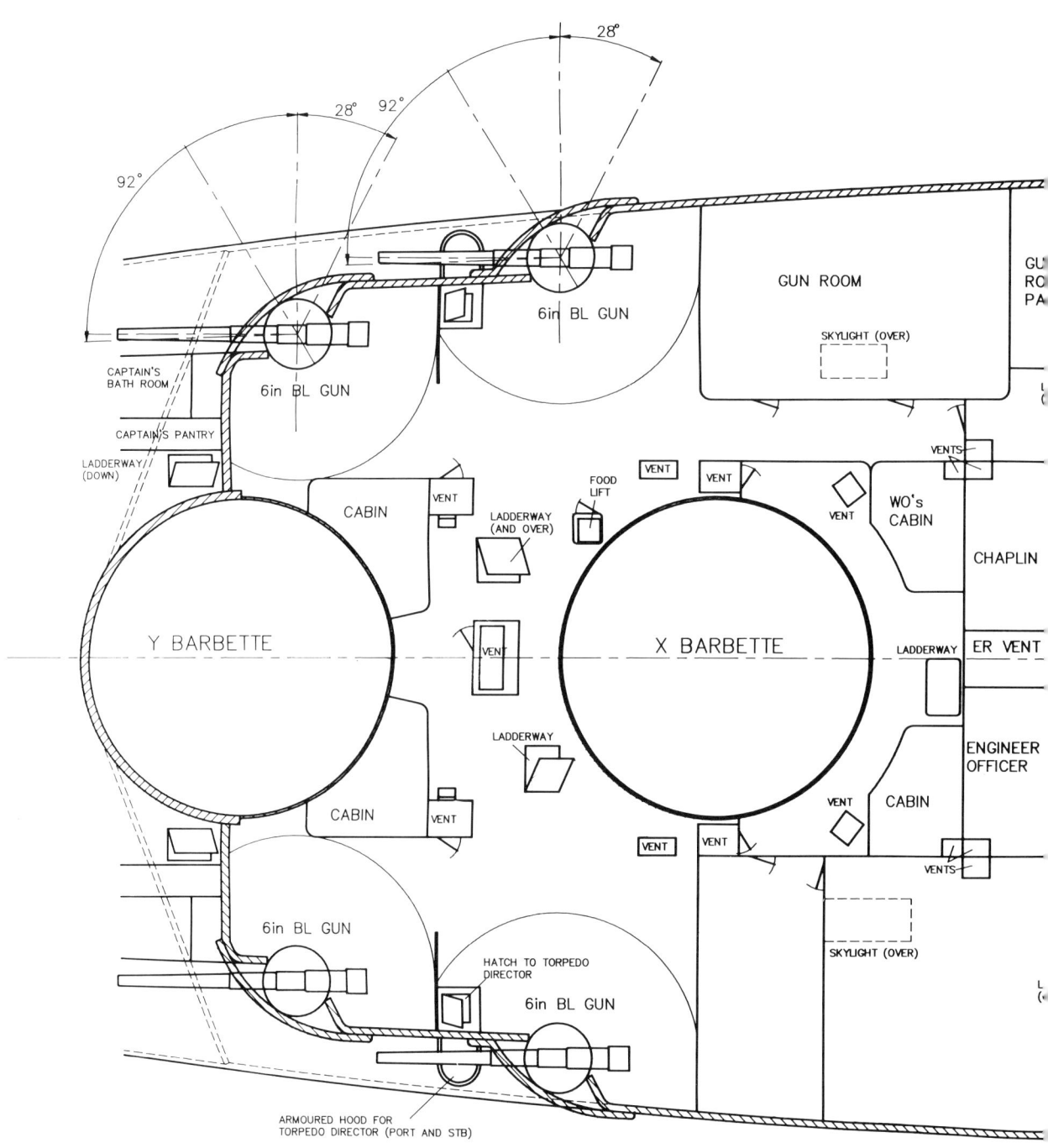

A plan of the after end and officers' accommodation which shows four *6in aft – since it has a midships turret it is clearly not a* Queen Elizabeth, *but nothing in the Covers matches it.*

SUPER-DREADNOUGHTS

'M' to 'MII' may not have existed, but 'MIII' had sixteen 6in behind 6in armour, the other armour between main and upper decks being reduced to 7in to reduce topweight, which meant a hull of 580ft × 90.5ft × 28ft; 25,000 tons. These were the largest dimensions which would go into the four docks at Devonport North and into No 15 at Portsmouth. She would cost £2,150,000. 'MIV' had only twelve 6in, but retained the 8in upper armour; cost and displacement were as for 'MIII'. 'MV' combined the reduced 6in battery with 7in armour, which brought her back to 90ft beam, 24,750 tons and £2,130,000.

Consideration was given to fitting small-tube boilers to the class and the parallel cruiser *Tiger*, but the latest DNC, Eustace Tennyson d'Eyncourt, could not carry engineering opinion with him; it would have saved 900 to 1000 tons in the battleships.

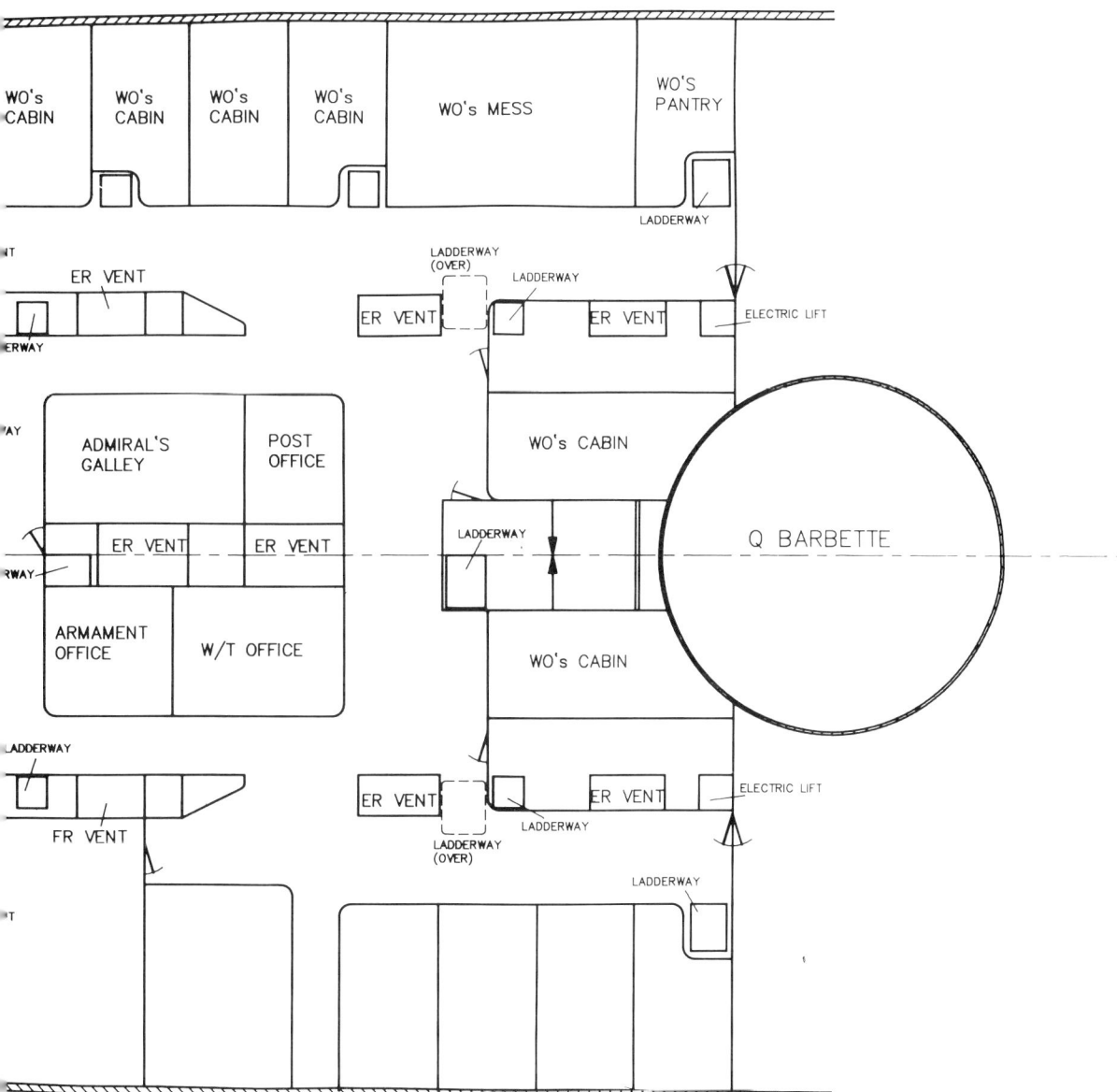

This well-known wartime view of Benbow *at speed is often used to point out the rangefinding baffles and their ineffectiveness; equally, it demonstrates the disadvantages of a low secondary battery, the 'white water' below the forward turrets being clear evidence.* (IWM)

The Board decided, about 1 June 1911, on 'MIV', specifying that the secondary armament must have good arcs forward and aft, that its protection must take preference over the upper belt, that two guns should be in gunhouses on the forecastle and, apparently, that a stern tube should be fitted as well as the four beam tubes. This last led to endless trouble in providing a route for reload torpedoes. It was feared that a door would have to be cut through the torpedo room bulkhead. By July, Mooney had managed to get the tube abolished. At Jutland, the *Lützow* was flooded through just such a door as was proposed.

The gun houses also led to trouble; their supports and hoists took the space for a pair of guns in the battery below, they needed power, which was not trusted for small guns, and they were very exposed to blast from 'A' and 'B' turrets. On 18 August, 'Tug' Wilson, the new First Sea Lord, proposed moving them to the main deck abreast 'Y' turret, and moving the fifth pair of 6in slightly forward, sacrificing their after arcs somewhat. A G H W Moore, the new DNO, somewhat sceptically had some 12 pdrs put into them and the *Queen Elizabeth*s as 'quick reaction' anti-torpedo weapons at the request of various officers, but Bridgeman had them omitted before his departure on the grounds that they were too small to be effective, and that 6in could react as quickly. 'If anti-aircraft guns are needed some rearrangement would be necessary, but I do not think our experiments have as yet gone far enough in this direction to warrant provision for this nature of armament.'[18]

Table 2: *BUILDING DATA*

Name	Builder	Laid down	Launched	Commissioned	Fate
Orion	Portsmouth DYd	29 Nov 1909	20 Aug 1910	Jan 1912	Scrapped 1923
Monarch	Armstrong	1 Apr 1910	30 Mar 1911	Mar 1912	Expended as a target 1925
Conqueror	Beardmore	5 Apr 1910	1 May 1911	Nov 1912	Scrapped 1923
Thunderer	Thames Iron Works	13 Apr 1910	1 Feb 1911	Jun 1912	Scrapped 1927
King George V	Portsmouth DYd	16 Jan 1911	9 Oct 1911	Nov 1912	Scrapped 1927
Ajax	Scotts	27 Feb 1911	21 Mar 1912	Mar 1913	Scrapped 1927
Centurion	Devonport DYd	16 Jan 1911	18 Nov 1911	May 1913	Blockship Jun 1944
Audacious	Cammell Laird	Feb 1911	14 Sep 1912	Oct 1913	Mined Oct 1914
Iron Duke	Portsmouth DYd	12 Jan 1912	12 Oct 1912	Mar 1914	Training ship 1932, bombed 1939, scrapped 1946
Marlborough	Devonport DYd	25 Jan 1912	24 Oct 1912	Jun 1914	Scrapped 1932
Benbow	Beardmore	30 May 1912	12 Nov 1913	Oct 1914	Scrapped 1931
Emperor of India	Vickers	31 May 1912	27 Nov 1913	Nov 1914	Scrapped 1931

More happily, Moore checked that the 6in were to be purely anti-torpedo weapons, and then got a separate control system put in for them; it had proved impossible to find a spot safe from blast for the usual sighting hoods. Finally, a pair of 3in AA guns were installed.

The larger hull permitted more bunker capacity, giving a theoretical 7780 miles at 10kts, against 6280. Speed was nominally 21kts, but the ships had to be pressed hard to attain it. There was a brisk argument between Winston Churchill and the King over names: the former wanted *Assiduous*, *Liberty* and *Oliver Cromwell*, the latter offered *Delhi*, *Marlborough* and *Wellington*. *Delhi* referred to the King's Durbar at Delhi in 1911, and was changed to *Emperor of India*; *Marlborough* survived, in more senses than one; and *Wellington* became *Iron Duke*. *Benbow* seems to have passed *nem con*.[9]

The second *Iron Duke* is inseparably linked with Jellicoe, but in fact he did not like her; according to him, she trimmed 1.5ft by the head, and water got into the main deck 6in casemates, despite much remedial work. The after pair of 6in were very wet, and had to be moved to the fore shelter deck, roughly where they started. Deep load draught was estimated at 33.5ft forward and 32ft aft against 31.5ft designed.

Service Experience

The 12in dreadnoughts had formed the 1st Battle Squadron of the Home Fleet; as the 13.5in ships were commissioned, they formed the 2nd, which by August 1914 formed an almost homogeneous squadron of eight. A little earlier, the *King George V*s visited Kiel, in what was intended to be the start of an Anglo-German *detente*. In their own eyes, the 2nd formed a *corps d'elite* and were evidently expected to form the van of the line of battle on deployment. The first two *Iron Duke*s were in service just before the war, and that class were distributed among the 1st Battle Squadron and the newly formed 4th to stiffen the older ships.[10]

At 0845 on 27 October, the *Audacious* was mined off Northern Ireland, as she was about to take part in a gunnery practice. A P Cole, RCNC was on board, and reported the explosion as being mild and apparently inboard of the port bilge keel just forward of the after end of the engine room; the most vulnerable spot. It took some time to shut watertight doors, and flooding, though slow, could not be stopped. Leaks where electric cables passed through bulkheads, and via a sanitary pipe, slowly destroyed her buoyancy. Every five hours or so, she lurched badly, as the swell, free water inside her and her righting force got into synchronism. Determined attempts to tow[11] or steam her inshore failed and at dusk she was abandoned, to capsize and blow up about two hours later. The only casualty was a Petty Officer aboard the *Liverpool*, hit by falling wreckage.

A replacement was soon obtained in the *Erin*, ex-Turkish *Reshadieh*. She was not unlike 'MIII', squeezed somewhat. She is generally attributed to Sir George Thurston, Vickers' designer, but, like the ex-Chilean

Marlborough, *late in the war, with flying-off platform on 'B' turret*. (CMP)

Jutland, 31 May 1916: Iron Duke *(centre) opens fire, following the four* Orions *(*Thunderer, *behind* Conqueror, *is next ahead).* Royal Sovereign *and* Superb *are visible in the foreground.* (IWM)

Canada, is believed to have been an Elswick design, almost certainly based on the five-turret designs '641' or '643', rejected by Brazil for the ship which ultimately became the *Agincourt*.[12] With her thin closely-spaced funnels, she looked quite unlike the Admiralty designs and was known as 'The Turkish Delight'. She proved cramped and insanitary, and her guns were not up to Admiralty standards.

As with most big British ships of the era, combat experience was limited to Jutland.[13] The super-dreadnoughts were at almost full strength, only the *Emperor of India* being away refitting. The battle proved something of a fiasco, especially for the 2nd Battle Squadron. On deployment, they were in the van, but their fire was largely masked by battlecruisers and light craft; the *Ajax* was only able to fire one salvo, which missed, and the *Erin* could not use her 13.5s at all. The *Orion*s did a bit better, *Orion* making one hit on the *Markgraf* and the *Monarch* one on the *König*, but the *Iron Duke*s, further back in the line, more than made up for this. Initially, the Fleet Flagship scored seven hits on the *König*, from 43 rounds, one of which let in 1000 tons of water and nearly blew her up. The *Marlborough* did more shooting than any other of the main fleet battleships and also survived a torpedo hit. During the second clash of battleships, she made three hits on the *Grosser Kurfurst*, as well as evading three more torpedoes. The *Conqueror*, *Thunderer* and *Benbow* were badly masked and were not able to fire effectively. The *Canada* fired 42 rounds without effect and could hardly make out her targets. As the Grand Fleet turned inwards by divisions, the 2nd Battle Squadron were left on the outside track.

The *Marlborough*'s torpedo hit abreast the forward boiler rooms and the 6in magazine; the actual hole was 25ft long, damage extending for 70ft. Two men were killed and the flooding rapidly put six boilers out of action, though the ship continued in action. By dusk, speed had fallen off to 15.6kts and she and her division fell astern. She was badly down by the head, and at dawn the flag had to be transferred. It was probably as well that the *Seydlitz*, which passed close in the night, was not engaged; the concussion might well have done further damage. The return home was long and difficult; bulkheads showed signs of weakness, the pumps clogged and had to be freed by a diver, she was attacked by submarines and the weather worsened; destroyers laid an oil track to smooth her passage. By the time she entered the Humber, she was drawing 40ft. Her survival may well have been due to the excellent pumping facilities of her class. The contrast with the *Audacious* is interesting, but the *Seydlitz*, of about the same size, survived a torpedo and a great deal of gunfire as well.

During the second battleship clash, the *Orion*[14] and *Monarch* made five hits between them on the already badly damaged *Lützow*, but that ship had already been fatally damaged by the *Invincible*.

One of the minor legends of Jutland arises from the misuse of the 'interceptor' – the equivalent of a safety catch – in an anonymous British battleship. She glimpsed the enemy and fired two salvoes. There was a mighty roar and five shells sped towards the foe. The second roar was not quite so tremendous, and only four shells went forth. The petty officer in the turret concerned investigated, and said 'I have waited for twenty years to fire this gun at the enemy, and you, you silly man, forgot to close the interceptor', or words to that effect, and then proceeded, clean contrary to King's Regulations, to commit assault, battery and grievous bodily harm on the wight responsible. The ship never got another chance. The *King George V* fired exactly nine 13.5in (heavy) shell at Jutland.

Although intended for use against heavy ships, the 13.5s were used against destroyers, the *Iron Duke* sinking *S-35* with all hands. Several destroyers were damaged by 6in fire, but 4in seems to have been quite ineffective.

As it fell out, Jutland was little more than a skirmish for the British battleships, but it might have been much more serious. At dusk, 2100, when the Grand Fleet was at last in its full battle formation and between the Germans and their base, the *King George V* and other ships in the van sighted three German ships to the west. They were probably the southern end of the High Seas Fleet battleships; German light cruisers and battlecruisers were in the vicinity. The light cruisers *Caroline* and *Royalist* made a torpedo attack, but the 2nd Battle Squadron withheld fire, thinking that the three ships were the British battle cruisers, which could have been in that position.

During the night, the *Seydlitz*, as already mentioned, passed very close to the *Marlborough* and her division, while the *Moltke* three times sighted the 2nd Battle Squadron. Fire was withheld in strict accordance with Grand Fleet orders. The *Moltke*'s torpedo officer wanted to have a go, but Admiral Hipper also had doubts.

The rest of the war was endless training, maintenance and preparation for a clash that never came again. By the end of the war, in spite of a lot of effort on entertainment and distraction, crews were nearly mad with boredom. When she paid off, the *Ajax*'s stokers threw the driver and fireman off the train taking them south, and drove it all the way themselves.

The *Orion*s, the *Erin* and two *King George V*s were scrapped after the Washington Treaty, three *Iron Duke*s in the early 1930s. The demilitarised name ship was crippled by bombing at Scapa Flow early in the Second World War, but the *Centurion* had an unlikely afterlife as a target ship and then as a dummy of the new *King George V*s, taking part in the 'Operation Vigorous' Malta convoy attempt, and ended her career as a blockship for one of the Mulberry harbours.

Like so many ships, they were never fully tested against the weapons of their own time; probably they were good but not outstanding, better than the French *Bretagne*s, but not as good as the German or American ships of their day. Nevertheless, their crews were very proud of them.

Sources
Based largely on the 'Ships' Covers', ADM 138/347 & 348 (*Orion* class), 138/338 & 339 (*King George V* class), and 138/268–270 (*Iron Duke* class).

Marlborough *after Jutland, down by the bows; apparently taken in the Humber*. (NMM)

A contrast in styles: an Iron Duke *passes the surrendered German* Kaiser *and* Kaiserin *at Scapa Flow; these ships were of a similar size to the British vessel but of a slightly earlier date.* (IWM)

Emperor of India *postwar, with plated-in after casemates.* (CMP)

Notes

1. The cost of the 1909–10 ships led to the Lloyd George Budget and the Parliament Act of 1911.
2. This shows, incidentally, that he was not the all-powerful naval dictator of legend.
3. The Japanese had the same idea many years later with their 24in.
4. Who had worked on the *Warrior* as a young apprentice.
5. See below.
6. Someone underlined this in red.
7. In October, 1910, the Ordnance Board reported that British AP shell were useless at impacts of more than 20 degrees to the normal; even half inch armour greatly reduced their effect.
8. This decision may have saved LII on the morning of 1 June 1916. The Germans carried 5.9in secondary and 3.5in tertiary guns in their big ships; the former were deadly; the latter were used, but does anyone know what effect they had? Moore noted prophetically that in a close range night attack it was essential to stop a destroyer with the first salvo. A 12pdr could not do this, but a 6in might, as the British 4th Flotilla found out.
9. In his own time, the Duke of Wellington was known simply as *the* Duke; a 131-gun ship bore his title. Shortly after his death, the iron collier *Duke of Wellington* was built and became known as 'the Iron Duke'. By 1867, this was acceptable as the name of a second class ironclad.
10. The 3rd were the *King Edward VII*s and had left the Grand Fleet by Jutland.
11. By the giant liner *Olympic* among others. It seems odd that *Audacious* had not gone to full action stations.
12. Described in David Topliss' article in *Warship International* (3/1988).
13. On 16 December, six of the 2nd Battle Squadron sighted light craft in the game of blind man's buff off Scarborough, but did not open fire – in fact, a narrow escape from the whole High Seas Fleet.
14. The writer's father was serving in *Orion*'s transmitting station.

THE FIRST HUNTER-KILLERS

British 'R' Class Submarines of 1917

The information that has been published about the 'R' class in the past has tended to be generalised and inaccurate. This article by David Miller, based entirely on primary sources, sets the record straight, describing the evolution of the design, the construction of the class and the service record of those boats that attained operational status before hostilities ended on 11 November 1918.

By early 1917 the Royal Navy was so concerned by the havoc wrought by the Imperial German Navy's U-boats that Admiral Jellicoe, the First Sea Lord, submitted an extremely gloomy paper to the War Cabinet forecasting defeat unless urgent action was taken.[1] Jellicoe's successor as Commander-in-Chief of the Grand Fleet, Admiral Beatty, was equally blunt in another letter written at the same time: 'It is acknowledged on every side that the enemy submarine is a threat of extreme gravity. *It actually threatens us with defeat.*'[2]

Jellicoe proposed several countermeasures at national level, but also listed steps being taken by the navy itself. These including more escorts, more decoy ('Q') ships, more air stations along the Channel coast and also mentioned that: '. . . schemes for using submarines to hunt submarines have been prepared and are being carried out', probably the first official mention of the use of submarines in an anti-submarine warfare (ASW) role at this high level.

Within the Admiralty one of the most important outcomes of those 'schemes' was the 'R' class submarine, a type which has attracted relatively little attention by naval historians, possibly because its operational life was so brief. This ASW submarine was, however, one of the most revolutionary naval concepts of the First World War, the precursor of the 'hunter-killers', which next appeared some thirty years later and which exist in such large numbers today.

R-7, built by Vickers, Barrow, and the first 'R' class boat to be commissioned. The lifebelt on the bridge was carried at Vickers' insistence for the safety of their staff during trials, much to the indignation of the commanding officer! (VSEL, Barrow)

Anti-Submarine Submarines

Some of the Royal Navy's ASW measures, such as more convoy surface escorts, were predictable, although some novel ideas were also considered. One naval officer, for example, proposed using seagulls to detect hostile submarines, while another scheme, which actually reached fruition, involved using specially constructed craft, based on dumb barges, to simulate British submarines in distress. It was hoped to lure a U-boat into attacking the dummy, whereupon it would itself be attacked by a lurking British submarine.[3]

The earliest recorded British trials of a submarine operating against another submarine took place in 1909 and further trials took place in the following year. This line of approach was, however, not pursued and when the First World War started other roles were given priority, although whenever the occasion offered British submarines did manage to sink some U-boats. In 1917, however, every possibility was re-examined and the Grand Fleet formed a special committee to investigate these 'submarine-versus-submarine' operations, which reported in August.[4]

One of the first factors to be considered was the record of such anti-submarine attacks (Table 1). Examination of submarine logs showed that there had been no shortage of periscope sightings of surfaced U-boats, but with an underwater speed of (at best) some 6kts, by the time the British submarine had manoeuvred into a firing position the target had usually either long since departed on the surface, or had submerged and thus, with the technology of the day, was no longer detectable. Indeed, between January and June 1917 there had been no less than 144 sightings of enemy submarines by British submarines, but of these only five resulted in hits, of which just three led to the sinking of the U-boat (Table 1, Serials 6–8).

Table 1: U-Boats sunk by British Submarines, 1914 to mid-1917

Serial	Date	German Submarine Sunk	Sunk By	Weapons Used	Where
1	23 June 15	U-40	C-24	Torpedoes	North Sea
2	20 Jul 15	U-23	C-27	Torpedoes	North Sea
3	15 Sep 15	U-6	E-16	Torpedoes	North Sea
4	14 Jul 16	U-51	H-5	Torpedoes	North Sea
5	21 Aug 16	UC-10	E-54	Torpedoes	North Sea
6	10 Mar 17	UC-43	G-13	Torpedoes	Atlantic
7	1 May 17	U-81	E-54	Torpedoes	Atlantic
8	7 Jul 17	U-99	J-2	Torpedoes	North Sea

R-2 *leaving harbour. The second of the class to be completed by Chatham Dockyard she has the original gun platforms fore and aft of the bridge, whose height is clearly shown by the sailors. Note the two 30ft periscopes and the extended hydraulic aerial/signal mast.* (IWM)

THE FIRST HUNTER-KILLERS

The committee made many recommendations, most of which were acted upon. They proposed that at least one gun of 4in (102mm) or greater calibre be carried by all submarines, with the mount being as high as possible to enable the gun to be brought into action as soon as possible after surfacing. The first such mount appeared on *L-3* and the practice continued well into the postwar era. A curious feature of this proposal, which seems to have been overlooked by the committee, is that not one of the recorded submarine-versus-submarine engagements had resulted in the target being destroyed by gunfire; on the contrary, in every case, the sinking had been due to a torpedo fired from a submerged submarine.

The committee also proposed that at least one periscope should be 30ft (9.14m) long, so that the submarine did not break surface in a seaway. Concerning torpedoes, the committee used a game-shooting analogy, which would have been well understood by the 'hunting-and-shooting' types among British naval officers of the day, when it stated that using torpedoes against another submarine '. . . can be likened to shooting grouse with a rifle'. It therefore recommended that '. . . as many torpedoes should be fired as possible'.

This led them to propose the use of modified Russian Drzewiecki drop-collars on the upper decks of existing submarines, although, as far as is known, this was never implemented. A more realistic proposal was to increase the number of torpedo tubes in new designs from the existing four to six bow tubes.

Other attempts to improve submarine capabilities included mounting a 7.5in (190mm) howitzer in one boat, while *J-1* was fitted with twenty depth charges in lieu of its beam torpedoes. Neither was a success!

It appeared, however, that a submarine with a very high underwater speed was the answer and, as a result, the Director of Naval Construction (DNC) sent a proposal to the Third Sea Lord for such a boat, which led directly to the 'R' class.

The Admiralty Board Proposal

The DNC's proposal was put to the Admiralty Board on 27 September 1917.[5] The preliminary sketch proposed a boat 180ft (54.86m) long and 15ft (4.57m) in diameter, displacing 370 tons on the surface and 480 tons submerged. It had a single propeller and was powered by a single 'H' class 240bhp engine on the surface and two 600bhp electric motors when submerged. It also introduced – for the first time in any submarine design – an auxiliary low-powered motor on the main shaft for 'economical submerged propulsion at low speeds and also for surface propulsion while charging-up batteries when under way'.[6]

The proposed armament fulfilled all the requirements of the Grand Fleet committee. There were six torpedo tubes, all in the bow, with six reloads, using a new design of 14in (357mm) torpedo to conserve space and weight.[7] The design also included two single 4in (102mm) guns, mounted on platforms, one before and one abaft the very high bridge. There were also twin 30ft (9.1m) periscopes.

The paper ended by posing the Admiralty Board with a number of questions and it is clear from the record that there was much discussion, especially concerning the

R-10 shows the smooth lines of this design and high standard of finish on the hull, accentuated by her all-over black paint. She has the low bridge of the later boats, but has retained slightly more of the forward gun platform. Note that a safety line has been rigged from the forward end of the bridge to the jumper wire to provide a handhold for the crew on the slippery foredeck. (RN Submarine Museum)

armament. Eventually this was referred to a sub-committee for further examination and, although no records have survived, the outcome can be assumed from subsequent actions. A revised drawing was initialled by the DNC in November 1917 in which the design had been developed in certain ways, with the six 14in (357mm) tubes being replaced by six standard 18in (457mm) tubes and a more powerful diesel engine being fitted, although the original two 4in (102mm) guns were still mounted.[8]

Final Design

The Admiralty Board was concerned from the start about the construction programme. At their 27 September meeting the Chief of Naval Staff (CNS) informed the Board that, provided the single-engine design was followed and assuming that there was no delay in producing the new 14in (357mm) tubes or in getting out the working drawings, then it would be possible to deliver the new class within twelve months of the Board's decision. After discussion the Admiralty Board directed that:

> a. The design as submitted to the Board was generally approved and be worked out in detail embodying a single engine.
> b. The construction of eight of the new type of submarine be approved in substitution for eight 'H' type submarines about to be constructed, the order for which shall be cancelled.
> c. The construction of an additional four of the new type of submarine be approved, these four to be built at Chatham.

The orders were then placed: HM Dockyard Chatham – *R-1*, *R-2*, *R-3* and *R-4*; HM Dockyard Devonport – *R-5* and *R-6*; Vickers, Barrow – *R-7* and *R-8*; Armstrong – *R-9* and *R-10*; Cammell Laird – *R-11* and *R-12*.[9] This was

Batteries and tanks in the 'R' class. Note the buoyancy tank under the fore end of the bow casing, designed to overcome fears of a sudden spontaneous dive when steaming at speed on the surface.

Table 2: *R Class Building Programme*

Name	Launched	Completed	Builder	Fate
R-1	25 May 18	14 Oct 18		Sold 20 Jan 23
R-2	25 May 18	20 Dec 18	HM Dockyard, Chatham	Sold 21 Feb 23
R-3	8 Jun 18	31 Mar 19		Sold 21 Feb 23
R-4	8 Jun 18	23 Aug 19		Sold 26 May 34
R-5			HM Dockyard, Pembroke	Cancelled 28 Sep 19
R-6				Cancelled 28 Sep 19
R-7	14 May 18	29 Jun 18	Vickers, Barrow	Sold 21 Feb 23
R-8	28 Jun 18	26 Jul 18		Sold 21 Feb 23
R-9	12 Sep 18	14 Oct 18	Armstrong Whitworth	Sold 21 Feb 23
R-10	5 Oct 18	12 Apr 19		Sold 19 Feb 29
R-11	16 Mar 18	8 Sep 18	Cammell Laird	Sold 21 Feb 23
R-12	9 Apr 18	29 Oct 18		Sold 21 Feb 23

quickly amended, however, and the construction of *R-5* and *R-6* was transferred to HM Dockyard Pembroke Dock.[10] The Admiralty Board was clearly enthusiastic about the 'R' class and in its subsequent consideration of the shipbuilding programme for 1919 (which was later cancelled due to the ending of the war) the Board stated a requirement for a further twenty-two 'R' class boats, suggesting a total of thirty-four by 1920, had the war continued.

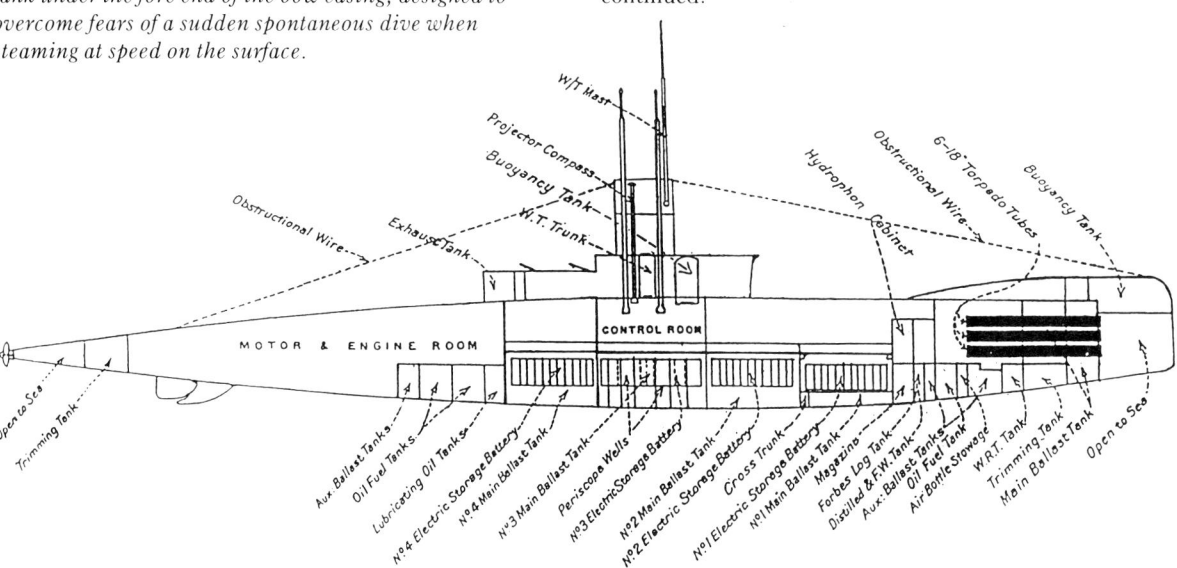

In the event the promised ten were not delivered within twelve months of the Board's decision, the actual dates achieved being shown in Table 2, while *R-5* and *R-6* were never completed at all, being cancelled in 1919.

Technical Details

The design was finalised in early 1918 and, despite some changes, it was still clearly recognisable as that presented to the Admiralty Board in 1917. Extensive tests were conducted at the Admiralty Experimental Establishment on various aspects of the design and Commodore (Submarines) [Cdre(S)], Commodore S S Hall, RN, paid several visits to inspect progress, reminding the staff on each occasion of the overriding need for high underwater speed.

The Hull. The pressure hull was 161.5ft (49.23m) long, 163.75ft (49.91m) overall. It was circular in cross-section for most of its length, forming a body-of-revolution (*ie*, it was perfectly symmetrical about its main axis), narrowing laterally into an elliptical shape from Frame 21 forwards. This gave it unusual strength for the time and the quoted operating depth of 250ft (76.2m) was the best of any First World War class.

The single-hull design minimised underwater drag and the free-flooding superstructure aft of the conning tower found on all other contemporary submarines was totally eliminated. The lines of the hull aft were very fine in order to achieve a combination of propulsive efficiency and manoeuvrability.

The most revolutionary feature of the 'R' class was its high underwater speed and the designers showed great awareness of the problems. In the words of the technical history:

> 'R' class boats were designed solely for speed and endurance submerged; they were meant to get 15kts and got 14.4. The submarine officers wanted higher bridges to prevent being washed down by every wave but Commodore Hall wanted streamlining. Vibrating wires, bollards, cleats and other obstructions when their individual effect is added up cause extraordinary loss in underwater efficiency and the streamline effect on submersibles is just as important as the streamlining of aircraft.[11]

The effect of drag was highlighted in a speed trial in 1919

Longitudinal section of the 'R' class.

Table 3: R CLASS SPECIFICATION

Displacement:	Surface – 415 tons (421.7 tonnes); Submerged – 503 tons (511 tonnes).
Dimensions:	Length overall – 163.75ft (49.91m); length of pressure hull – 151.5ft (46.18m); beam – 15.54ft (4.74m); mean surface draught – 11.5ft (3.51m).
Main Engine:	One 8-cylinder diesel; 240bhp at 380rpm; fuel – 13.25 tons (13,463kg); total weight of engine – 10.5 tons (10,668kg).
Electric Motors:	Two main motors; 1200bhp; 600rpm; total weight – 20.2 tons (20,524kg). One auxiliary motor; 25bhp.
Batteries:	Type – Chloride E4400; number of cells – 220 in 4 tanks; weight – 93.5 tons (95,001kg).
Performance:	Surface speed – 9½kts; surface radius of action – 2000nm at 9kts. Submerged speed – 15kts (design), 14.4kts (trials); range 15nm at full power; 240nm at 4kts on auxiliary motor.
Design Diving Depth:	250ft (76.2m).
Armament:	Six 18in torpedo tubes; 12 torpedoes carried (war), 7 (peace). [NOTE: The original plans show two 4in guns. Amended plans show one 4in gun. As far as is known no submarine of this class ever actually had a gun mounted.]
Complement:	Officers – 3; petty officers – 3; ratings – 16.

when *R-3* performed rather worse than had been expected. She was taken out of the water and found to be encrusted with barnacles; when they had been removed the top speed increased 31 per cent (Table 4).

The exceptional design speed of 15kts, however, caused great concern, as was described some years later by the former captain of *R-7*, the first of the class to run trials.

> There was considerable speculation as to what would happen when an R-boat went full speed for the first time – some thought they would dive uncontrolled to the bottom and stick in the mud, never to rise again, but the really gloomy prophets

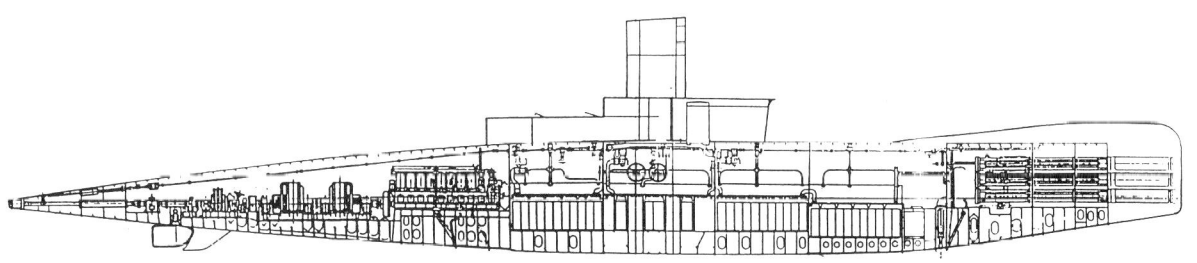

Table 4: SPEED TRIALS ON R-3. (2) (3)

2 July 1919			24 July 1919		
Current	Revolutions	Speed	Current	Revolutions	Speed
(amps)	(rpm)	(knots)	(amps)	(rpm)	(knots)
300	288	4.7	300	290	4.7
1000	440	7.4	1000	450	8.4
1500	508	8.6	1500	540	10.8
2800	616	10.5	3000	709	13.8

Notes:
The result of the 2 July trials were so disappointing that the boat was docked and the hull found to be very foul with weed and barnacles. She was thoroughly cleaned and three weeks later achieved the results shown in the right-hand columns. Speeds are submerged; periscopes down.
Source: Ship's Cover, Folio 74.

said that the propeller at full power would gradually come to rest and the submarine revolve around it![12]

There was also a fear that the boats could get into a spontaneous dive either when running on the surface or at high speed when submerged, which led to lengthy tests at the Admiralty Experimental Works. To overcome the anticipated surface problem a small bow superstructure was fitted, whose primary function was to carry a buoyancy tank high in the nose, while two additional main ballast tanks were installed well forward inside the hull. These, coupled with an instruction to captains to trim well down by the stern were intended to overcome the surface problem. It was also planned at one stage to fit 'emergency hydroplanes' amidships to counter the anticipated submerged problems, but these were not fitted, since blowing the forward tanks was much more effective.

It is not usually appreciated when looking at side elevations of the 'R' class that there were, in fact, two rudders either side of a fixed skeg. In a postwar Minute the DNC stated that

> In preparing the design of the 'R' class it was essential to cut down to a minimum all unnecessary resistances and to enable this to be done the rudders were placed forward of the propeller with all the actuating gear inboard, instead of behind the propeller with supporting brackets and the greater part of the operating gear outboard as in the 'B' and 'C' classes. It was recognised at the time that this would make control difficult but was accepted to achieve maximum submerged speed.[13]

According to the DNC the fitting of Kitchen rudders was examined at the request of Cdre (S) to overcome the control problems. But it would have required much redesign to fit these hemispherical devices and they would also have markedly reduced speed, so the idea was dropped. After the first boats to be completed showed poor directional control while submerged an extra 4in (102mm) was added to the after edge of the rudders, adding 1sq ft (929cm^2) in area to each.

Sketch of rudder and skeg arrangement. (Courtesy of J Cheetham)

Armament was originally planned to be six (14in (357mm) torpedo tubes to an entirely new design. This proposal was dropped since there was insufficient time to develop them and six standard 18in (457mm) tubes were installed instead, all in the bow. The torpedoes were embarked through the hatch above the petty officers' mess tail first, drawn aft and then pushed forward directly into the tubes. The war complement included six reloads, which are shown in the plans lying in the POs' mess, which would have been highly unpopular and in peacetime it was rare for even one spare torpedo to be carried.

The original intention was to mount two 4in guns (102mm) in open mounts, one forward of the bridge, the other abaft. This was amended to a single 4in gun forward of the bridge and several contemporary documents show such a gun mounted, with a small magazine sited immediately forward of No 1 Electric Storage Battery.[14] As far as is known, however, no boat actually entered service fitted with such a gun.

Small arms were also carried, with a pistol for officers and a rifle for petty officers and ratings. At least one boat also carried a Lewis machine gun, presumably for defence against attacking aircraft or airships. One curious feature was a central strengthening plate in the bows, made of 0.625in (1.6cm) steel plate, which (according to the DNC) was intended to enable hostile submarines to be rammed, although it seems unlikely that any captain would have used such a tactic.

Superstructure. The original bridge was exceptionally high, both to provide maximum visibility and to provide good protection for the watchkeepers when running on the surface, and the first seven boats to be completed (*R-1, R-2, R-4, R-7, R-8, R-9* and *R-11*) all had this high bridge.

THE FIRST HUNTER-KILLERS

R-3 and *R-12*, however, were completed with a different bridge, which was lower, longer and more streamlined, while *R-10* also had a lower bridge, but to a slightly different design.

The main trunk gave access to the bridge, while a second trunk was intended to permit rapid access to the forward gun position, but was never used as such and was initially simply closed off. In the boats with the revised bridge, however, it was used as an extra ballast tank.

Propulsion. Of the internal space 35 per cent was allocated to the machinery, which drove a single shaft and a three-bladed propeller. The main engine was built to the same American design as those fitted in the 'H' class, with eight cylinders and an output of 240bhp at 380rpm. Of the ten engines, four were built at Chatham (*R-1* to *R-4*), one by the North British Diesel Co (*R-10*) and the remainder by Ruston & Hornsby. These proved effective in service and the captain of *R-9*, for example, reported that on his acceptance trials the main engines had run at 380rpm for seven hours 'very satisfactorily'.

Directly aft of the main engine the shaft carried two main electric motors giving a total of 1200bhp. Further aft on the shaft was an auxiliary propulsion motor, with a power output of 25bhp. The Chloride E4400 battery consisted of four sections of 55 cells each, similar to those used in the 'J' class.

Tanks. There were four main ballast tanks, one abreast of each battery tank, with a total capacity of 76.3 tons. There were also two main tanks forward (A – 6.8 tons and B – 10 tons) and five auxiliary tanks with a total capacity of 9.2 tons. There were six oil tanks with a capacity of 13.25 tons and a distilled water tank was fitted for the first time in a British submarine.

Hydrophones. Other contemporary British submarines were fitted with three hydrophones, but the 'R' class had five, of which four were Hydrophone Plate Receivers,

Mark V, fitted forward in the pressure hull, with 3in (7.62cm) diaphragms flush with the outside of the hull plating and in direct contact with the water. There was also a single Revolving Directional Hydrophone (RDH), which was mounted under a dome on top of the forward casing. This consisted of a 4ft 6in (1.37m) long teak bar, fitted with a microphone at either end, which was rotated

The Development of the 'R' Class Design.
A. *The original design with tall bridge and two single 4in (102mm) guns, as approved by the DNC in November 1917. (Source: NMM, Ships Plans)*

B. *The 'final' design of early 1918, with tall bridge and single 4in mounted before the bridge. It is possible that some boats may have been completed with one gun as shown here, but, if so, it was removed before they entered service. (Source: DNC Notebook)*

C. *The first six boats as they entered service, with tall bridge and gun platforms plated over.*

D. *The final four boats were built with totally redesigned bridges, which were lower, long and considerably more elegant. In some the front edge was cut back to just before the access trunk originally intended to serve the forward gun, as shown here (see R-12 in photograph.) R-10, however, had a similar low bridge, but retained part of the forward gun platform (see photograph).*
[In all cases the periscopes and wireless masts are not shown at full height.]

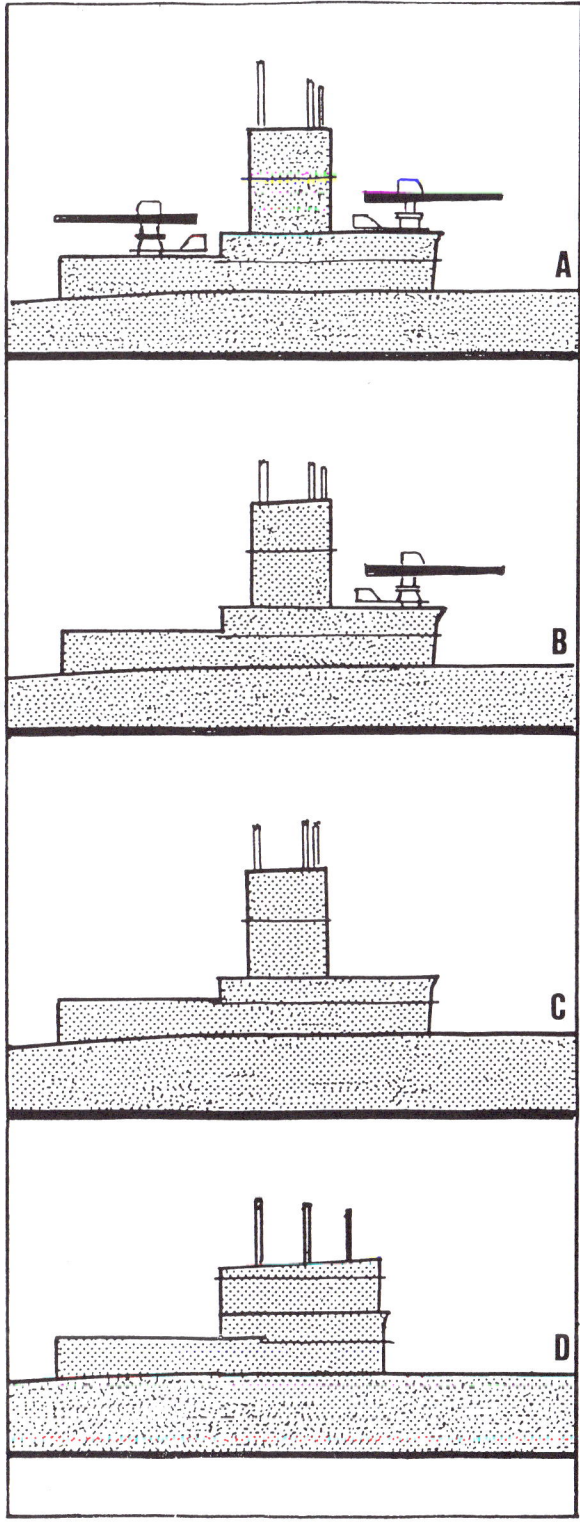

71

by means of a telemotor. (Postwar photographs of *R-4* show a device mounted on the plate covering the disused forward gun mounting, which is assumed to be a type of RDM.) In all cases the hydrophones used the standard Pattern 6400 Microphone.

The Hydrophone Cabinet was situated forward on the port side. The operator was highly trained and with his equipment was expected to achieve an accuracy of roughly ±5° up to 5000yds (4600m), but much less beyond that. The equipment, however, was only intended for use against surface targets and the standard textbook of the time makes no mention of the use of such equipment against submerged submarines.[15]

A report on the effectiveness of these devices states that:

> PO Telegraphist Swatton recalled an incident off Portland when he was keeping a hydrophone watch. The CO asked if it was clear to surface and Swatton replied he was sure there was a surface vessel in the vicinity, but he was unable to get a bearing or be more precise. The CO waited a little longer and then asked Swatton again, who said he still felt there was a ship nearby – possibly lying almost stopped – and listening for *R-4*. After considering this the CO said he would come up to periscope depth. No sooner had he raised the periscope than he immediately ordered the boat deep. 'By God, you were right, Swatton, I actually saw the whites of his eyes!

The launch of the R-11 *from Cammell Laird's Birkenhead shipyard, 16 March 1918. The advanced hull form is readily apparent.* (Williamson Art Gallery & Museum)

Wireless. The 'R' class was fitted with a 3kW Poulsen wireless. This set transmitted at 1400–3980m and received at 350–5000m. Under moderate conditions this gave a range of some 200nm (322km) when working submarine-to-submarine and up to about 400nm (644km) when working to the shore.[16] Fixed aerials were mounted fore and aft, with a three-section, hydraulic mast mounted on the bridge, which could be used when surfaced to obtain greater range. The wireless operator's position was on the starboard side of the control room.

In Service

Submerged handling. One of the frequently-repeated charges levelled at the 'R' boats is that they became unstable at high underwater speed, a parallel often being drawn with the much later German Type XXI and subsequent high-speed 'long-thin' hull designs. No evidence whatsoever has been found of such a phenomenon; indeed, all contemporary comments are most complimentary about the type's underwater performance, especially at depth. The captain of *R-9*, for example, said that his boat was '. . . quite the steadiest at depth-keeping I have ever seen, very exceptionally good'.[17] In the same report he referred to a small tendency to 'get light forward on speeding up' but Cdre(S) spent several days aboard this boat in October 1918 where he found that she could be controlled to ±2ft (0.6m) even at full speed.

Handling at periscope depth – 25ft (7.62m) – it was not quite so good and several captains complained of the tendency to break surface, especially in bad weather. As mentioned above, initial trials showed poor directional

control when submerged but this was improved by a slight enlargement of the rudders.

Surface handling was never good and in a following sea there was a very bad tendency for the boats to be 'pooped'. However, Cdre (S) found that by judicious adjustments to the ballast tanks it was possible to raise the stern by some 2ft (0.6m), which greatly improved the handling without actually bringing the propeller out of the water.

The boats also tended to roll very badly in a beam sea, but again it was found that careful attention to ballasting reduced this somewhat. In addition, deeper bilge keels, fitted soon after completion, further alleviated the problem.

Head-on to a sea the boats handled very well and a letter by the Assistant DNC suggested that some captains found the handling easier than others.

> Have received a letter from the captain of *R-12* stating that he has done 600 miles at 300revs without a hitch. It was very rough and while *Albacore* [a tender] had a very bad time of it, they [*R-12*] were alright, the enlarged bilge keels made all the difference and he expressed himself delighted with the boat and no hitch anywhere.[18]

At slow speed, and especially when manoeuvring in harbour, the boats were a real handful and the rudders became virtually useless. The boats in the Blyth Flotilla all insisted on the use of a tug to dock, but, interestingly, those in the Campbeltown Flotilla never seemed to need such assistance.

The boats were relatively slow on the surface and in the 1920s *R-4*, then part of the 6th (Training) Flotilla, was given the nickname of 'The Slug'. This had nothing to do with her handling, as is sometimes assumed, but was a comment on her slow speed when the training boats returned to harbour at the end of a day's training, since she was always the last in.[19]

War Service

Despite the urgency accorded to the programme, only six boats joined the fleet prior to the Armistice. Of these, *R-7*, *R-8*, *R-11* and *R-12* were based on the depot ship HMS *Platypus*, which throughout this period was moored in Campbeltown harbour. The other two boats – *R-1* and *R-9* – were part of the 14th Submarine Flotilla, based on HMS *Vulcan*.

The Vulcan flotilla was based at Blyth, having arrived there from Stornoway on 8 October 1918. The pace of life seems to have been slow and the logs of both 'R' class boats give a feeling of a lack of urgency in getting them to operational status.

The first of the Chatham boats to be completed, *R-1* was taken over from the builders on 14 October 1918 and

Four 'R' class boats lie alongside HMS Platypus *at Campbeltown. The marked contrast between the low bridge on* R-12 *and the high bridge on* R-11 *is very apparent. The two inboard boats are almost certainly* R-7 *and* R-8, *but the purpose of the long casing on the stern of the nearest cannot be identified.* (RN Submarine Museum)

sailed on the 19th, arriving in Blyth on the 22nd, the only problem having been 'great difficulty in steering the vessel in a heavy swell'. Diving trials were carried out on the 24th and the crew is shown as 'painting' or 'cleaning' from 25 October to 7 November, when she was moved to a drydock for new bilge keels to be fitted. The most noteworthy part of her wartime log is the record of the extraordinary number of losses overboard, including a bucket, a pair of binoculars and four torpedo charging hoses!

The first of two Armstrong's boats, R-9 was taken over by the Navy on 2 November and sailed for Blyth the following day. She carried out torpedo firing trials on 5 November and again on the 9th, the remainder of the time being spent cleaning.

The Platypus Flotilla, based on HMS *Platypus*, moved from Killybegs, Ireland to Campbeltown, Scotland in February 1918 with three 'D' class and two 'E' class submarines. The first 'R' class boat joined the flotilla in August and it is very clear from the logs of these boats that, despite their newness, they were hard at work from the very start.

R-8 was taken over from her builders, Vickers, on 26 July 1918, sailing from Barrow to Campbeltown on 3 August, where, on the 6th, the captain instituted a practice which was to appear in the log for every day of his command: '6.30am – Divisional Prayers'. The remainder of the month was devoted to tests, as for any new warship: compass swinging, hydrophones, torpedo firing and diving. Then on 3 September she set off on her first war patrol, although she had to return to harbour shortly afterwards, her engine clutch bolts having sheared. The problem resolved, she put to sea again at 1400 hours on the 4th and, having reached her operational area, submerged off Barra Head at 1740 hours on the 5th. She operated quite happily until the 11th when she rendezvoused with R-7 at 1800 hours and the two boats returned to base.

R-8 carried out two further war patrols, one from 25 September to 6 October and the other from 16 October to the 19th. The latter was brought to a premature end because the captain and four of the crew were laid low by the influenza which was sweeping Western Europe at that time. She was in quarantine from the 19th to the 28th and was working-up for more patrols when the war ended.

One of the Cammell Laird boats, R-11 was commissioned on 8 September 1918, but did not join the flotilla until 18 October. She carried out one (uneventful) war patrol, between 23 and 26 October. R-12 was commissioned at Birkenhead on 29 October and sailed to join the flotilla in early November. She was forced to put into Belfast by bad weather and did not reach the flotilla before the Armistice.

The Log of R-7

A detailed analysis has been made of the log of R-7 as she served longer than any other in the class and was the only boat to attack a U-boat. She was taken over from the builders, (Vickers, Barrow) on 29 June and joined *Platypus* shortly afterwards. She left on her first patrol on 17 August, but was forced back the next day by bad weather, sailing again on the 19th, only to return a second time on the 20th with a minor defect; she eventually managed a three-day patrol between the 23rd and 25th. Her second patrol was much more successful, lasting from 4 to 12 September, following which she was sent to Harland & Wolff at Belfast to have deeper bilge keels fitted. She was back in Campbeltown on the 22nd and, following some training and maintenance, sailed again on 4 October.

A summary of her patrol is given in Table 5, from which it can be seen that during this ten-day operation she travelled some 381nm, of which 138nm (107 hours 5 minutes) were on the surface in transit to and from the operational area, and 243nm (124 hours 40 minutes) were submerged. Her captain followed the general instructions given for this type of patrol by surfacing at night and lying stationary watching for enemy submarines while the batteries were charged, the average charging time being 5 hours 25 minutes. There were no mechanical defects during the patrol and the only problem that Lieutenant Lockhart reported was that on the third and fourth days

Table 5: *PATROL BY HM SUBMARINE R-7 4–13 OCTOBER 1918*

Date (all October 1918)	On Surface		Submerged		Charge Hours
	Time	Distance (nm)	Time	Distance (nm)	
4	9hr 20min	58.0	6hr 25min	10.0	4hr 50min
5	11hr 10min	–	12hr 50min	19.75	4hr 00min
6	6hr 05min	–	17hr 55min	33.0	7hr 10min
7	8hr 55min	–	15hr 05min	30.5	4hr 20min
8	7hr 20min	–	16hr 40min	34.0	5hr 30min
9	10hr 50min	–	13hr 10min	27.0	5hr 30min
10	8hr 00min	–	16hr 00min	32.0	5hr 25min
11	13hr 10min	–	10hr 50min	26.0	5hr 20min
12	14hr 15min	–	9hr 45min	18.5	–
13	18hr 00min	80	6hr 00min	18.5	–
TOTAL	107hr 05min	138	124hr 40min	242.75	42hr 05min

Source: Captain's Log (PRO ADM 173/12627)

he had difficulty in maintaining periscope depth due to stormy weather and so he went to 65ft (19.8m).

It was during this patrol that the only 'R' class contact with a U-boat took place, which is best described in the captain's own words:

> On 11 October at 0705 when at 27ft [8.23m] sighted submarine, apparently just risen, bearing 285 degrees, distant about 2 miles. Enemy appeared to be steering Southward at slow speed, zigzagging. Proceeded at about 11 knots for 20 minutes and at 0730 found herself [ie, R-7] to be about 2500 yards, 3 points on Port bow of enemy, continued attack at low speed, but shortly afterwards enemy appeared to turn to Westwards and then disappeared. Observed smoke of steamer, Northbound, to westward, closed latter but nothing further of enemy. Sea calm, visibility good, variable later. Enemy class unknown but 2 guns visible and large conning tower. On 12 October at 0650 when at 25ft [7.62m] boat was felt to lurch as if passing over jumping wire of submarine. No blow or damage and nothing heard or seen.[20]

R-7 returned to base on 14 October and on 20 October the log records that all hands were engaged in painting the boat black, although whether this was an idea of the captain's or ordered by the flotilla is not clear. R-7 was out again from 21 to 26 October, the only event of note on this patrol being when a 'friendly' motor vessel tried to ram her and, having failed in that, fired at the submarine with her stern gun, a not infrequent occurrence for British submarines. R-7 exercised with hydrophone trawlers from 1 to 6 November and then suffered a defect in her air compressor, which required repair at Rathmullen, which was where she was when the war ended.

R-7's brief war record shows that most of the stories about the 'R' class are untrue. She was the first of the class to become operational and suffered a few inevitable problems, but they were all very minor. Her diesel engine was fully able to charge her batteries and she was able to carry out perfectly normal patrols. Finally, she was able to react exactly as she had been designed to do, sighting the enemy at a distance and then descending to 60ft (18.3m) to proceed at full speed to an attacking position. It was simply bad luck that when Lockhart was creeping in for the attack the enemy was frightened by the appearance of a small steamer and dived, and, from the incident the following day, it would appear that she remained in the area.

Postwar Service

Once the war ended the Royal Navy could scarcely wait to rid itself of the 'R' class. Eight of the class were taken out of service in 1922, while R-10 survived until 1929 and R-4 until 1934. The attitude of the Royal Navy is indicated by an assessment of the submarine types which would be needed in a possible war against Japan. At a conference in May 1922 it was concluded that: 'No special design of submarine is necessary for local defence or anti-submarine work, and these duties should be carried out by normal types no longer suitable for their primary duties.' This led to a statement in 1924 by Rear-Admiral (Submarines) (the post of commodore having been upranked as the number of boats declined!) that '. . . no mention is made of special anti-submarine submarines as practically all classes of submarine can carry out this work and, unless a large submarine campaign were embarked upon by the enemy, special development of this type is not considered necessary'.[21]

Thus, the careful analysis carried out by the Grand Fleet committee was totally ignored, and even the lessons learned about streamlining were quickly forgotten. The 'R' class had shown what could be achieved by reducing

The high quality of external finish can be seen in this launching photograph of R-11; great attention was paid to reducing surface resistance in hull of this class. (Williamson Art Gallery & Museum)

underwater drag to a minimum, but the designers of the postwar boats seem to have paid scant attention, with topside fittings such as guns, gunshields, aerials, masts, bollards, cleats and stays proliferating, which reduced underwater speed to a meagre 3–4kts accordingly.

There was, however, an astonishing postscript to the 'R' class's story. In 1947 Western navies were faced by yet another submarine threat, this time from the Soviet Union, one answer to which seemed to be small, fast and highly manoeuvrable 'fighter' submarines and the US Navy constructed three boats of the 'K' (*Barracuda*) class, with a submerged displacement of 1160 tons and a speed of 13kts. The Admiralty appears to have come up with a similar requirement and someone with a good memory recalled the 'R' boats. As a result an officer was despatched to the Public Record Office to examine the 'R' class plans and it is quite clear that very serious consideration was given to placing the thirty year old design back in production. However, after further thought it was seen that such a proposal was impracticable, especially as the DNC calculated that the smallest hull to accommodate modern sensors would have a surface displacement of 1680 tons.[22] So it was, perhaps with a feeling of regret, that the plans were returned once more to the archives!

Conclusion

It can thus be seen that many of the stories which have affected the 'R' class's reputation for many years are incorrect. Their submerged performance was excellent, their surface performance, while not good, was certainly not as bad as has been stated, and their wartime captains – particularly those operating from Campbeltown – were very happy with them. The 'R' boats were fully capable of charging their batteries, carried out patrols of up to ten days duration and were, by a very great margin, the fastest submarines of their day.

Table 6: GERMAN U-BOATS SUNK BY THE ROYAL NAVY, 1914 TO 1918

Platform	Weapon	Number	Totals
Surface Ships	Depth Charges	26	
	Rammed	21	
	Guns	17	
	High-speed paravanes	2	67 (38%)
	Explosive sweeps	1	
	Torpedoes	0	
Mines	–	36	36 (20%)
Submarines	Torpedoes	18	18 (10%)
	Guns	0	
Aircraft	Bombs	1	1 (0.6%)
Unknown	–	37	37 (21%)
Accident	–	19	19 (10%)
TOTALS		178	178

The concept of a specialised ASW submarine was sound and, as Table 6 shows, submarines accounted for 10 per cent of all U-boats losses. The 'R' class never really had chance to prove itself due to its late entry into service, but the indications are that it would have been most successful.

The Royal Navy was, of course, under very stringent financial restraints in the 1920s and cuts had to be made somewhere, but it was most unfortunate that the admirals failed to see what a remarkable advance had been made in the 'R' class. So, had the effort and money that was poured into other classes been devoted instead to developing the 'R' class the Royal Navy might well have had submarines in the 1930s which outperformed every other submarine in the world.

Notes

[PRO = Public Record Office; NMM = National Maritime Museum; SubMus = Royal Navy Submarine Museum.]

1. PRO ADM 1/8480/36. Paper dated 21 February 1917.
2. Letter, Beatty to Captain (Submarines) Tees and Captain (Submarines) Blyth, dated January 1917 (PRO ADM 137/1926).
3. The scheme is described and two photographs included in PRO ADM 1/8498/208.
4. PRO ADM 137/1126.
5. Admiralty Board memorandum PRO ADM 167/54, pp62–64. Several sources refer to the DNC submitting a proposal to the Board in early 1917. There is no mention of any such proposal in the Admiralty Board Minutes until 27 September, although it may be that this was simply the time taken, even in wartime, for a proposal to work its way through 'the system'.
6. Some previous British classes had been fitted with auxiliary electric motors but these were connected to the main shaft by gears, and not mounted directly on the shaft as in the 'R' class.
7. This is *not* the old design, as the Admiralty Board paper (Note 5) refers specifically to '6–14in torpedoes' and to 'the design of the *new* torpedo tubes' [author's italics].
8. 'R' class 'Ship's Cover' held at the National Maritime Museum, Folio 18.
9. Admiralty Minute dated 29 October 1917 in Ship's Cover.
10. Dockyard Branch Minute dated 13 December 1917 in Ship's Cover.
11. Technical History No 21, p7.
12. Lecture by Captain C B Barry, DSO, RN to the RUSI on 23 January 1935 (SubMus A1983/32).
13. DNC Minute dated 14 June 1919 in Ship's Cover.
14. This manuscript handbook is held in the NMM at DEY/102. Although undated it appears to have been prepared after the war.
15. *Handbook of Hydrophones in Submarines, 1924* (CB.1644), PRO ADM 186/440.
16. Technical History Number 21, page 17.
17. Report to Cdre (S) by Lt C B Barry, 25 July 1918, in Ship's Cover.
18. Ship's Cover, Folio 80.
19. *One Man Band*, R-Adm Ben Bryant, William Kimber (London 1958).
20. PRO ADM 137/2077.
21. All these letters are in PRO ADM 116/3164.
22. Submarine Museum file SubMus 1947/1.

REVOLUTION MANQUÉ

Technical Change in the Royal Navy at the end of the First World War

War concentrates the mind wonderfully and, in the last years of the First World War and the following few years of peace, British naval staff, constructors and scientists were full of bright ideas. Major improvements or radical changes were in hand in virtually every aspect of the design of warships and their equipment though the benefits of many were to be lost in the era of peacetime economies. Though most of these changes have been described individually, the full scope of technical development is not generally appreciated and in the following survey David K Brown RCNC pulls together these separate strands.

The effectiveness of conventional designs, in some cases including existing ships, was much enhanced by a series of improvements, many of which were initiated by the lessons of Jutland.[1] The chance of hitting was improved by the extension of director firing to the main armament and later to the secondary armament of all battleships and by 1918 modern cruisers had been so equipped. Some ships were given longer rangefinders. In 1917 the Henderson gyro equipment was fitted to ensure that guns only fired at zero angle of roll (though this implied maximum roll velocity) and improved Dreyer fire control tables were introduced.

The new shell greatly increased the chance of penetrating armour, particularly at oblique impact, and their effectiveness was clearly shown in 1919 when *Erebus* and *Terror* fired thirty-one 15in shells of various design into the *Baden*. A 14in turret face plate was penetrated at a striking velocity equivalent to 15,500yds, at 18½° to the normal, by an APC shell while an SAPC penetrated the 6¾in upper belt and then broke up on the 6¾in 'B' barbette (42° to normal) knocking out a piece of armour 4ft × 3ft and jamming the turret.[2]

Temporary measures to improve flash protection were improvised within days of Jutland and tests of these and more elaborate precautions were carried out using the fore turret of *Vengeance* in August 1917, successfully resisting the explosion of two full 15in charges. Later tests, with more charges, showed a weakness in the flash doors but further trials in *Prince of Wales* in July 1919 showed that this problem, too, had been overcome. A number of other minor, but potentially serious, hazards were also identified and remedied, such as that of molten lead dropping from electric cables and igniting charges.[3]

New design mountings, such as the 16in and 8in turrets were given elaborate flash precautions, depending on complicated interlocks, which contributed much to the initial unreliability of the mountings.

Improved quality control during manufacture reduced the amount of impurity in cordite, making it somewhat safer. It took much longer to introduce the much safer solventless propellant known as SC which only entered service in 1927.

There were many other detailed but often important improvements in other aspects of warships, several of which are described in later sections.

New Battleships

The numerous studies for battleships of the 1913–14 programme, none of which were built, included fast options based on *Queen Elizabeth* but with a raised protective deck as well as slow options based on *Royal Sovereign*. Surprisingly, all options carried both coal and oil fuel. There was a more novel study in July 1914 for a 'High Speed Battleship', a 30kt *Queen Elizabeth* of 31,350 tons, foreshadowing the merger of the fast battleship and the battlecruiser. A further range of studies was produced in early 1916, with speeds ranging from 22–27kts. They had a thinner belt, only 10in thick, whose effectiveness was enhanced by sloping it inwards from top to bottom. The protective deck was the traditional 1–2in thick.

The postwar designs of 1920–21 were radically different,[4] much bigger and with heavier guns (16in for battlecruisers, 18in for battleships). The guns were not particularly novel. The existing 18in were 40-calibre and three 45-calibre were ordered from RGF (wire wound), Vickers (half wire wound) and without wire from

Baden *in 1919 during firing trials in which* Erebus *and* Terror *fired thirty-one 15in shells into her*. (CMP)

Prince of Wales*: two 15in charges exploding in the forward barbette during successful trials of magazine flash protection*. (Author)

Elswick. They weighed from 130 tons upwards but the shell was light at 2916lbs following some misleading trials. The conservative Ordnance Board chose the wire wound version before any of the guns was finished. The 16in was also wire wound and fired a light shell.

The early battleship and battlecruiser designs, under Attwood, resembled in appearance *Royal Sovereign* and *Hood* respectively but incorporated major new features. In particular, the armour deck was typically 8in thick over magazines and 5–6in over machinery. There was a price to pay: the G3's deck weighed 4430 tons, more than the belt. The battlecruisers aimed for 32–33kts while most battleship studies were for 23kts.

Later designs had a radically different configuration when Attwood, himself an innovator, was joined by Goodall with even more advanced ideas. The turrets were grouped forward to reduce the length of the heavily armoured magazines while the machinery, better protected by far than in previous ships was aft. For their day, they had an exceptionally heavy AA armament of 6–4.7in and 32–2pdr pom-poms (4×8).

Campbell has described these ships in detail and shown their superiority over all contemporary and most later designs. Much of the thinking went into the designs of *Nelson*, started in December 1921, which, as details were gradually revealed, was seen by other countries as an exceptionally effective battleship.

Air Power

The Royal Navy with the Royal Naval Air Service (RNAS) had pioneered most of the technology and operational doctrine of the air war at sea. It had carried out the first air strikes from sea against land targets, the first torpedo bomber attack and, when the war ended, had the world's only true aircraft carrier, *Argus*, with a squadron of Cuckoo torpedo bombers embarked, ready to attack the High Seas Fleet at anchor.

Many of the later developments derived from an imaginative plan put forward in 1916 by Commodore

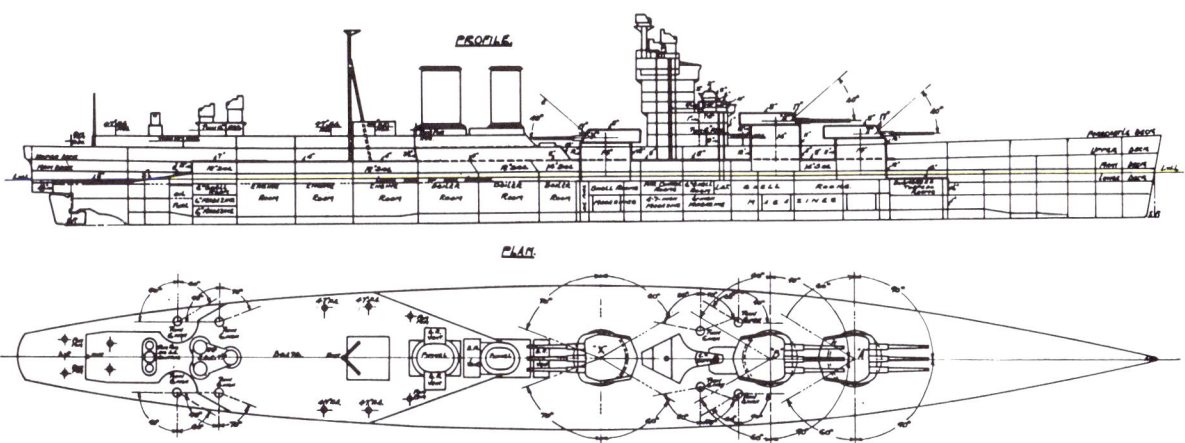

The ultimate battlecruiser, the G3 design, in which Atwood and Goodall grouped all the turrets forward reducing the length requiring heavy armour which was 16in thick on the side and 8in on deck.

▼ *There was to have been a battleship equivalent (N3) of the G3 – 48,500 tons, 23kts and 9–18in guns.*

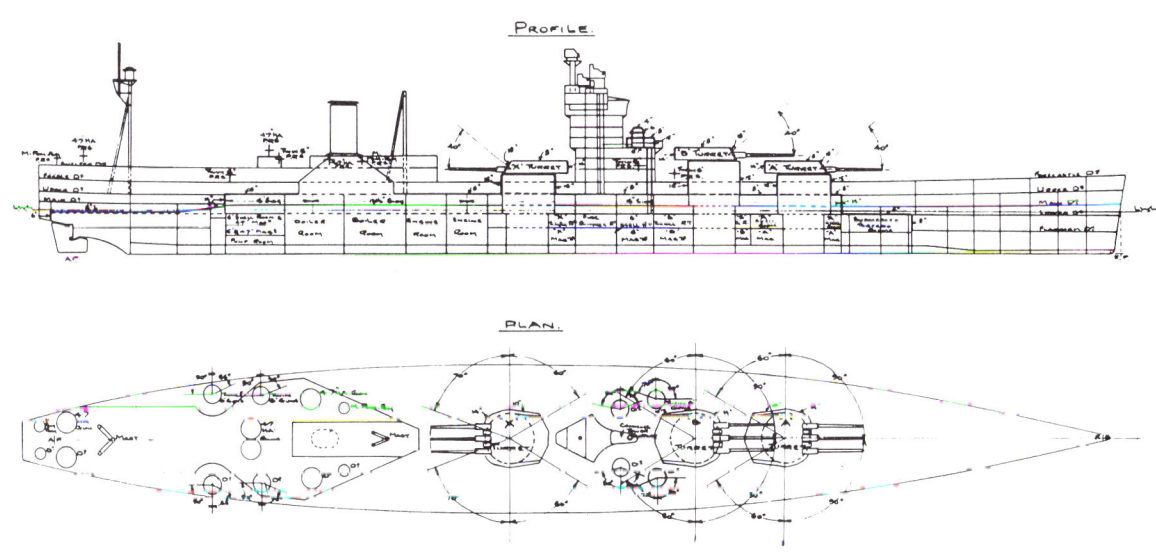

A Cuckoo *dropping a torpedo.* Argus' Cuckoo *squadron was working up when the end of the war prevented a torpedo attack on the High Seas Fleet at anchor.* (By courtesy of Owen Thetford)

Sueter for an attack on the German fleet at anchor by 121 torpedo bombers launched from eight aircraft carriers. This plan was quickly endorsed by Beatty, commanding the Grand Fleet, and implemented in large part (although, limited by shipyard capacity) by the Board under Jellicoe. The carriers *Argus*, *Eagle* and *Hermes* owe their origin, in part at least, to this plan, as did the Cuckoo torpedo bomber.

The RNAS was a very large air force and in 1918, when it was subsumed into the new Royal Air Force, it comprised some 2000 aeroplanes, 650 seaplanes, 150 flying boats and 100 airships. Neither the formation of the RAF nor the impact of peace led to any immediate loss of interest in naval aviation. Thinking on carriers culminated in the 1923 designs of which Design B (submitted on 27 November 1923) was favoured. On a displacement of 16,500 tons (standard) she would have carried 32 Dart or Bison aircraft, themselves postwar designs, with a speed of 34.5kts and an armament of 6–4.7in AA guns. There was a bigger variant with 36 aircraft and 6–8in guns.

At the end of the war there were 103 aircraft embarked on ships (besides those on the carriers) of the Grand Fleet which could take off from flying-off platforms carried above turrets.[5] These planes could not be recovered at sea and if far from land they had to come down on the water and were lost. Carriage of planes on battleships and cruisers was temporarily abandoned after the war as the new planes were too big to take off from platforms and too expensive to be abandoned after one flight. They were not yet strong enough to land on floats in anything other than a dead calm sea. However, catapult development was in hand for launching and the stronger seaplanes from about 1930 could land in a moderate sea or even on a carrier's deck.

Much of the RNAS strength was shore based for ASW operations in the North Sea and other areas adjoining the United Kingdom.[6] Aircraft on these duties included: 75 airships, non-rigid; 190 landplanes; 216 seaplanes, floats; 85 large flying boats.

The effectiveness of their operation is somewhat uncertain as only one sinking of a U-boat was recorded. On the other hand, it is claimed, almost certainly correctly, that submarines were forced to dive and remain submerged in daylight, much reducing their mobility and affecting their morale. The flying boats were very impressive aircraft indeed, perhaps the largest military aircraft operated in any number, but the Kangaroo, a land plane of half their weight carried twice the payload.

Shore based ASW probably suffered more than any other branch during the years of peace and Coastal Command of 1939 was a pale shadow of the 1918 force – its most numerous aircraft, the Anson was inferior to the Kangaroo. The RAF had devoted most of the small resources available to the flying boat rather than to the more effective land plane. They, too, were possibly affected by the idea that Asdic had rendered the submarine impotent.

The Air Ministry applied quite considerable resources to the development of the rigid airship which in British hands had proved expensive and ineffective. One may also doubt if the German airship programme really justified its cost. The cheap and effective non-rigid airforce was scrapped as soon as the war was over.[7] Perhaps a small part of the funds wasted on rigid airships should have been applied to a small development unit of non-rigids though the USN non-rigid airships of the Second World War do not seem to have been very effective.

About 170 RN ships (and some USN) were fitted to operate kite balloons and five special ships were fitted as depot ships with hydrogen generators. The balloons were used for gunnery spotting in big ships and looking for submarines. It was thought that the advent of sufficient 'spotters' from aircraft carriers obviated the need for balloons though one may query if there were ever enough carrier aircraft.

There were plans for even more novel weapons. By 1921 a primitive radio controlled 'cruise missile', the 'RAE 1921 Target' (a small, propeller driven aeroplane) was tried with a 200lb warhead and by 1927 this had evolved into the Larynx with a 250lb warhead.[8] These missiles were ahead of their day: radio control was uncertain – it took many years to develop a radio controlled AA target aircraft – and the warhead was not worth the effort.[9]

The contrast between the RNAS of 1918 – numerous, ahead in technology and operational thinking – with the Fleet Air Arm and Coastal Command of 1939 is pathetic.

Anti-Aircraft

The Admiralty believed that a powerful anti-aircraft gun armament could keep attacking aircraft at a height and range from which hits would be few. There was debate over the choice of gun with the 4.7in Mark VIII, which had quite a good performance for the day, as the choice for the first postwar ships.[10]

The close range weapon was far more radical and at its inception well ahead of thinking in other navies. It was based on the 2pdr pom-pom, Mark VIII, and an extemporary six-barrel mount was tried in *Dragon* in 1921. The mock-up of the final eight-barrel version was approved in 1923 but its production was delayed by the Treasury who refused to make funds available (possibly inspired by RAF objections), perhaps the only real instance of Treasury delay.[11] In consequence, ship fitting did not take place until the early 1930s by which time the pom-pom was approaching obsolescence.

The new Admiralty Research Laboratory worked on a number of AA control systems, including a tachymetric version, stopped by the Board when on the point of success in 1928. Lord Chatfield says it was the single most disastrous error between the wars – perhaps their Lordships were obsessed with game bird shooting.

The Admiralty was not seriously concerned over the effects of bombs of the early 1920s against battleships.

The Kangaroo *shore based anti-submarine aircraft of 1918 was superior in most aspects of performance to the* Anson *of 1939.* (By courtesy of Owen Thetford)

During the 1920s attempts were made to develop a guided missile, culminating in the Larynx *of 1927, seen being catapulted off from the* Stronghold. (By courtesy of R D Layman)

There were 2000lb bombs in existence but they could only be carried by large, slow and unmanoeuvrable aircraft, easy targets, and their striking velocity was less than that of plunging fire from 16in APC which armoured decks were intended to resist. However, a considerable number of full scale trials were carried out both to test the effectiveness of bombs and to ensure that there was no Achilles heel in the battleships' protection.

The first problem was that of hitting an undefended, stationary target at all with bombs dropped from aircraft. For example, in the 1924 trials against *Monarch*, 57 bombs were dropped for 11 hits even after two or three practice runs using 9lb practice bombs. To obtain a hit on a specified part of the ship was clearly impossible. The solution was ingenious: it was found possible to fire bombs of all sizes, up to 2000lbs, from howitzers against targets representing the horizontal structure of a ship, but arranged vertically, at Shoeburyness. The penetration through these structures was measured for different bombs and replicas were placed in realistic positions which could cause serious damage on board target ships and detonated.

Such trials were carried out in 1921 in the ex-German *Baden* with six bombs from 520lbs to 1800lbs. The following year *Gorgon* was used to study the effect of near-miss bombs exploding in the water close to the hull following the USN trials against *Ostfriesland*. Damage, even from a 2000lb bomb was slight.

In 1923 a replica of a 60ft length of *Nelson*'s side protection was built on to *Superb* and tested. The biggest charge was of 2082lbs TNT (representing a 4000lb bomb) and was exploded 7ft 6ins from the side, at a depth of 40ft. The protective bulkhead was distorted but watertight; the turrets still functioned as did the machinery. There were leaks in the circulating water inlets so new designs were prepared and tested.

Several tests in the *Monarch* trials of 1924 were aimed at studying the effects of bombs exploding in the uptakes. As a result, stronger armour gratings were introduced. The effect of the blast on a working boiler was considerable and improvements were considered. It was noted that none of the bombs dropped from aircraft succeeded in penetrating the 1½in upper deck in a fit condition to explode. Bombs which were positioned and exploded in

The Admiralty was keen to develop the capability of the Fleet Air Arm and, in 1923, a class of advanced carriers were designed but not built. (Norman Friedman)

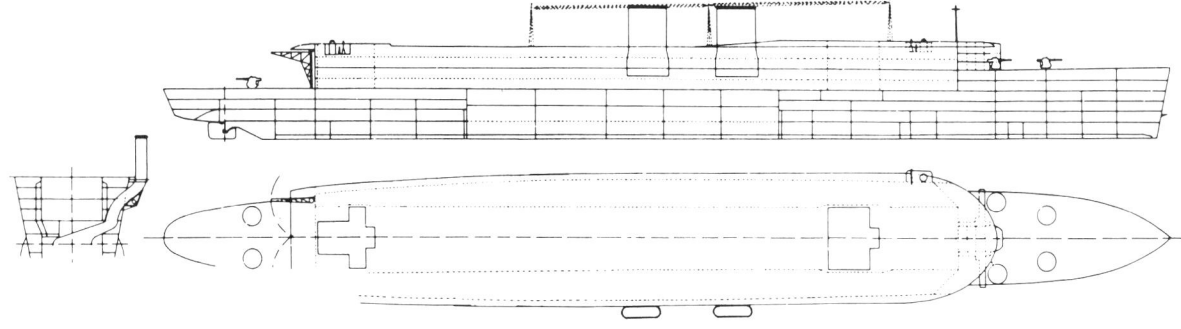

unprotected areas caused extensive devastation to structure and systems over two to three main compartments.

The difficulty in hitting a stationary target was seen to justify the view that bombing could not hit a small ship manoeuvring at speed. While this view was probably correct in the 1920s, it ceased to be so with the introduction of dive bombing.

Torpedo Protection

The RNAS had pioneered the air launched torpedo attack and countermeasures for RN ships were not forgotten. Since submarine torpedoes were bigger than those which could be carried by aircraft, the design of protection was directed towards the larger weapons. Tennyson d'Eyncourt's bulge had proved very effective in resisting torpedoes during the war and most large warships were 'bulged' during the 1920s, though often these were too shallow to be effective against bigger torpedoes. Even a small bulge, projecting from the ship, could reduce speed by 1–1½kts with a greater loss if the bulge was enlarged to protect against bigger torpedoes.

Attention was focused on internal protection, with a series of longitudinal bulkheads forming compartments, some filled with water to stop splinters, others empty to dissipate blast. There would be an inner bulkhead, up to 2in thick and, since even this might leak, there would either be a cofferdam or a small watertight compartment used for a non-vital function arranged on the inboard side. The inboard, sloping belt armour was arranged so that torpedo explosions would vent outboard.

A large pontoon, 80ft long, was built at Chatham in 1915 and for many years was used to test protection against underwater explosions.[12] In 1920 a replica of *Hood*'s bulge was successfully tested against a 500lb charge but it was recognised that a modern protection scheme must resist 750–1000lbs. Scale tests representing a 1000lb charge were held in 1921 against both RN and USN systems in parallel with similar US tests – with contradictory results, each country showing that its scheme was the better.

The *Nelson*'s side protection was tested with a 1000lb charge which caused the protective bulkhead to distort through 3½ins. Tests of under-bottom explosions were carried out in July 1921 when a 350lb charge was exploded under the float. As a result, the new capital ship designs were given a double bottom 7ft deep though it was later realised that this stiff structure would apply a severe load to the base of the transverse bulkheads, rupturing them over a greater length of the ship.

As a result of these realistic full scale tests, the Admiralty was convinced that modern large warships could be protected against all contemporary weapons.

The multiple pom-pom was tried as a mock up in 1921 and the eight-barrel design was approved in 1923. Few were fitted until the mid 1930s due to financial restriction. (CMP)

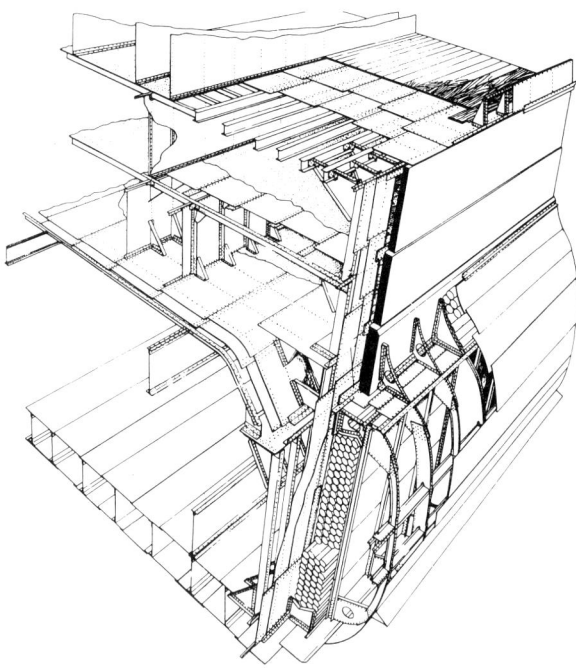

Testing of torpedo protection systems both on models and full scale continued energetically. The diagram shows that fitted to Hood, *a fairly early form of the bulge. Many of the sealed steel tubes forming the inner layer were seen floating after the* Hood *sank.* (John Roberts)

Anti-Submarine Warfare

From early 1917 onwards, experiments had been carried out into the detection of submerged submarines using echo ranging with high frequency sound.[13] By early 1918, success had been achieved, with French assistance, using the piezo electric effect on quartz to generate the sound and, in June 1918, the first production sets of Asdic (later active sonar)[14] were ordered. By 1919, one of these sets, Type 112, mounted in *P-59* was achieving detection ranges of 2500–3000yds, in ideal conditions, with the ship steaming at 15kts.

The various *ad hoc* establishments set up during the war to study different aspects of ASW were rationalised, ultimately forming an experimental seagoing unit at Portland (HMS *Osprey*) and the Admiralty Research Laboratory at Teddington for more fundamental work. This establishment, ARL, was given a wide range of tasks in the weapon area. Relationships between the scientific staff and naval officers were not good as they came from widely differing cultures, though Hackmann[15] seems to overstate the problem.

At the risk of oversimplification, the scientist wished to delay production until the perfect unit had been tested whilst the naval officer would rush an imperfect but usable set into service to get some operational benefit – and then complain that all the bugs had not been removed. Clearly this was a real dilemma with no perfect solution then, or even today.

The introduction of Asdic, (parallelled in USN work), was an enormous step forward in ASW. By 1918, the U-boat offensive had been defeated by a combination of the convoy system and escorts dropping depth charges almost by guess work. It was not unreasonable to assume that Asdic, with a capability of detecting and locating a submarine at well over 1000yds, had almost eliminated the submarine threat.

It does not seem to have been appreciated that detection ranges of 1000–3000yds still implied the need for a large number of escort vessels and the use of night surface attack was ignored. Such thinking was encouraged as most AS training was held in waters where conditions were favourable to the use of Asdic. Over-confidence in the value of Asdic and hence the belief that the submarine threat had almost vanished led to a number of decisions some of which, with hindsight, must be regretted.

Though the British delegation to the Washington Treaty negotiations (and later Treaties) argued for a total ban on submarines, it appears that this was a formal position and that it was always recognised that it had little chance of success.

One major casualty of Asdic's success was the silent propulsion programme. Early attempts to locate submerged submarines depended on listening for the noise they made using a hydrophone. This could only be done with the hunting vessel stopped as, when under way, the noise from the submarine was drowned by noise from the ship's own machinery and propellers. From early 1917, various attempts were made to silence the surface ship sufficiently for it to use hydrophones – today's passive sonar – when moving.

Several individuals suggested surrounding the hull outside the machinery with a screen of air bubbles to absorb the sound. Model tests showed promise and a trial was arranged on *ML-497* using air at 100psi from a reservoir. There was some reduction in noise radiated into the sea but it was clear that the air supply was inadequate.

Further trials were carried out in May 1918 with *PC-43* but there was no suitable air compressor to meet the requirement of 15,000cu ft/min at 20psi. Eventually two 90hp water pumps delivering through an aerator had to be used. The trials were a failure, *PC-43* being noisier with the screen than without. It was recognised that the principle was sound and that the problem lay in the air supply. Air bubble screening of machinery was rediscovered after the Second World War as 'Masker' and proved successful once continuing difficulties in supplying air were overcome.

Another approach to reducing underwater noise involved replacing the propeller with jet propulsion using internal water pumps and two of the 'Strath' class trawlers were converted before completion. The *George Ireland* (later *Teviot*) and *Henry Jennings* (later *Ure*) were given jet drive designed by Major Gill and were compared with a propeller driven sister, the *Thomas Ansell*. The jet was very inefficient.

EFFICIENCY OF JET DRIVE

Ship	Propulsion	Coal consumption (cwts/hr)	Speed (kts)
Henry Jennings	Jet	5.25	7.91
Thomas Ansell	Screw	5.0	10.3

It was reported that the jet drive was much quieter but it needed 395ihp for 8kts compared with 165 for the screw. The overall efficiency was 0.218 which was worse than an earlier jet driven craft, *TB-98*, which achieved 0.254. Astern thrust was provided by a reversing bucket but *Jennings* failed to go astern at all.[16]

Both bubble screening and jet propulsion were abandoned and with them, passive sonar, as it was believed that Asdic had eliminated the need for 'silent propulsion'. It is ironic that Asdic, which was one of the great successes of the period should, indirectly, lead to several failures.

Submarines

There was some muddled thinking on future submarine policy which may, in part, have been due to the impression that the role of the submarine had been diminished. The 'R' class submarine had been designed for anti-submarine work with the then very high under water speed of 15kts. They were soon disposed of for reasons which are not entirely clear. The feeling that their role was now unnecessary may have been one reason but it may also have been appreciated that their combination of high speed, limited diving depth and crude control operation was hazardous in peacetime.

Wartime lessons were misread by major navies, all of whom built large cruiser submarines which were unsuccessful. The RN built an improved, steam driven submarine, *K-26*, and then *X-1*, armed with four 5.2in guns. Both were failures. On the other hand, the Admiralty realised that in a Pacific war with Japan, the RN would be the inferior power, at least for a time.[17] The 'O' class were designed, based on the 'L's with much longer endurance and a 500ft diving depth – though no one seems to have been brave enough to try it. As part of this Pacific strategy, the Admiralty and the Post Office developed long range, low frequency radio communications from the UK to submarines.

Torpedoes

Much effort was devoted to more powerful torpedo propulsion systems while, if possible, reducing the bubble track left behind.[18] From 1920 these efforts were concentrated on enriched oxygen using 'air' with up to 57 per cent oxygen in the Mark VII of 1928. There were fears of explosion with such an oxidant and real concern over the rapid corrosion which it caused in the air vessel. By 1923, some work was in hand on a high test peroxide plant but all development on propulsion came to an end in the late 1920s with the introduction of the Brotherhood Burner Cycle engine. This was so compact and powerful, so far ahead of other navy's plant that, for a decade or so, there seemed no need for further improvement.

There was some interest in bigger torpedoes but the 24.5in used in *Nelson* was adopted so that a shorter weapon resulted – one not quite so difficult to transport to the torpedo flat, low in the ship and right forward. The magnetic influence exploder, used unsuccessfully by the Germans during the war, was studied but the difficulties were great and even the partial solution lies outside the period discussed here.

Air bubble screen of machinery and propellers was tried on ML-497, the air being distributed from belts round the hull, seen above. The principle was sound and was developed after the Second World War but practical problems were not overcome before the scheme was abandoned. (Author)

Despite the RN's successful use of magnetic mines in the war and tests of a prototype acoustic mine, there seems to have been little effort made on mines or on countermeasures.

Cruisers and Destroyers

The 'C', 'D' and 'E' class light cruisers and the 'V & W' destroyers were seen as excellent and, since they were available in considerable numbers, little thought was given to these categories. There was a clear need for larger, trade protection cruisers but there is no record of design studies between the *Hawkins* and *Kent* classes.

The *Kents*, the first post Washington design, were extremely advanced ships for their day. Their 8in guns were intended for a high rate of fire: 5 rounds per minute, with 70° elevation and quite fast training and elevation so that they had some AA capability. They also had 4in AA and two of the new eight-barrel pom-poms, an AA capability far in excess of their contemporaries. Tennyson d'Eyncourt gave them a novel, deep hull to save structural weight under the Treaty which, incidentally, gave them a high freeboard and airy mess decks.

Tennyson d'Eyncourt was also much interested in the

Big, 'cruiser' submarines were built by most major navies despite the failure of German cruisers in the war. K-26 was a much improved steam driven boat whilst X-1 had a heavy gun armament of four–5.2in (with director). Neither was successful. (CMP)

idea of diesel engines for cruisers and corresponded with Armstrongs on tentative schemes. Nothing came of this: available diesels were too big and heavy, but a diesel-electric cruising plant was put into the minelayer *Adventure*, which served to demonstrate that current diesel reliability left much to be desired. The Washington Treaty excluded fuel from the standard displacement making the use of diesels, with their low fuel consumption, less attractive.

The big battlecruisers had been designed with transom sterns to reduce their length and so permit existing docks to be used. Model tests at the Admiralty Experiment Works showed that this stern shape led to a small increase in top speed and it was incorporated into most new designs including, unfortunately, *Adventure*. The eddies behind the transom caused mines to break their horns on the transom as they were laid and *Adventure* had to be rebuilt with a new stern, putting back the general adoption of transoms for many years.[19] *Adventure*'s bridge design, high freeboard and even her three funnels most probably influenced the *Kents*.

In passing, one may note that CMBs, fast motor torpedo boats, were largely abandoned, perhaps justifiably in the light of the destruction of a flotilla by German seaplanes in April 1918.

Technology

During the war there were rapid developments in electric arc welding of both structure and equipment and there was a very active co-operation with the USN, fostered by Goodall, the constructor attaché in Washington. It was expected that the all-welded ship would appear shortly but there was intense opposition from shipbuilders, both management and labour which, in some cases endured

REVOLUTION MANQUÉ

Canberra; *similar to the Royal Navy's* Kent *class but with taller funnels. These ships were the most advanced of the many 'Treaty' cruiser designs. To reduce hull weight, the depth of the hull, keel to deck, was increased, leading to the high freeboard seen in this photo.* (CMP)

until the end of the Second World War. One must exempt Cammell Laird from this criticism as they built the cargo ship *Fullagar*, the world's first all-welded ship in 1921. The RN's all-welded ship, *Seagull*, did not appear until the late 1930s and had to be built in a Royal Dockyard because of opposition in the industry. There were real problems in developing suitable steels for welding and in the supply of welding equipment but these could have been solved.

The first generation of postwar machinery developed from wartime plants and all were geared turbines with small-tube boilers but using superheated steam making them more efficient. There were few later developments in machinery as Board policy had become hostile to naval engineers and their work.[20]

The newly formed Admiralty Research Laboratory was to do much for weapon system development. In 1919, R E Froude ended his forty years as Superintendent of the Admiralty Experiment Works (hydrodynamic research and testing) and considerable resources were devoted to updating the equipment, culminating in the commissioning of a new, much larger ship tank in 1928.

Why?

There are two questions to answer: why was there this flood of innovation from about 1917 to 1923 and why did it almost stop? During the good years, new ideas usually came from the middle ranks, navy, constructor and scientist, both temporary and permanent. Leading from the top, d'Eyncourt was an exception, with the bulge and *Kent*'s novel structure to his personal credit. However, new ideas were generally supported by the most senior officers as instanced by the backing given to Sueter's torpedo bomber plans by Beatty and Jellicoe, neither known as innovators.

It must be realised that the navy was accustomed to technical change. Since the turn of the century there had been continual changes in gunnery, machinery, radio and ship design, etc. Change was perhaps seen as natural in 1917. During the war there was a clearly perceived need to improve many aspects, particularly ASW, technical skills were available and funding was not difficult. It is no surprise that enthusiasm so generated led to the innovations in the later war years and there was sufficient impetus to carry it forward for a few more years.

At the end of the war, many temporary officers and civilians left the service and many of those remaining were inclined to accept the general view that the peace would be lasting. A specific case was the loss of the RNAS, with so many inventive officers, to the RAF.

Resources of all kinds were scarce and it was no longer possible to 'have a go' with every bright idea and all such schemes had to be justified in tedious detail before priority and finance committees and, even if it was accepted, it

Adventure, *a minelayer, introduced a number of advanced technical features including a transom stern and diesel-electric cruising propulsion. Eddies behind the transom led to mines hitting the ship as they were laid and the transom was rebuilt in* Adventure *and not used in other classes where it would have been more appropriate.* (CMP)

was at the expense of another proposal of comparable value. It is almost impossible to write a formal justification for the most far out schemes and limited improvements stood a better chance of approval.

The Board seems to have thought that the war had shown that too much thought had been given to technology in earlier years and not enough to the philosophy of strategy, tactics and planning. While this conclusion may well be right, the solution which they adopted – to cut technical work, rather than extend strategic planning – was wrong. Reaction was not universal as shown by Haslar's new tank in 1928 but it required a great deal of hard work to overcome the obstacles. Any reaction within the Admiralty was negligable compared with that in the marine industries, as shown by their opposition to welding and later to longitudinally framed small ships. Comparison with the years after the Second World War cannot be made since the threat of the Third World War was clearly perceived.

Overall, the navy did change remarkably in this period with aircraft carriers, Asdic, *Nelson* and *Kent* to name but a few examples; but much more could have been done, at modest cost, in new technology between 1923 and rearmament in the late 1930s. The generation of senior officers who grew up under Fisher seem to have rejected his vision of technology.

Notes

1. David K Brown, 'The Surface Fleets of World War I', *Les Marines de Guerre du Dreadnought au nucleaire*, Service Historique de la Marine (Paris 1991).
2. John Campbell, *Naval Weapons of World War Two* (London 1988).
3. David K Brown, *op cit*.
4. John Campbell, 'Washington's Cherrytrees', *Warship* 1–4 (London 1977).
5. Roger Nailer, 'Flying off Platforms', *Warship Supplement* 86–88 (Kendal 1986–87).
6. Alfred Price, *Aircraft versus Submarines* (London 1980).
7. Patrick Abbott, *The British Airship at War 1914–1918* (Lavenham 1989).
8. Roger Nailer, 'Larynx – The First Deterrent?', *Warship Supplement* 100 (Kendal 1990).
9. Richard D Layman, 'Catapults', *Warship International* VIII/3 (1970).
10. John Campbell, *Naval Weapons*.
11. G C Peden, *British Armament and the Treasury 1932–1939* (Edinburgh 1979).
12. David K Brown, 'Attack and Defence, Part 3', *Warship* 24 (London 1982).
13. Willem Hackmann, *Seek and Strike* (London 1984).
14. The derivation of the word Asdic is considered by Hackmann, *op cit*. While it is clear that ASD stands for Anti-Submarine Detection, it seems likely that IC does not have any meaning.
15. Willem Hackmann, *op cit*.
16. David K Brown, 'Jet Propulsion in the Royal Navy', *Marine Propulsion* 3/80 (Redhill 1980).
17. David Henry, 'British Submarine Policy 1918–39', in *Technical Change and British Naval Policy*, B Ranft (ed) (London 1977).
18. This section is based on unpublished work by G J Kirby.
19. David K Brown, 'The Transom Stern', *Warship* 5 (London 1978).
20. Sir Louis Le Bailly, *From Fisher to the Falklands* (London 1991).

THE FRENCH FLOTILLA PROGRAMME OF 1922

John Jordan describes the design origins of French inter-war flotilla craft, tracing the thinking behind both the standard destroyers and the large contre-torpilleurs, *and concludes with an evaluation of the resulting* Bourrasque *and* Jaguar *classes.*

The *Statut Naval* of 1912, which was intended to remedy the decline of the *Marine Nationale* in relation to the other major European powers, established the force structures to be achieved by 1920 as: 28 battleships, 10 scout cruisers, 52 fleet torpedo boats, 94 submarines, and 10 vessels for distant stations.

The imbalance between the number of capital ships and the number of torpedo boats intended to accompany them is striking. The low priority accorded to flotilla craft in the programme would become alarmingly apparent in the following months. The year 1912 saw the authorisation of no fewer than twelve 'super-dreadnought' battleships of the *Bretagne*, *Normandie* and *Lyon* classes. In contrast, the number of fleet torpedo boats ordered in 1913 was cut to three (in previous years an average of six or seven boats had been ordered).

Last of the twenty-one 800-tonne torpedo boats designed in the years which preceded the First World War, Enseigne Gabolde *was modified while on the stocks to serve as a trials ship for the geared steam turbines to be installed in the next generation of destroyers.* Enseigne Gabolde *was the only ship of her class to have an additional superimposed 100mm gun forward – a feature which was to prove unsatisfactory because of the concentration of weights in the bow. Note the gunshields on 'A' and 'X' mountings, which are similar in configuration to the early model fitted in the ships of the 1922 Programme.* (Marius Bar, 1921)

Even more ominously, a ministerial decree of March 1913 reclassified 53 elderly torpedo boats of 300/330 tonnes as *'torpilleurs d'escadre'*. It was becoming increasingly clear that the figure of 52 fleet torpedo boats would be achieved not by new construction but by reclassifying elderly vessels with inadequate speed, endurance and weaponry for the purpose. These fears were confirmed when budget proposals for 1914 included no provisions for new torpedo boat construction; nor was there any indication that new vessels would be authorised in the following year. Meanwhile, the battleship programme continued to roll, with four ships of the *Normandie* class being laid down in 1913, and a fifth in January 1914. The only fleet torpedo boats capable of accompanying them were the twenty-one 800-tonne ships of the *Bouclier*, *Bisson* and *Enseigne Roux* classes laid down from 1909 onwards.

This potentially disastrous situation prompted a series of detailed submissions directed to the Minister by the Etat-Major Général (the French Naval Staff), in which the case for an increase in both the quantity and the quality of torpedo boat provision was forcibly argued. A submission dated 24 April 1914 made the following proposals:

- that 52 fleet torpedo boats of 800/1500 tonnes (the classification *'torpilleurs de haute mer'* – 'ocean-going' or 'high seas' torpedo boats – was used to distinguish these from the elderly 300/330-tonne types recently reclassified as *'Torpilleurs d'escadre'*) should be built at the rate of 9/10 per year for work with the battle fleet.
- that the 300/350-tonne and 450-tonne torpedo boats of earlier programmes be relegated to anti-submarine duties, and attached to the *Flotille des Défenses Sous-Marines*.
- that the 52 *torpilleurs de haute mer* be organised as six *escadrilles* each of eight units (four of the six *escadrilles* would be new-build 1500-tonne vessels) and attached to the battle fleet (the *1ère Armée Navale*).
- that the four remaining units (to be used for replacement) be attached to the *2e Escadre Légère*, which would comprise three *escadrilles* each of eight torpedo boats (predominantly 450-tonne and the elderly 300/330-tonne types).
- that the 800-tonne type could be regarded only as an intermediate type, as its limitations for fleet work were already apparent, and that it should therefore be relegated to the *Flotille des Défenses Sous-Marines* as new 1500-tonne torpedo boats were completed.

In order to deliver the above the following construction programme was proposed:

- nine 1500-tonne torpedo boats to be laid down in 1915, ten in 1916, six in 1917, and seven in 1918 (a total of thirty-two, to which would be added the twenty 800-tonne torpedo boats already completed or building – the third unit of the 1913 series, the *Enseigne Gabolde*, was to be completed as an experimental ship for geared steam turbines.

Bourrasque, *the name ship of the first series of* 1500 tonnes, *photographed in January 1927 shortly after completion. Note the height of the funnels, soon to be reduced, and the single 75mm HA gun on its platform amidships between the triple banks of torpedo tubes.* (Marius Bar, 21 January 1927)

– a follow-on programme 1919–25 to replace older torpedo boats, with three 1500-tonne ships plus five of a smaller type intended for the *Flotille des Défenses Sous-Marines* per year (in 1923 the figure would be two 1500-tonne + six smaller)

This programme could hardly be described as over-ambitious when compared with the destroyer programmes of other major European navies or indeed with the French battleship programme which was to run concurrently. It did not however receive official approval, and in the event proved incapable of fruition given the difficult military and industrial circumstances in which France found herself following the outbreak of the First World War.

The Genesis of the '1500 Tonnes'

The twelve 800-tonne torpedo boats of the *Bouclier* class, completed from 1911 to 1913, proved unsatisfactory in a number of respects: they were criticised for their mix of 100mm (3.9in) and 65mm (2.6in) gun calibres, for the fragility of their hulls and machinery, and for their lack of endurance. Although these privately-built boats largely exceeded their designed speed of 30kts on acceptance trials, they were found in service to be capable of only 25kts in formation. This was acceptable when France was building dreadnoughts with a maximum speed of 20–21kts. However, the super-dreadnoughts of the *Lyon* class due to be laid down in 1914 had a maximum speed of 23kts; the battlecruisers under consideration in 1913 had designed speeds of 26–28kts; and 'fast battleships' modelled on the British *Queen Elizabeth* class were projected, also with a speed of 26–27kts.

In order to provide the necessary speed margin over the new capital ships the design speed of the next generation of destroyers would have to be at least 33kts. This implied an increase in displacement to 1000/1200 tonnes, making these ships comparable to the lastest British destroyers. Such was the torpedo boat design proposed by the *Conseil Supérieur de la Marine* in April 1913. Hull strength and endurance (2600nm at 10kts) would be improved, and the increase in size would enable a much more powerful gun and torpedo armament to be carried. After considering a uniform gun armament of three/four 100mm disposed on the centreline, the *Conseil* opted instead for two light-weight 140mm (5.5in) mountings. There would be eight 450mm (17.7in) torpedo tubes (double the number in the *800 tonnes*), disposed as two singles abeam the bridge structure (*cf* contemporary German torpedo boats) and two triple mountings on the centreline.

The 140mm calibre effectively outgunned all contemporary foreign destroyers and torpedo boats, and had been first considered in 1912, when the *Service des Travaux* estimated that a short 140mm gun firing a 36.5kg (80lb) shell at a muzzle velocity of 514m/s (1686f/s) could be built for a weight of only 4.8 tonnes, as an alternative to a 100mm firing a 14kg (31lb) shell at an initial velocity of 830m/s (2723f/s).

In June 1913 a detailed study was commissioned from Schneider, who in February 1914 submitted a proposal for a semi-automatic 140mm/25 with a horizontal sliding breech – a novel feature which would not be adopted by the *Marine Nationale* until the late 1920s, following trials with the German 15cm KL/45. The proposed 140mm/25 would have a maximum elevation of 20°, and would fire a 36kg shell at a rate of 15 rounds per minute with a muzzle velocity of 550m/s (1800f/s). The total weight of the mounting would be 6.65 tonnes, somewhat heavier than the *Service des Travaux* had initially anticipated, but still significantly less than that of comparable 45-calibre mountings designed for cruisers and battleships. The only reservation expressed by the *Marine Nationale* concerned the wide dispersion of the gun (a consequence of low muzzle velocity), which meant that it could be considered effective only out to 6000–6500m (6500–7100yds).

It was proposed that two *torpilleurs d'escadre* with the characteristics outlined above, designated M89 and M90, be laid down at Rochefort under the 1915 Programme for completion in 1917. In the meantime the Naval Staff decided that the geared steam turbines proposed for these ships (a further novelty for the *Marine Nationale*) would be trialled aboard the third unit of the *Enseigne Roux* class, the *Enseigne Gabolde*. The ministry refused to sanction the construction of M89 and M90, even though orders had already been placed for their torpedoes. However, studies of the new torpedo boats continued until the outbreak of the First World War, when a moratorium was placed on all new construction. By this time they had grown in size to 1530 tonnes, with a complement of 155.

In April 1917, with the U-boat war in the Atlantic at its peak and the Western Front stabilised, the project was revived. The lessons of wartime operations were incorporated in an amended proposal for a vessel with at least three 140mm guns (two forward of the bridge, one aft) and two AA guns. The single torpedo tubes of the 1914 design were to be suppressed, but this would be compensated for by increasing the calibre of the triple centreline tubes to 550mm (21.7in) and providing centralised fire control from the bridge. Endurance would be increased to 3000nm at 16kts.

This proposal was further refined in December 1917, when the Naval Staff amended the *torpilleur d'escadre* requirement to a 35kt vessel with an endurance of 3000nm at 14kts, and established a parallel requirement for a 'flotilla leader' (*conducteur d'escadrille*), capable of the same speed but with increased endurance (3500nm at 14kts) and a fourth 140mm mounting.

Although a study for two 1700/1800-tonne vessels designated '*contre-torpilleurs*' (literally 'torpedo-boat destroyers') was commissioned from the Chantiers Normand in the same month, these ships must be regarded as only indirect antecedents of the revolutionary type of the same name authorised in 1922. The EMG submission of 1917, together with subsequent notes dating from 1918, makes it clear that the inspiration for both the above types was British, the 'V' class being specifically mentioned as a model for the new *torpilleurs d'escadre*. The larger ships would therefore simply have been flotilla leaders in the mould of the British *Shakespeare* and *Scott* classes, three of which had been launched in 1917.

Tigre: *Inboard Profile*

Key:
1. Fwd 130mm magazine
2. Fwd magazine hoists (P&S)
3. Navigation bridge
4. W/T office (fwd)
5. W/T office (aft)
6. After magazine hoists (P&S)
7. After 130mm magazine
8. Depth charge magazine
9. Emergency steering position
10. Depth charge chutes

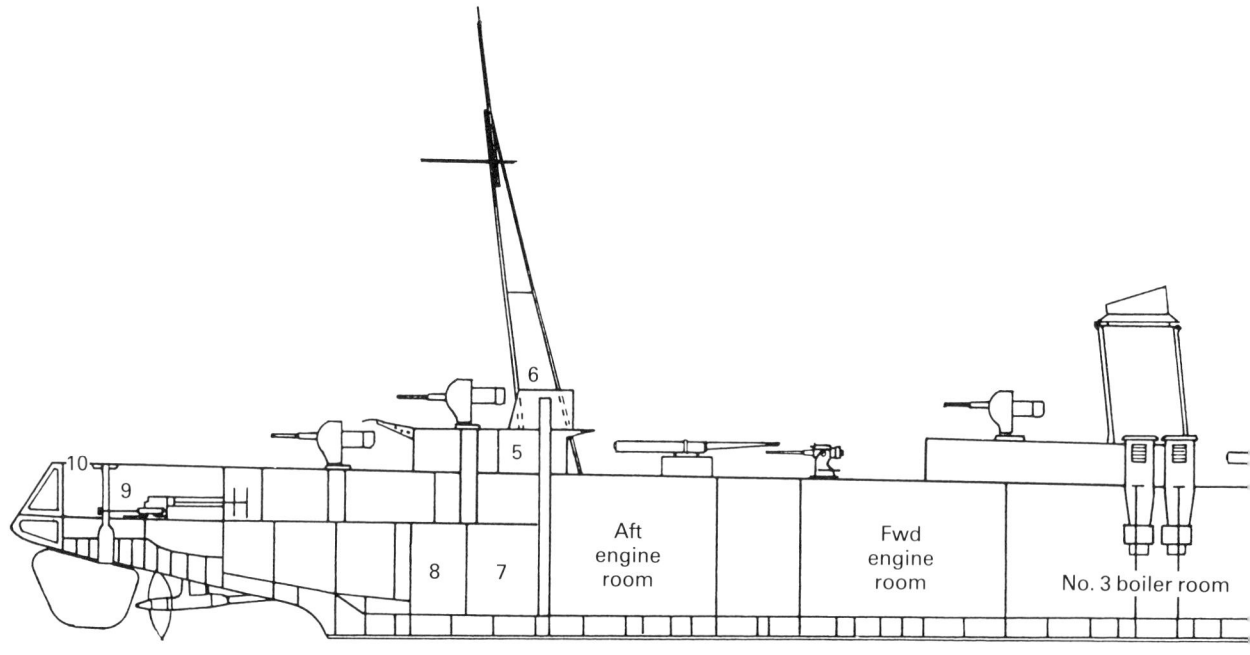

Bourrasque: *Inboard Profile*

Key:
1. Fwd 130mm magazine
2. Fwd magazine hoists (P&S)
3. Navigation bridge
4. W/T office (fwd)
5. W/T office (aft)
6. After magazine hoists (P&S)
7. After 130mm magazine
8. Depth charge magazine
9. Emergency steering position
10. Depth charge chutes

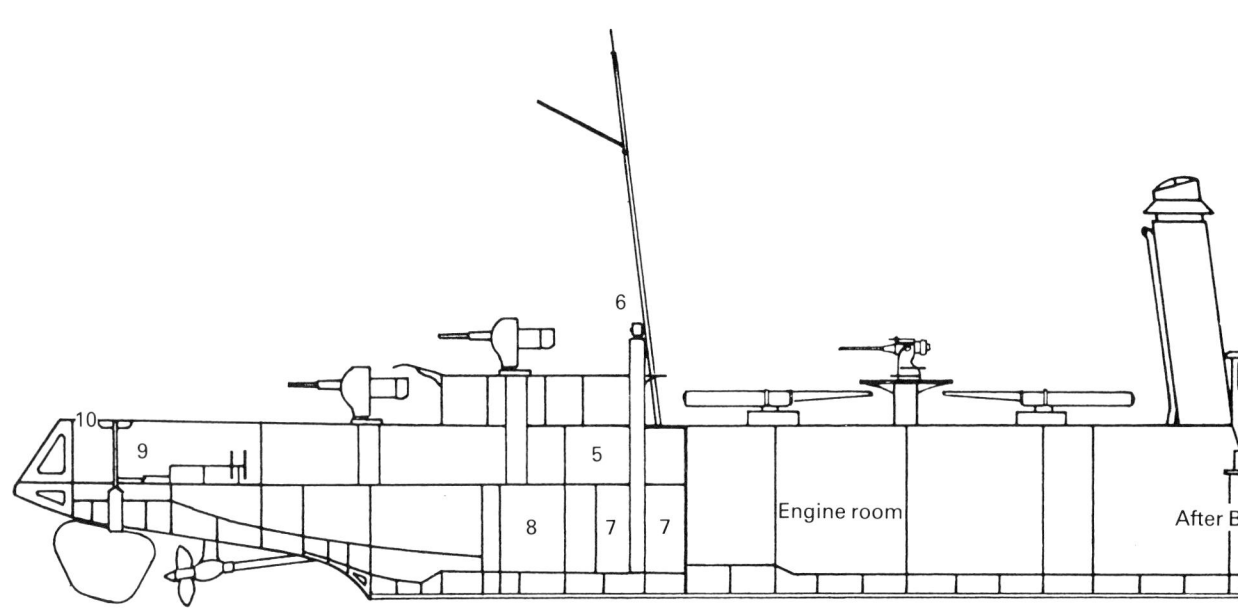

THE FRENCH FLOTILLA PROGRAMME OF 1922

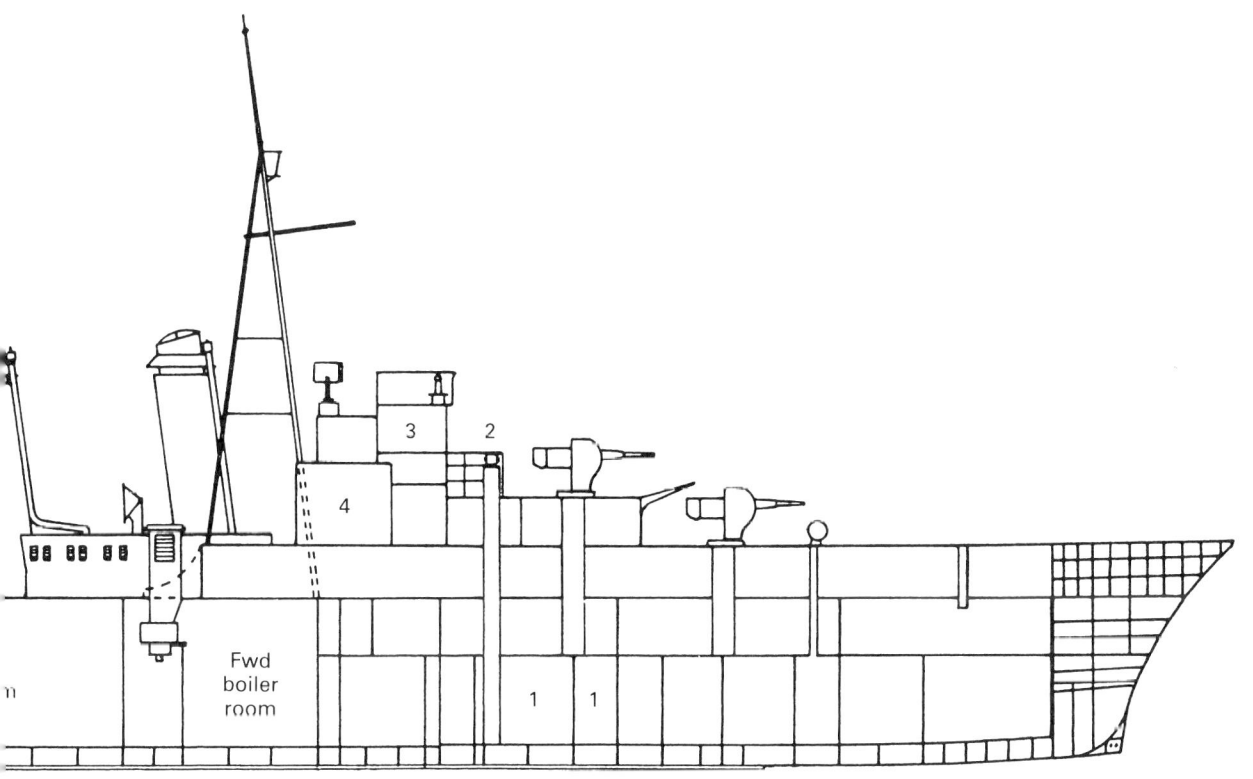

Inboard profiles of Bourrasque *and* Tigre. *The similarities between the bow and stern sections are striking; the layout of the main guns, magazines and hoists fore and aft is virtually identical, as is the size, configuration and layout of the bridge structure. The major disparity between the two types is the relative space accorded to the propulsion machinery. The length of the machinery spaces in* Bourrasque *is just over 40m, or 40 per cent of the ship's length between perpendiculars; the corresponding figures for* Tigre *are 60m and 47 per cent. In the* 1500 tonnes *the steam turbines were located in a single engine room, which also housed the auxiliary machinery. There were two engine rooms in the* contre-torpilleurs, *the auxiliary machinery being distributed between the two compartments. (Author)*

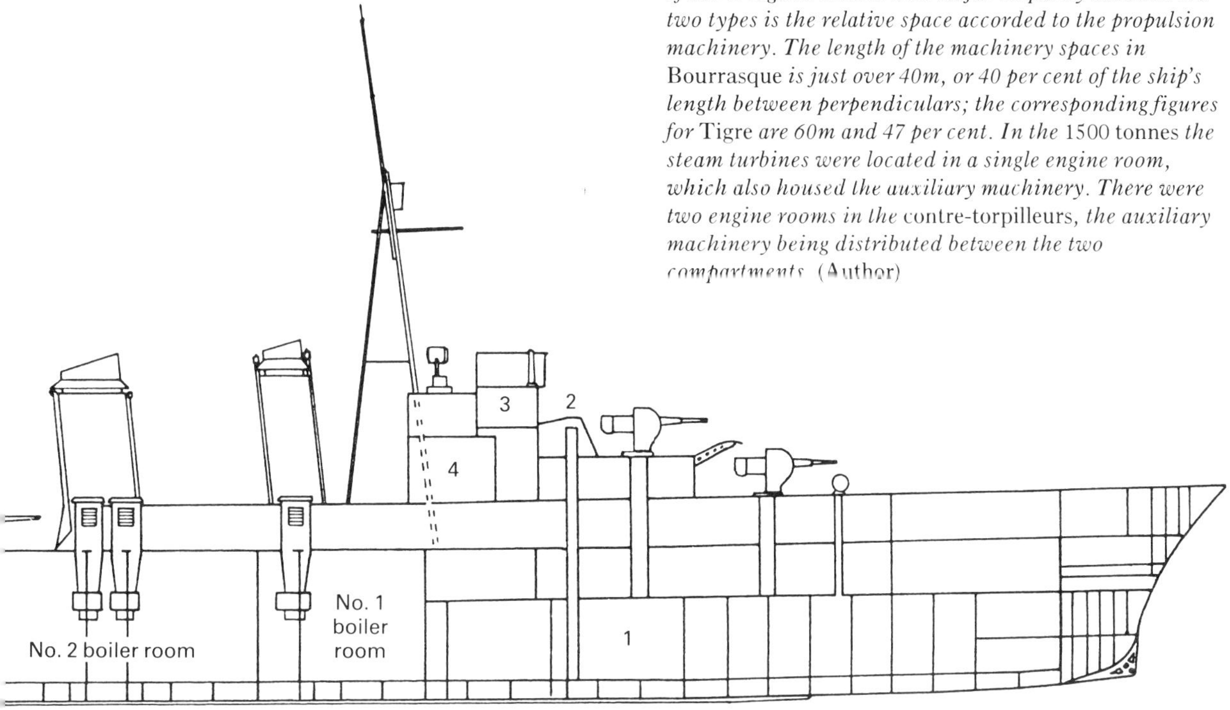

Table 1: *Large destroyers ordered or under construction at the end of the First World War*

	S-113 (Ger)	Scott (UK)	Leone (It)
Ordered:	1916	1916–17	1917–20
Laid down:	1917	1917–18	1921–22
Completed:	1918–19	1918–19	1924
Displacement:	2415t full load	2050t full load	2290t full load
Dimensions:	106m × 10.2m × 3.4m	101m × 9.7m × 3.8m	113m × 10.3m × 3.2m
	350ft × 33.5ft × 11.2ft	332.5ft × 31.8ft × 12.5ft	372ft × 33.8ft × 10.5ft
Machinery:	4 boilers; 2-shaft	4 boilers; 2-shaft	4 boilers; 2-shaft
	geared steam turbines;	geared steam turbines;	geared steam turbines;
	45,000shp = 36kts	40,000shp = 36kts	42,000shp = 34kts
Oil fuel:	720t	500t	400t
Armament:	4 single 150mm/45 (5.9in)	5 single 120mm/45 (4.7in)	4 twin 120mm/45 (4.7in)
		1 single 76mm AA (3in)	2 single 76mm AA (3in)
	2 twin 600mm TT (23.6in)	2 triple 533mm TT (21in)	2 triple 450mm TT (17.7in)
	(40 mines)		(60 mines)
Complement:	176	164	204

The Genesis of the 'Contre-Torpilleur'

With the conclusion of the Great War of 1914–18 the *Marine Nationale* at last had the opportunity to take stock of its current situation and to plan for the future. Although German revanchism could not be altogether discounted in the longer term, attention was focused for the moment on Italy, a potential rival in the Mediterranean with ambitions for colonies in North Africa.

In a note dated 25 February 1919 Admiral de Bon, Chief of the Naval General Staff, pointed out to the Minister that the Italian Navy had completed or laid down no fewer than twelve flotilla leaders and forty torpedo boats during the 1914–18 period; the corresponding figure for the *Marine Nationale* was three (to which should be added four torpedo boats building for Argentina requisitioned in August 1914, and twelve *Kaba* class destroyers purchased directly from Japan in 1917 to compensate for the forced inactivity of the French shipbuilding yards). Admiral de Bon proposed that the new priority for construction be as follows: 'destroyers' (British terminology was adopted to denote the entire category of flotilla craft), followed by light cruisers, and finally capital ships.

These suggestions were immediately approved by the new Minister, Georges Leygues, who also approved a more detailed *note sur les destroyers* only a few days later, on 12 March. This was the most significant note yet, as it attempted to define the missions and capabilities of the *contre-torpilleur*, as opposed to the *torpilleur d'escadre*.

The double rôle of the latter had been clearly set out in a study dating from 1914, which summarised the experience of the Russo-Japanese War together with the theories for the employment of flotilla craft which had been developed and refined in the build-up to the First World War. The primary function of the *torpilleur d'escadre* was to attack the enemy battle line with torpedoes; its secondary rôle was to disrupt by torpedo and gunfire the attacks of enemy flotilla craft against the French line.

The *contre-torpilleur* as defined in the note of March 1919, however, had a triple rôle, with a markedly different set of priorities. Its primary rôle was scouting, followed by the protection of its own battle line against enemy flotilla craft. Torpedo attacks against the enemy line of battle were relegated to third place, and were circumscribed by constraints on approaching the enemy ships too closely.

The qualities required for the first two rôles were stated to be: high speed, endurance, a large radius of action, and a powerful armament. When scouting for the battle line they would be expected to hold a contact and to be capable of engaging not only destroyers but *small cruisers* (author's italics). This implied a speed and armament superior to current flotilla craft, light protection, and a displacement of at least 2000 tonnes.

The ability to engage small cruisers is particularly significant. Of the major European naval powers only France had failed to build light cruisers capable of scouting for the battle line prior to or during the First World War. Ten *éclaireurs d'escadre* had featured in the *Statut Naval* of 1912, but although a prototype (the *Lamotte-Piquet*) had been ordered from the Arsenal de Toulon in 1914, the orders for this ship and for two other units contracted to private yards were cancelled in 1915, and the project was finally dropped.

The Italians, in contrast, had completed three 3500-tonne scout cruisers of the *Quarto* and *Bixio* classes in 1913–14, and the *Marine Nationale* would have been familiar with their capabilities from liaison duties with the Italian Fleet during the war. These ships had only light protection (a 38mm deck plus 100mm plating around the conning tower), and were armed with six 120mm (4.7in) and six 76mm (3in) guns. The *contre-torpilleurs* proposed by the Naval Staff would therefore be quite capable of holding their own against such ships, and their comparative lack of protection would have been compensated for by a significant advantage in speed (35kts as compared with only 26–28kts for the Italian cruisers).

The Naval Staff considered that torpedo attacks on the

enemy battle line with torpedoes would no longer have to be launched at the close ranges accepted before and during the war. Torpedo technology had progressed to the point at which attacks could be launched at 12,000–15,000m (13,000–16,400yds). While accepting that the percentage of hits obtained at such distances would be small, the solution was seen to lie in combining multiple torpedo firings with superior fire control. The visual image conjured up by the Naval Staff submission is not that of a Royal Navy destroyer of 1914, tearing into action against the enemy battle line with the spray flying from its bows and smoke pouring from its funnels, intent on torpedoing or ramming its opponent. Rather it is that of the Japanese cruisers at the Battle of Java Sea, manoeuvring in formation for a favourable firing position, then launching salvoes of 'Long Lance' torpedoes against the enemy line.

Interestingly, the French considered torpedo reloads impractical under action conditions, even at these longer ranges. This clearly reflects operational requirements in the European theatre, where ships would normally return to their home port after each engagement; the Pacific powers were faced with longer campaigns over vast expanses of ocean, and therefore viewed things from a different perspective.

Close combat was envisaged by the report as being most likely at night, when hostile forces might stumble into one another. In these conditions ramming might still be possible (ramming was also considered important for the effective prosecution of submarine contacts), so the bow would need to be reinforced. Both the scouting rôle and night combat would require propulsion machinery which was flexible and responsive.

In the context of the above considerations the following detailed recommendations were made:

Hull: to be designed for strength and speed; greater draught to ensure good seakeeping (earlier French flotilla craft had been designed with shallow draught in order to minimise the threat from mines and torpedoes); reinforced bow; anchors to be carried in hawsepipes.

Armament: need for compromise between weight of shell and rate of fire; largest practical calibre 138.6mm or 5.5in (need stressed for high level of reliability of loading mechanisms and fire control to achieve acceptable rate of fire); four 138.6mm proposed, disposed as superimposed single mountings fore and aft; all guns to have shields to protect their crews from splinters and spray; each gun to be provided with 150 rounds, including some 'ready rounds' stowed close to the gun (an innovation attributed to British practice), and star shell; anti-aircraft protection to be provided by one heavy 75mm HA gun plus four machine-guns; two triple mountings for 550mm long range torpedoes with good arcs, especially forward; although not primarily anti-submarine vessels, might need to use depth charges in the event of a hostile submarine being present, or against enemy flotilla craft in the event of a failed ramming or close encounter(!); eight 100kg (220lb) *grenades* to be carried for this purpose; study proposed for laying mines in the path of the enemy battle line, ten mines being carried by each destroyer.

Fire control: fire control for main artillery to comprise a director incorporating a rangefinder; torpedo sights to be located on bridge, which must be protected against both sea *and wind*; depth-charging and minelaying also to be controlled from bridge; searchlight projectors to be replaced in their traditional rôle of illuminating enemy ships at night by star shell in order not to provide a point of aim for opponent); one 60cm (24in) projector to be fitted for long range signalling, plus a 30cm (12in) projector for signalling in formation (the stress laid on power and

The contre-torpilleur Tigre *running trials in 1926. The bridge structure is virtually identical to that of the* 1500 tonnes, *and she has the early-model gunshields. The fifth 130mm gun, installed atop a deckhouse immediately abaft the third funnel, had to be served from the after magazines.* (Marius Bar, 1926)

reliability of the electrical circuits suggest that these were particular failings of earlier French models); main W/T office to be close to the bridge with direct communication.

Speed and endurance: to be capable of 40kts at full load displacement for six hours in order to ensure a comfortable 35kts in normal service; boilers and auxiliaries to be robust and reliable (a rough weather trial was proposed); must be able to accompany the battle line in all circumstances.

Protection and damage control: boiler and machinery rooms to be protected by a belt of 5cm (2in) grilles with a height of 3m (1m above, 2m below waterline), with transverse and bulkheads of same composition; protective deck of 4cm (1.5in) plated grilles over the same area; protective mattressing around the bridge, torpedo tubes and guns to absorb splinters; paravanes against mines; comprehensive damage control arrangements employing hand-operated steam pumps, and powerful ventilators to disperse gases; ship to be divided into three sections for damage control purposes, of which central section to comprise boiler and machinery rooms; ship to be able to steam with either of outer compartments flooded; each of the three sections to have powerful independent pumps per compartment; counter-flooding to be possible in the centre section.

The above proposals are reproduced in detail because of the insight they provide into French technical and tactical thinking of the period. Some of that thinking is retrospective and analytical, reflecting lessons based on hard wartime experience (*cf* the observations on hull form, the reliability of machinery, and on damage control). However, the document also shows an acute awareness of the tactical possibilities opened up by new technology (lightweight large calibre guns, long range torpedoes, star shell, high-pressure boilers). Of particular note is the emphasis on superior battle control from a capacious, relatively comfortable bridge, with centralised director fire control and long range torpedo sights.

Such was the philosophy which was to underpin not only the *contre-torpilleur*, but also the new *torpilleur d'escadre*, the design of which would also incorporate many of the above features.

The 1922 Programme

Unfortunately the concepts elaborated in the 1919 submission could not as yet be translated into orders for new ships. The parlous state of the French economy in the immediate postwar period, the disorganised state of the established shipyards and the naval infrastructure, and the need to develop the new weapons and machinery required for the proposed flotilla craft, imposed a hiatus on construction which the *Marine Nationale* could ill afford.

In 1919 it tinkered with the idea of purchasing surplus 'Modified W' hulls offered by British shipbuilders, and in April 1920 contemplated ordering two ships of an intermediate *contre-torpilleur* type of 1780 tonnes armed with five 100mm guns. However, the purchase of modern destroyers in Britain was rendered impractical by an alarming fall of the franc against sterling, while the *contre-torpilleur* proposal was rejected because the 100mm calibre was considered inadequate at a time when both the British and the Italians were moving to the 120mm (4.7in) calibre. In the meantime successful trials were conducted with a new 5.1in 130mm/40 gun (Model 1919), which subsequently served as the basis for a larger 2400-tonne design, and aggressive attempts were made to secure former German and Austro–Hungarian torpedo boats as war reparations.

The inadequacies of the 100mm gun were also an important factor in the design of the new *torpilleurs d'escadre*. The staff requirements for the new ships as approved in April 1920 stipulated four single 100mm mountings plus two 75mm AA, but these were quickly followed by a recommendation that a higher calibre be adopted, either a new model 120mm or the lightweight 130mm Model 1919. Without a 120mm gun on the drawing board there was little alternative to the 130mm,

Table 2: *Fleet destroyers ordered or under construction at the end of the First World War*

	H-145 (Ger)	Modified 'W' (UK)	Curtatone (It)
Ordered:	1916	1918	1915–20
Laid down:	1917	1918	1920–21
Completed:	1918–20	1919–20	1923–24
Displacement:	1145t full load	1510t full load	1215t full load
Dimensions:	85m × 8.4m × 3.4m	95m × 9.0m × 2.9m	85m × 8.0m × 3.0m
	272ft × 27.6ft × 11.2ft	312ft × 29.5ft × 10.5ft	278.7ft × 26.2ft × 9.8ft
Machinery:	3 boilers; 2-shaft	3 boilers; 2-shaft	4 boilers; 2-shaft
	geared steam turbines;	geared steam turbines;	geared steam turbines;
	24,000shp = 34kts	27,000shp = 34kts	22,000shp = 32kts
Oil fuel:	332t	367t	
Armament:	3 single 105mm/45 (4.1in)	4 single 120mm/45 (4.7in)	2 twin 102mm/45 (4in)
		1 single 76mm AA (3in)	2 single 76mm AA (3in)
	2 twin, 2 single 500mm TT (19.7in)	2 triple 533mm TT (21in)	2 triple 450mm TT (17.7in)
	(24 mines)		(16 mines)
Complement:	105	127	117

THE FRENCH FLOTILLA PROGRAMME OF 1922

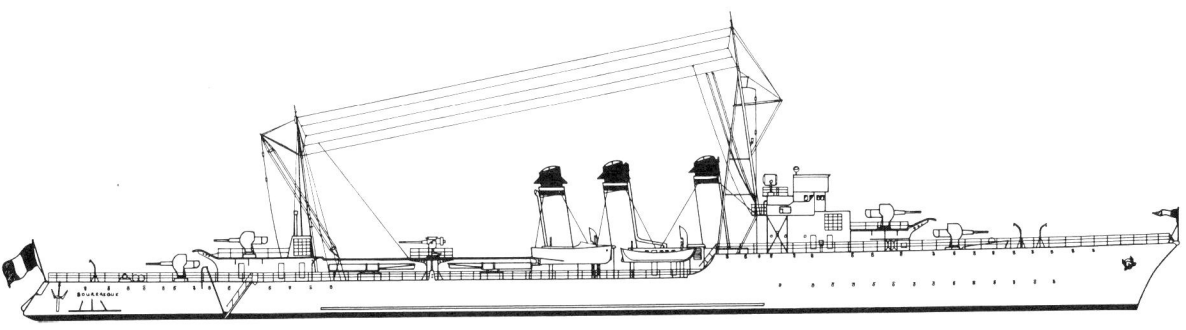

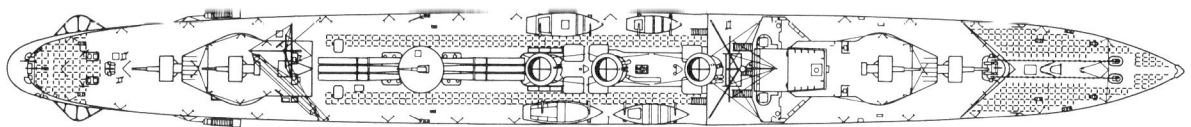

Profile and plan views of Bourrasque *as initially completed, with the taller funnels and half-gunshields. All units of the class appear to have received modifications to both within two years of completion.* (Author)

and installation of the latter aboard the new ships was approved in June 1921. Design displacement was increased by 75 tonnes to accommodate the heavier mounting, but the latter would be largely responsible for the problems which would be experienced with topweight and seakeeping.

The final characteristics of the *1500 tonnes* were approved in December 1922, and construction of twelve ships of the *Bourrasque* class was authorised under the 1922 Programme, together with six *contre-torpilleurs* of the *Jaguar* class (originally designated *2400 tonnes*).

Hull and Superstructure

In accordance with the recommendations of 1919 the hull of both types was designed for speed, habitability and good seakeeping. High freeboard, and a long forecastle incorporating a distinctive raked bow would be features of all French flotilla craft built during the inter-war period. Nevertheless the fine lines of the bow section, adopted for high speed, were to be a problem for the *1500 tonnes*. The concentration of weight forward resulting from the weight of the two single 130mm mountings and the layout of the fuel tanks caused them to bury their heads in a seaway.

The *Jaguar* class do not appear to have suffered the same problem. However, whereas it seems generally agreed that the *1500 tonnes* were of robust construction, the *contre-torpilleurs* have been criticised as too lightly built, and operations in heavy weather resulted in some hull deformation. This appears to be the classic engineering problem involved in 'scaling up', and it should not be forgotten that the largest flotilla craft previously built by the *Marine Nationale* were the *800 tonnes* of the 1908–13 Programmes, with maximum dimensions of 82.6m/8.6m,

as compared with the 126.78m/11.32m of the *Jaguar*. Nevertheless, there was also a powerful incentive to the STCN designers to keep hull weight to the minimum, given the premium placed on high sustained speed by the staff requirements. The 4/5cm grilles covering the machinery spaces stipulated in the EMG submission of 1919 were presumably an early casualty of these weight-reduction measures, leaving these ships entirely without protection.

The adoption of superimposed guns forward, together with the emphasis on a unified and centralised control of combat operations outlined in the 1919 document, resulted in a high unstreamlined bridge. A full enclosed navigation bridge was topped by an open bridge with a single 3m (10ft) rangefinder at its after end. At the after corners of the bridge structure were two 75cm (30in) searchlight projectors. The lower level of the bridge structure housed the main W/T office (see inboard profile).

Both the *1500 tonnes* and the *Jaguar* were three-funnelled ships, although there were important differences in the configuration and spacing of the funnels resulting from the number and layout of the boiler rooms. The *1500 tonnes*, which had three boilers in two boiler rooms (there was a single boiler in the forward room, hence the larger gap between the first and second funnels), had three slim funnels of circular configuration. The *Jaguar*, on the hand, had five boilers in three boiler rooms; the first funnel was similar to those of the *1500 tonnes*, but the uptakes from each of the paired boilers in the after boiler rooms were combined into broad, flat-sided funnels. Both types were fitted with a high tripod mast forward, but the main mast was a tripod in the *Jaguar* and a pole mast in the *1500 tonnes*.

The relatively high silhouette of these ships was subsequently much criticised. It presented an excellent target for enemy rangefinders and made them instantly recognisable at a distance. The combination of a high silhouette and a lightweight hull made them lively in a seaway and difficult to steer in high winds, and both types rolled badly (although the *Jaguar*, unlike the overloaded

Table 3: BOURRASQUE CLASS (1500 TONNES)

Name	Builder	Laid down	Launched	In service
1922 Programme				
Bourrasque	A Ch de France	Apr 23	5 Aug 25	1926
Cyclone	F Ch Méditerranée	Apr 23	24 Jan 25	1927
Mistral	F Ch Méditerranée	Dec 23	6 Jun 25	1927
Orage	C N F Caen	Apr 23	30 Aug 24	Dec 26
Ouragan	C N F Caen	Aug 23	6 Dec 24	Jan 27
Simoun	Penhoët	Apr 23	3 June 24	Jan 26
Siroco	Penhoët	Apr 23	3 Oct 25	May 27
Tempête	Dubigeon	Jun 23	21 Feb 25	Jul 26
Tornade	Dyle et Bacalan	Aug 23	12 Mar 25	1928
Tramontane	F Ch Gironde	Sep 23	29 Nov 24	1927
Trombe	F Ch Gironde	Nov 23	29 Nov 25	1927
Typhon	F Ch Gironde	Sep 23	22 May 24	1928

Characteristics (as designed)

Displacement:	1319 tons standard
	1500 tonnes normal
	1800/2000 tonnes full load
Length:	99.33m (325.9ft) pp, 105.77m (347.0ft) oa
Beam:	9.64m (31.6ft)
Draught:	4.3m (14.1ft)
Complement:	7 officers; 131–135 ratings
Machinery:	Three Du Temple small-tube boilers, 18kg/cm^2 (216°C)
	Two-shaft Parsons (Rateau-Bretagne in *Orage*, *Ouragan*, Zoelly in *Tornade*, *Tramontane*, *Trombe*, *Typhon*) geared steam turbines for 33,000shp; speed 33kts
	Oil fuel 345 tonnes; radius 3000nm at 15kts
Armament:	Four 130mm (5.1in)/40 Model 1919 semi-automatic guns in single mountings (110 AP shells + 16 star shell per gun, plus 70 practice rounds)
	One 75mm (3in)/50 Model 1922, two single 8mm MG
	Six tubes in two triple mountings for 550mm (21.7in) Model 1919D torpedoes
	Two racks each for ten 200kg (440lb) depth charges; two Thornycroft mortars for 100kg (220lb) depth charges; sweep gear
Fire Control:	One 3-metre rangefinder; torpedo sights on bridge

1500 tonnes, at least had a good margin of stability). Interestingly, their seakeeping was considered superior in the Atlantic, with its longer swell, than in the Mediterranean.

In 1929, only three years after the completion of *Bourrasque*, it was decided to reduce both the silhouette and topweight of the *1500 tonnes* by cutting down the tall funnels. This modification was incorporated from the outset into the follow-on design, the *L'Adroit* class.

Convoy duties in 1939–40 led to criticisms from the ships' commanding officers that the turning circle was too great for effective anti-submarine manoeuvres, although there were successes (notably the depth-charging and ramming of *U-54* by *Simoun* in February 1940). The *contre-torpilleurs* suffered badly in the narrow waters off Dunkerque and Boulogne in May 1940, their large size and lack of manoeuvrability being largely responsible for the losses of *Jaguar* and *Chacal* to MTBs and Stuka dive-bombers respectively. However, the latter ships, which had good stability and a high reserve of buoyancy, appear to have resisted damage well enough, *Jaguar* surviving a serious collision in January 1940 when she was rammed amidships at speed by the British flotilla leader *Keppel*.

Armament

The 130mm/40 Model 1919 lightweight gun did not live up to expectations. It was a powerful weapon for destroyer-sized ships, firing a 32kg (70lb) shell out to a distance of 18,900m (20,700yds) at its maximum elevation of 36°, and capable of piercing 80mm of armour at 10,000m (11,000yds). These figures compare favourably with the contemporary 4.7in/120mm mounted in the latest British and Italian destroyers, which fired shells weighing 22.5–25kg (50–55lb) at a maximum range of about 15,000m (16,400yds).

Table 4: JAGUAR CLASS (2400 TONNES)

Name	Builder	Laid down	Launched	In service
1922 Programme				
Jaguar	Lorient	22 Aug 22	17 Nov 23	Jan 1927
Panthère	Lorient	23 Dec 22	27 Oct 24	Jun 1927
Léopard	A Ch Loire	Aug 23	29 Sep 24	15 Nov 1927
Lynx	A Ch Loire	1923	24 Feb 24	15 Nov 1927
Chacal	Penhoët	16 Aug 23	27 Sep 24	1 Dec 1926
Tigre	A Ch de Bretagne	15 Sep 23	2 Aug 24	1 Dec 1926

CHARACTERISTICS (as designed)

Displacement:	2126 tons standard
	2400/2500 tonnes normal
	2950/3050 tonnes full load
Length:	119.70m (392.7ft) pp, 126.78m (415.9ft) oa
Beam:	11.32m (37.1ft)
Draught:	4.10m (13.5ft)
Complement:	8 officers; 187 ratings
Machinery:	Five Du Temple small-tube boilers, 18kg/cm² (216°C)
	Two-shaft Rateau-Bretagne (Bréguet in *Léopard*, *Lynx*) geared steam turbines for 50,000shp; speed 35.5kts
	Oil fuel 530 tonnes; radius 3500nm at 15kts
Armament:	Five 130mm (5.1in)/40 Model 1919 semi-automatic guns in single mountings (130 rounds per gun)
	Two 75mm (3in)/50 Model 1922 (300 rounds + 120 star shell per gun)
	Six tubes in two triple mountings for 550mm (21.7in) Model 1919D torpedoes
	Two racks each for eight 200kg (440lb) depth charges; four Thornycroft mortars for 100kg (220lb) depth charges
Fire Control:	One 3-metre rangefinder; torpedo sights on bridge

However the superiority of the French 130mm was more theoretical than real. The design and construction of the gun, which had a Welin screw breech block, was traditional, and compared unfavourably with the German quick-firing 15cm KL/45 with which the *Marine Nationale* would shortly become familiar through trials with the former *S-113* (renamed *Admiral Sénès*), The high

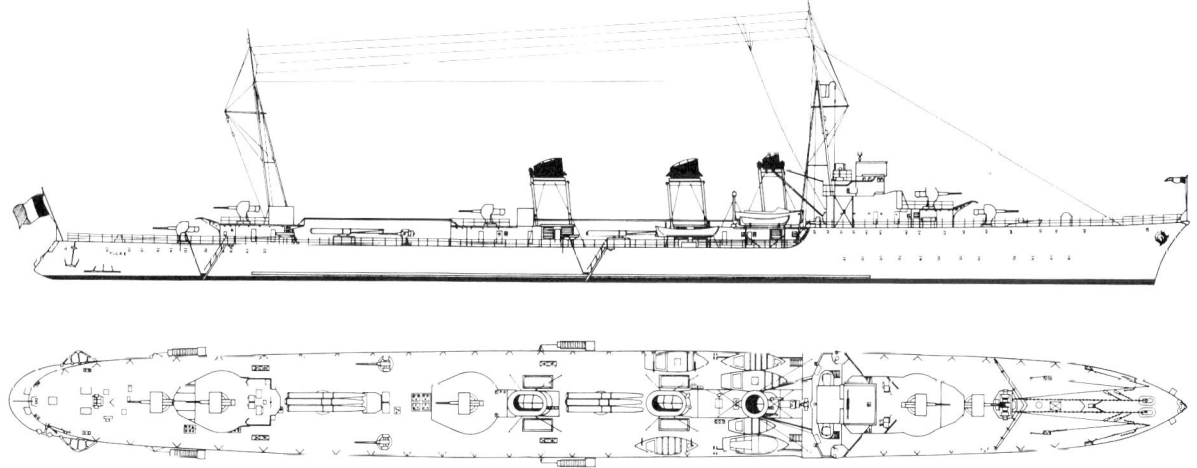

Profile and plan views of Tigre *as completed. These ships would remain virtually unmodified until the mid 1930s, when the single 75mm Model 1922 AA guns were removed and replaced by four twin Hotchkiss 13.2mm Model 1929 mountings, and the single 3-metre rangefinder was replaced by a 5-metre type of more modern design.* (Author)

angle of elevation had been achieved only at the expense of excessive height of the trunnions (1.5m), which made the gun difficult to load at any angle. This effectively reduced the rate of fire to four/five rounds per minute in even the most favourable conditions.

Moreover, at its maximum range of 18,900m, achieved at such high cost, the 130mm Model 1919 was unlikely to hit anything. The short 40-calibre barrel, with fire control provided initially by a single 3m coincidence rangefinder, was found in trials to have an effective range of only 10,000/11,000m. Disappointed with this aspect of the ships' performance the General Staff in a note dated 16 June 1933 recommended that the 3m rangefinder be replaced by two 4m (13ft) stereoscopic rangefinders. Subsequent photographs show the *1500 tonnes* with a single rangefinder (possibly 4m) incorporated into a circular housing, and the *Jaguars* with a single 5m (16ft) model.

The original gunshield fitted in both types proved inadequate to protect the crew from the elements, and within two or three years of completion all units were retro-fitted with a more substantial shield.

Initially both the *Bourrasques* and the *Jaguars* were fitted with the 75mm (3in) Model 1922 AA gun, the two mountings in the *Jaguars* being provided with star shell rounds in order to maximise the number of HE/AP rounds available for the 130mm guns. The 75mm gun proved relatively ineffectual, and in the second *1500 tonnes* series it was replaced by two single 37mm Model 1925, fitted at upper deck level abeam the after deckhouse. This modification was subsequently extended to the *Bourrasque* class. The 75mm AA guns of the *Jaguar* class were disembarked around 1934, to be eventually replaced by four twin 13.2mm Hotchkiss Model 1929 (two abeam the bridge structure, and two in place of the original AA guns).

The decision to develop a 550mm torpedo to supersede the 450mm model in service aboard prewar French flotilla craft quickly produced the results looked for by the General Staff in their anticipation of long range firings. The Model 1919D, 8.2m (26.9ft) long with a 238kg (525lb) picric acid warhead, had a range of 6000m at 35kts and 14,000m at 25kts. Its much-improved successor, the Model 1923DT, which had an alcohol-fuelled four-cylinder radial engine, was 8.3m long with a 310kg (683lb) TNT warhead, and had a range of 9000m at 39kts and 13,000m at 35kts. This latter model would be carried by all the French flotilla craft built between the wars. Both types constituted a major improvement on the prewar 450mm Model 1909, which had a warhead of 144kg and a range of only 3000m at 30kts.

The *contre-torpilleurs*, as evidenced by the 1919 submission, were not intended as anti-submarine vessels. They carried sixteen 200kg (440lb) depth charges in twin tunnels beneath the quarterdeck, discharging them via angled rails over the stern. Depth settings for 30m, 50m and 100m were possible and the fusing was hydrostatic. Initially all six ships were fitted with four Thornycroft mortars, but these were removed shortly after completion; two were reinstated during the 1930s. The mortar, which fired a 100kg (220lb) charge, had a range of 60m with a flight time of four seconds.

The *1500 tonnes*, which were intended to furnish more direct protection to the battle fleet, were provided with twenty depth charges but only two Thornycroft mortars. Neither they nor the *Jaguars* were fitted with ASW

Orage *as completed. She has the original tall funnels and the early gunshields, which provided adequate protection from splinters but not from the weather. Note the high fully enclosed bridge.* (Marius Bar, 15 March 1927)

THE FRENCH FLOTILLA PROGRAMME OF 1922

detection equipment prior to the Second World War; both could therefore respond to submarine attack only by saturating the surrounding area with ordnance, a course of action for which their limited anti-submarine weapon load was poorly suited. However, the depth charge 'tunnel' beneath the quarterdeck had one significant benefit: it permitted the *1500 tonnes* to carry the minesweeping paravanes and associated winch gear standard on fleet destroyers of the period without impinging on the provision of above-decks depth charge racks.

Machinery

The *Jaguars* and the *1500 tonnes* were powered by a combination of Du Temple small-tube boilers and single-reduction geared steam turbines. The Du Temple boilers were of relatively conservative design, and operated at a steam pressure of $18kg/cm^2$ with a temperature of 216°, figures comparable to the Yarrow boilers of contemporary British destroyers. These were the first French flotilla craft designed from the outset with geared turbines, the prototype turbines having been installed in the *Enseigne Gabolde* postwar for trials which began only in 1923. The six *Jaguars* had either Rateau-Bretagne or Bréguet turbines, while the *Bourrasques* had Parsons, Rateau-Bretagne or Zoelly turbines.

In the circumstances the propulsion machinery proved

This photograph, taken in March 1928, shows Ouragan *with the modified gunshield which would become standard for French flotilla craft of the period. The new gunshield provided greater protection for the gun crews than the earlier model, resulting in a higher rate of fire in adverse sea conditions.* (Marius Bar, 26 March 1928)

remarkably successful. For a navy which had not laid down a new destroyer type since the 800-tonne *Bouclier* of 1909 the flotilla craft of the 1922 Programme constituted a considerable technical risk. The *1500 tonnes* required twice the horsepower installed in the *Enseigne Gabolde*, the *Jaguars* three times that figure. Moreover the 50,000shp (designed) of the latter was not only unprecedented in France (the machinery of the *Bretagne* class battleships was rated at a mere 29,000shp), but exceeded the installed horsepower of any contemporary foreign vessel built or building in the destroyer/cruiser category.

Speed trials with the *Bourrasques* proved disappointing. Few reached their design speed of 34kts, the average for the eight-hour trial being 32.8kts. In service it was rare for an individual ship to exceed 30kts, and speed in formation (divisions of three ships) was only 28/29kts, barely sufficient for fleet operations with the elderly prewar dreadnoughts and totally inadequate for the fast battleships which would be laid down in the 1930s.

French inexperience with geared turbines has been blamed by certain authorities, but it seems more likely that the problem lay with the design of the *1500 tonnes* itself, which following the decision to increase the calibre of the main armament to 130mm was seriously overweight (GM was measured at 0.36m as compared with a designed GM of 0.58m!). This is borne out by the contrasting performance of the *Jaguar* class, which achieved trial speeds which rocked Europe with boilers and turbine machinery of similar design. The fastest, *Tigre*, maintained 35.93kts with 55,200shp on her eight-hour trial at a mean displacement of 2380 tonnes, and reached 36.7kts with 57,200shp in the following hour.

In their youth the *Jaguars* could comfortably maintain 34kts in formation. By 1939, some 150/200 tonnes heavier, they could still operate at 29kts in formation, and individual ships could sustain 30/31kts with ease. Some

Panthère *in 1931, with the modified gunshield. The* contre-torpilleurs *proved to be better sea-boats than their 1500-tonne counterparts, and remained essentially unmodified until the mid-1930s.* (Marius Bar, 1931)

Bourrasque *in 1929, with cut-down funnels and the new gunshields. Shortening the funnels served to reduce both topweight and the ship's distinctive silhouette, and this modification was applied to the second group of* 1500 tonnes (L'Adroit *class) from the outset.* (Marius Bar, 1929)

commentators have pointed to their frequent periods of unavailability with machinery problems under wartime conditions. However, it should be remembered that by 1940 these lightly-built, high-performance ships were thirteen to fourteen years old. Furthermore it appears that the problems experienced were generally related to breakdowns in the auxiliary machinery (in particular the ventilators and fuel pumps) and the poor state of the electrical cabling rather than the propulsion machinery proper, which continued to operate reliably.

A less satisfactory feature of the machinery installed in the two classes was its high fuel consumption, which

resulted in a much lower action radius than had been anticipated. The problem appears to have been more acute at low (patrol/cruising) speeds, as the fuel consumption of 758kg (1671lb) per nautical mile recorded during the record-breaking speed trials of the *Tigre* was regarded as a matter for self-congratulation on the part of the shipbuilders.

Following trials with the first of the *Bourrasque*s the designed radius of 3000nm at 15kts was officially adjusted downwards to 2300nm at 14kts. In war service 1500nm was found to be a more realistic figure: during early 1940 *torpilleurs* of the *1500 tonnes* type charged with escorting

The contre-torpilleur Lynx *in 1936. The original single 75mm AA guns have been landed, but the Hotchkiss twin 13.2mm mountings Model 1929 have yet to be installed. These ships appear to have operated without any anti-aircraft armament during the mid-1930s.* (Marius Bar, 1936)

The 1500-tonnes Typhon *photographed in December 1934. By this time the single 75mm AA gun had been removed, and single 37mm Model 1925 guns had been added abeam the forward end of the after deckhouse. Note the range clocks fore and aft (not present in the early photos).* (Marius Bar, 27 December 1934)

Tigre *in April 1939, shortly before the outbreak of the Second World War. She now has twin Hotchkiss 13.2mm AA mountings abeam the bridge structure at forecastle deck level, and in the positions formerly occupied by the single 75mm AA. The new 5-metre rangefinder, partially concealed by the ensign at the fore mast halyard, has a small circular housing. The new-style pennant number marks her out as the lead ship of the 4th division of* contre-torpilleurs *(DCT).* (Marius Bar, 28 April 1939)

small merchant convoys from Casablanca to Brest – a distance of barely 1000nm – arrived at their destination with fuel bunkers virtually exhausted.

The *Jaguars*, confidently expected to achieve 3500/3750nm at 18kts, were found in peacetime service to be capable of 3000nm at 13/14kts only in the most favourable conditions. When *Léopard* took part in Atlantic convoy operations in 1940–41 maximum endurance was estimated at 2400nm at 13kts, and the ship had to break off to refuel in Iceland. As a result, it was decided to remove No 1 boiler and funnel in order to increase bunkerage from 530t to 780t.

In mitigation it should be said that neither the *1500 tonnes* nor the *Jaguars* were designed to escort Atlantic convoys. Indeed nothing could have been further from the thinking behind the *contre-torpilleurs*, encapsulated in the *Théories Stratégiques* of the influential Vice-Admiral Castex:

> ... it is certain that light, fast units, with a *moderate radius of action* [author's italics], will find better employment there [in the Mediterranean] than in other regions, given the short distances to negotiate and the proximity of numerous bases; it is the ideal operational area for the flotillas ...

Nevertheless commanding officers of the *1500 tonnes* criticised the design as better suited to transit between bases than to staying at sea, and reproached the designers for confusing theoretical radius with operational radius.

The theoretical radius envisaged for these ships was calculated on the basis of the key transit distances between the major bases of France and its North African colonies: Brest–Dakar, Brest–Toulon, Toulon–Bizerta and Toulon–Oran. It failed to take into account the weight of equipment added during the ships' service lives, hull fouling, the state of the machinery, unfavourable sea conditions, or the constant need to manoeuvre to avoid or prosecute submarine contacts.

Conclusion

The 1922 Flotilla Programme was a mixed success. It re-established the long-neglected construction of flotilla craft for the *Marine Nationale*, and laid the foundations for a major rebuilding of the fleet between the wars. It was courageous in that it aimed not merely to catch up with developments in the other major European navies but to jump ahead, to anticipate a new era of naval battles fought by new types of ships employing state-of-the-art technology.

Clearly risks were taken, only some of which paid off. Both the *1500 tonnes* and the *Jaguars* were expensive to build, especially the latter, for which an additional 1000 tonnes displacement bought only a fifth 130mm gun, a second 75mm AA gun, and three knots extra speed. However, the *Jaguars* were to prove by far the more successful of the two designs, and their large size was to make them more amenable to later modification when the early months of the Second World War served to highlight their deficiencies.

Sources
Henri Le Masson, *Histoire du Torpilleur en France 1872–1940*, Académie de la Marine (Paris 1963).
Robert Dumas, 'Les contre-torpilleurs français type Jaguar', *Cols Bleus* (18 June 1988).
Official plans of *Bourrasque* and *Tigre*, Centre d'Archives de l'Armement.
John Campbell, *Naval Weapons of World War Two*, Conway Maritime Press (London 1988).

SEETAKT

P F Wright describes the early history of German seaborne radar, and in particular the set fitted to *Admiral Graf Spee*. After the battle of the River Plate the British made intensive efforts to recover or at least to investigate this equipment, including purchasing the wreck of the ship via a third party in order to facilitate the work.

Over a year before the start of the Second World War, the German navy had in operation a radar gunlaying system for ranging ships' and coastal guns and locating vessels. It is said that the Royal Navy did not have anything similar until the end of 1940.[1] The German generic code name for their surface-to-surface radars was *Seetakt* and the cover identity for the whole spread of their radar was *Dezimeter Telegraphie*, shortened to *De Te* or *De Te-Gerät*.[2]

Early research by the German Air Force on very high frequency radio waves started about 1937–38 with an experimental station on the Brocken in the Harz Mountains.[3] This research led to further tests using reflected radio waves, a system similar to the radar idea which was to come. The French liner *Normandie* had an elementary form using reflected radio waves to detect icebergs.[4] British tests of this system culminated on 26 February 1935 when, using radio waves from the Daventry transmitter, a co-operating Heyford bomber had been detected flying at 6000ft.[5] The aircraft from Farnborough was flown by Squadron Leader R S Blucke. However, despite proving the feasibility of it, measuring the return of reflected radio waves was *not* radar.

Seetakt

The German scientists had been working on primitive radar at about the same time as those in the United Kingdom, but in a number of aspects they were ahead, especially in getting their experimental systems into production and deploying them in service. Moreover, they were working with centimetric wavelength radar. The quality of their units and work was of a very high order and by the outbreak of war in September 1939, they had established in service eight stations for *Freya*, an aircraft early warning radar and also *Seetakt*,[6] the subject of this article.

In one of the early RAF raids on the German fleet at Wilhelmshaven in December 1939, the bombers had been detected by German radar at a range of 120km (75 miles) from their target – this was *Freya* at its early best.[7] British losses on this raid were considerable. Early British radar had been codified under the cover of 'RDF', Radio Direction Finding. The name 'Radiolocation' was also in use, but the actual term 'Radar' came from the United States Office of Scientific Research and Development, as a contraction of Radio Detection And Ranging.[8] Britain thereafter adopted the American nomenclature.

By the Treaty of Versailles the size and strength of the German navy had been strictly controlled but on gaining power and declaring the creation of the Third Reich, Adolf Hitler disregarded the Treaty like he did many others. A build up and expansion of the navy was put in hand. One of the new types of ship laid down, was the

The tower structure on Admiral Graf Spee *showing, on the rangefinder, the* Seetakt *gun-ranging radar antenna closely covered while the ship was at Kiel on 22 August 1938.* Spee *was attending the launch of* Prinz Eugen. *Note battle honour 'Coronel' celebrating Admiral Maximilian Graf von Spee's 1914 victory against the Royal Navy off Chile.* (Erwin Sieche)

Panzerschiff, which came to be described as a 'pocket battleship' but in fact was more of an armoured cruiser. The third to be laid down was the *Admiral Graf Spee* of 12,100 tons displacement, launched from the Wilhelmshaven Yard in June 1934 and completed in January 1936.

By 1938, she was fitted with gunnery radar for her main armament which consisted of two triple 11in turrets, one forward, one aft, but the Admiralty was not aware of the radar. Although photographs of the ship had been examined, the *Seetakt* antenna, similar to a wire bed mattress, had not been identified as it was closely shrouded with covers. In fact the German Navy did call the antenna *Matratze* (mattress) or *Matratzen Gerät*. *Graf Spee* carried her 1.4m × 2.9m (4.6ft × 9.5ft) antennas above her rangefinder on the fore top along with a radar hut containing the equipment. This layout enabled results obtained from the optical rangefinder and the radar to be readily compared. The antenna was in two sections: the lower was the pulse-transmitter while the upper received the return signal.

Seetakt was an active gun-ranging radar operating on a frequency of 500MHz, 60cms wavelength with a range of about 15km (9 miles) and a beam width of roughly 15 degrees, while the range accuracy was of the order of 70m. It was identified by the *Kriegsmarine* as FMG 38G (O): FMG = *Funkmess-Gerät* (radio measuring apparatus); 38 = year of introduction; G = GEMA (manufacturer); 0 = antenna position (mounted on rangefinder tower).[9] The rotational rate of this was 360 degrees in 2 minutes. Other enemy ships carried later radar typified by types FuMO,

Relative locations of Seetakt *radar antenna on the 'Pocket Battleships'* Admiral Graf Spee *(A – 1938–1939)*, Lützow *(B – January 1942 – March 1944)*, Admiral Scheer *(C – aft rangefinder mid 1941 – April 1945; D – mid 1940 – April 1945)*. (Erwin Sieche)

Funkmess-Ortung (radio measuring – location) 22, 25 and 26 on *Admiral Hipper*, *Scharnhorst*, *Prinz Eugen*, etc.

Seetakt had been designed and developed by GEMA (*Gesellschaft für Elektroakustische und Mechanische Apparate*).[10] The very first German radar transmissions picked up by the British near Dover on 28 September 1940 came from *Seetakt* land-based radar, operating with the enemy's coastal guns around Calais, shelling a convoy in the Channel. The British did not know it was *Seetakt* at the time; in fact they were trying to track *Freya* transmissions.[11] 'Long' version of 'Window' (chaff) was used against *Seetakt* ground radar sets on the Channel coast prior to D-Day, since it matched the longer wavelength of such sets.[12]

Battle of the River Plate

Graf Spee, under Captain Hans Langsdorff, with 1134 officers and men, sailed from Wilhelmshaven on 21 August 1939, well before the outbreak of war. She was refuelled by her supply ship *Altmark* (Captain Dahl) in mid-Atlantic on the 28th, then continued sailing south. *Spee* remained under radio silence and displayed the name *Admiral Scheer* to confuse any visual contact. Starting on 30 September the surface raider set about her work in the South Atlantic and Indian Ocean, sinking in all nine merchantmen totalling over 50,000 tons. To enable her to continue her raiding, she transferred the crews, amounting to 229 prisoners, to the *Altmark*, but retained the ships' masters and first officers.

There followed, on 13 December 1939, the famous Battle of the River Plate. The British cruisers *Exeter* (the main threat) and *Ajax* with the New Zealand cruiser *Achilles*, fought the action to a brilliant conclusion. *Exeter* and *Ajax* were both damaged, the former having to

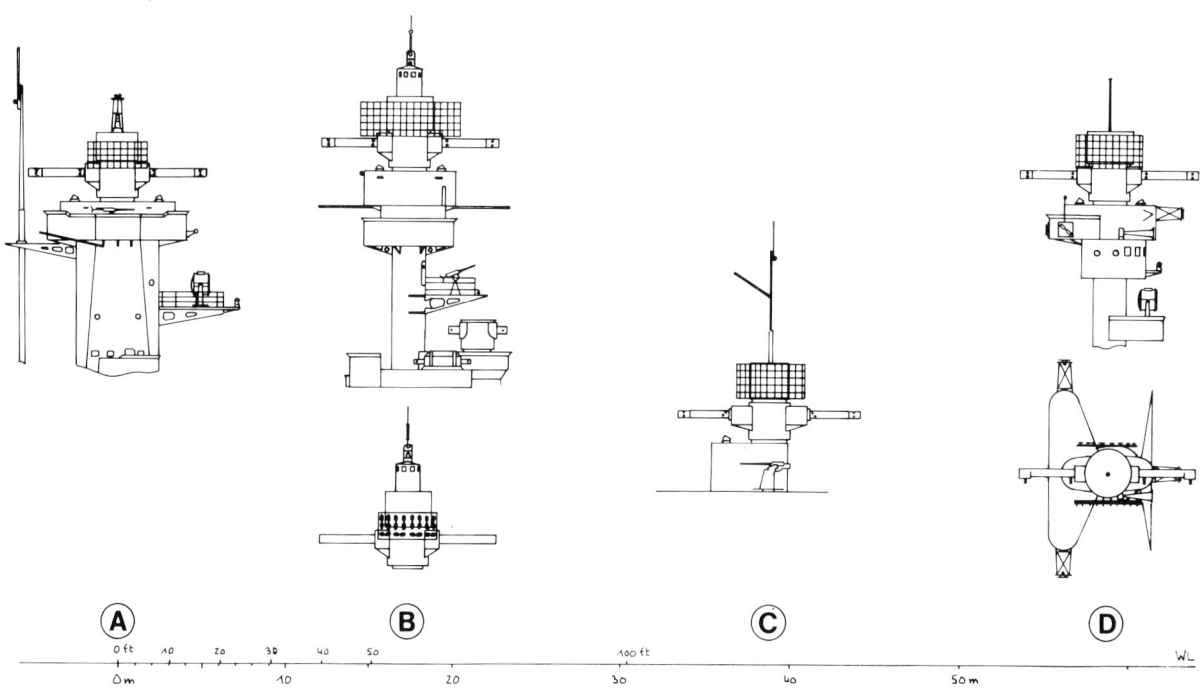

Typical of later Seetakt *equipment, this picture shows the light cruiser* Leipzig *with her FuMO 24/25 radar antenna. She was anchored at Kiel with a British guard aboard in July 1945. The RN Admiral's Barge passes and the German crew pay their respects.* (British Official)

withdraw, but *Graf Spee* had received twenty hits and took refuge at Montevideo in neutral Uruguay. She requested a stay of 14 days to put ashore her wounded and repair damage, but the Uruguayan authorities finally gave her only 96 hours to do what she could and leave. One of her urgent requests was to allow her to remove and land 'four or five military secrets',[13] no doubt including the *Seetakt* sets. This was refused.

On 17 December 1939, *Graf Spee* sailed from her anchorage in the outer harbour at 1820 hours with a skeleton crew, and at sunset was scuttled in shallow water 5 miles out and then blown up. She burned for six days. Two days after the scuttling, Captain Langsdorff lay on a German naval flag in his quarters ashore and shot himself in the head. After the ship's self-destruction, several photographs of the wreck were taken by the Press and others, when the *Matratze* antenna was clearly apparent. It seems odd that no attempt had been taken to destroy this aerial. The British Naval Attaché in Buenos Aires, Captain McCall, RN, who had moved to Montevideo during the relevant days, sent long and detailed reports to the Admiralty covering the whole episode and his actions during the period.[14] Luckily he had a fairly close relationship with the Uruguayan authorities which was advantageous. Some extracts from his reports are worth quoting:[15]

> ... the captured British seamen aboard *Spee* reported that the 'rangefinder' on top of the superstructure was continually in use at sea and kept revolving by motor. [The rangefinder of course, carried *Seetakt*.]

> ... the Germans believed that a local team of workers under their direction, could take part in breaking-up the wreck to remove four or five military secrets ...

> ... was His Majesty's Government interested in purchase of the wreck in order to make a careful examination of it with a view to finding one particular naval secret? ...

The Germans had already destroyed some items of 'secret equipment', ship's papers, the Ship's Log and the War Diary dated after 10 December 1939.[16] This of course included the battle details. However, her *Seetakt Matratze* was still *in situ* on the fore top and this required urgent investigation by a British radar expert. Naval Intelligence Division (NID9) in London, decided to fly out such a man.

The scuttled Graf Spee *burns, 17 December 1939* (CMP).

Scuttled and blown-up five miles outside the Uruguayan port of Montevideo, the Spee *settles into the estuary mud of the River Plate. The tower at this stage has slightly heeled to starboard.*

Bainbridge-Bell

During the First World War, Flight Lieutenant (Temp Captain) Labouchère Hillyer Bainbridge-Bell (born 17 August 1893), had served with the Royal Flying Corps and the Royal Air Force as an Engineering Officer (Wireless).[17] While with 34 Squadron RFC in 1916, he was awarded the Military Cross on 20 October. By March 1918, he was in the Middle East with 113 Squadron RFC as a Wireless Officer and stayed in the Service after the end of hostilities, in March 1919 moving to 111 Squadron RAF at Ramleh in Palestine. He left the RAF at the end of that year and during the 1930s became a civilian Scientific Officer employed by the National Physical Laboratory, at the Radio Research Station at Slough, working on radar receiver development and instrumentation.[18] He was co-author with Robert Watson-Watt and James Fleming Herd of *Applications of the Cathode Ray Oscillograph in Radio Research* (HMSO 1933).[19]

In April 1935, he was one of three scientists assigned to work on RDF (radar), the other two being A F Wilkins and E G Bowen.[20] Bainbridge-Bell has been described as 'the most talented circuit designer at Slough'.[21] In May, they moved to Orfordness (known to the radar team as 'The Island'), working for Robert Watson-Watt as part of his 'frighteningly small' group on long range early warning radar.[22] In 1938, he was working at Martlesham on 'Cathode Ray Tube and DF tests', flying in specially-equipped Anson aircraft with screened harnesses.[23] His pilot was Flying Officer D C Smith who was Flight Commander Radar Flight at Martlesham and nicknamed 'Blood Orange' from his complexion.

but it meant either first flying across the Atlantic to North America and then taking the long haul down the eastern seaboard of the United States and onwards through Venezuela and Brazil, or flying from the United Kingdom through France to North Africa, then down the Atlantic seaboard to West Africa and across the South Atlantic to South America. Bainbridge-Bell did the latter – it was somewhat more secure and fewer questions could be asked about his mission. Early in February 1940, he crossed to France and travelled down to Marseilles, and in relatively short hops via Oran, Casablanca, Agadir, Villa Cisneros (Rio de Oro, Spanish Sahara) to Dakar in Senegal, French West Africa. On 6 February 1940, Bainbridge-Bell left Dakar for his transatlantic flight to Natal in Brazil, changed plane at Natal for the hops to Recife and Bahia (Sao Salvador) and the following day, left Bahia for Caravellas, Rio de Janeiro, and the final leg to Montevideo, arriving on 7 February.[24] He was to stay for the next five weeks during which, he had to try and arrange by devious means, his access to the wreck of *Graf Spee*.

Investigating the Wreck

The fact that the *Graf Spee* lay in a neutral country's territorial waters raised a number of interesting and fairly complicated issues. Who owned the ship? (the Germans). Who had jurisdiction over access to her? (the Uruguayans). Who had a great technical interest in her? (the British – and to a lesser extent, the Americans). The local naval and harbour authorities put a patrol vessel on guard, not allowing any approach to the wreck nearer than 200 metres without a permit. Some sailors from the visiting United States cruiser *Helena* did get aboard and so did three local young men who swam to her from a boat while the guard ship was temporarily absent. They took ten or more photographs and in time, supplied prints to the British Legation in Montevideo,[25] who sent them on to London. The men were rewarded 'a modest compensation'.

In addition to the requirement for access to the wreck, the Admiralty was also fairly interested in the possible purchase of it, but if this were done, then it had to be a low-key almost clandestine operation.[26] It would hardly go down well with the British public if it appeared that the War Cabinet was paying the enemy for something that could be considered a prize of war! Negotiations were opened with a local entrepreneur to buy the wreck for £14,000, with expenses pushing this figure up to £20,000, and in fact it were done. The plan was that the Foreign Office through a local British businessman via the Banco Commercial, would secretly provide the cash, to enable Senor Julio Vega Helguera to buy the wreck as though he was the owner. However, a Deed of Sale contained the vital phrase '... purchased ... for the British Government'.[27] Vega would then pass the equivalent in pesos to the German Minister at their Embassy. It was hoped that the eventual sale of scrap metal, etc from the ship, would refund the British outlay.

Having now legally but secretly acquired the wreck, the British experts had to get on board without the authorities

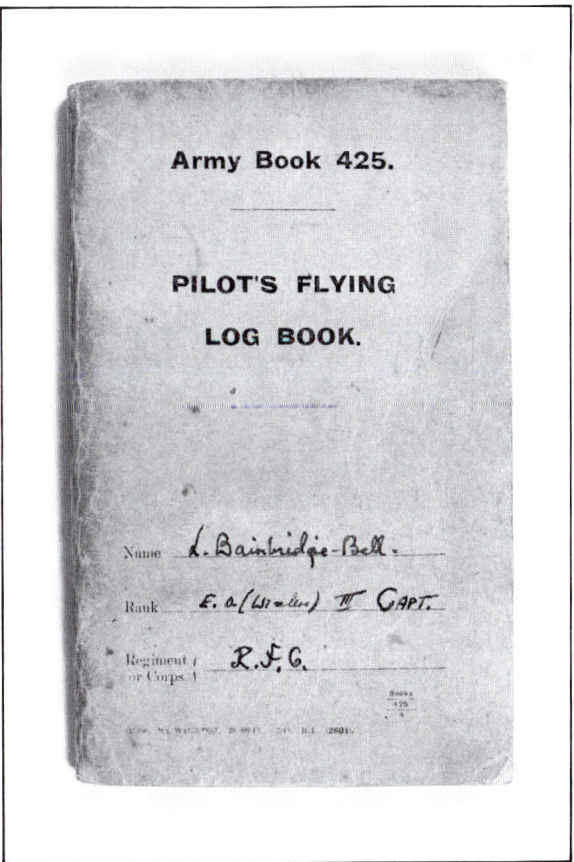

L H Bainbridge-Bell's RFC Log Book in which he recorded all his flights from the Great War until after 1956. His epic journeys to and from Montevideo in 1940 are included. (P F Wright)

realising what had actually happened. Ward & Co the shipbreakers were consulted and it was agreed that Bainbridge-Bell would be able to get aboard 'camouflaged' as a Ward adviser;[28] likewise the other two examiners the Admiralty later sent out, Lieutenant C P Kilroy, RN and Mr M K Purvis of the Royal Corps of Naval Constructors. These two flew to New York and onwards to Buenos Aires where their passports were replaced by new ones provided by the British Embassy, to avoid revealing their true occupations. They also flew the last stage of their journey from Rio, in an American-registered airliner, to arouse less suspicion.

Since Bainbridge-Bell's arrival in Montevideo, it had taken the best part of four weeks to get him on to the *Graf Spee* wreck; in fact at one stage, it was expected that he might have to compile a simple non-technical questionnaire covering the information required, to enable a local official to take aboard and complete. However, Bainbridge-Bell did get aboard on 6 March 1940 and followed this with two further visits. Time was of the essence because the wreck was in 30ft of water on a bed of soft mud and by February 1940 had already sunk into this to a depth of 8ft. Further hazards to survival of the wreck were in the offing. The local Uruguayan chief of their

A starboard bow view of the burnt-out wreck; the radar antenna is turned towards the camera and clearly visible on the rangefinder. (CMP)

The Spee's *fore top optical rangefinder with the* Seetakt Matratze *radar antenna carried forward. The antenna has two sections, the lower for pulse-transmission and the upper for receipt of return signal.*
(National Maritime Museum)

Naval Arsenal recommended blowing up the ship intentionally to avert the chance of possible internal explosions – *Spee* still had ammunition aboard. An attempt to blow up the bows had been made, but failed when the charges did not explode. The Uruguayans then wanted to dismantle the fore top cupola.[29]

Widespread pilfering from the wreck had already taken place and it was important that Bainbridge-Bell got into the bridge tower quickly to inspect the radar 'grey boxes' and the mattress antenna. This he did, sending a total of five reports to the Admiralty, four to the DSD (Director of Signals Department) and one to Naval Intelligence (NID 01192/40). Sadly, despite repeated searches at the Public Record Office at Kew, help from the Ministry of Defence Directorate of Naval Staff Duties, and from the National Maritime Museum library, the author has not been able to locate any of these particular reports. Bainbridge-Bell did confirm that at least one British shell had penetrated the bridge tower structure. He also gave his estimate that the radar operated on a wavelength of 57 or 114cms. It was eventually confirmed at 60cms.

A preliminary report from Kilroy and Purvis (NID 01465/40) had a limited circulation followed by their main report (NID 01601/40) sometime later.[30] The Director of Signals Department at the Admiralty gave his view that: 'The examination of the *Graf Spee* by an Admiralty Signals Establishment representative [Bainbridge-Bell], was most valuable in establishing the use of RDF by the

SEETAKT

enemy.³¹ Sir Henry Tizard in a report dated 8 April 1940 circulated by the DTSDD, surprisingly made the point that: 'He did not think *Graf Spee* had used RDF for gun-ranging during the battle, but used it for locating ships and aircraft. Shooting at our spotting aircraft at altitude 3000ft, had been very accurate at considerable range.' Vice-Admiral Sir James Somerville in an appended note considered: '*Spee* had used her excellent optical rangefinders during the battle'.³²

The question must be asked, 'did she use her radar for gun-ranging during the battle?' Herr F W Rasenack who, as a *Leutnant*, was the ship's Artillery Technical Officer, seems to imply that radar was not an important factor during the River Plate action. However, opening fire against *Exeter* was short by only 600yds and she soon hit the British cruiser at 9.5 miles (the maximum range incidentally of her radar).

L H Bainbridge-Bell left Montevideo for his journey home on 14 March, partly by sea this time, arriving on 8 April. During a storm soon after, the fore top tower on *Graf Spee* fell into the sea taking with it the rangefinder and the vital antenna.

Results of the Investigation

Despite the suggestion that Bainbridge-Bell's report was pigeon-holed by the Air Ministry and that it took at least a further year before the Royal Navy had their gun-ranging

A severe storm blew up in April 1940 with the result that the Spee's rangefinder tower with her Seetakt antenna and radar hut fell off into the estuary and was lost.
(National Maritime Museum)

Graf Spee *settles lower into the River Plate mud, for by April 1940, she had heeled-over to 50 degrees.*
(P F Wright)

111

radar, the document did reach high places. The amazing thing is, that having sent Bainbridge-Bell half way round the world to inspect and report on the equipment and aerials, the powers-that-be seem to have done little about it. Some people in authority were still asking 'Do the Germans have radar?'

Needless to say after reading the Bainbridge-Bell report, R V Jones of Air Ministry Air Intelligence in his Air Scientific Intelligence Report No 7 dated 17 July 1940, and aptly titled 'The Edda Revived', clearly stated 'It is safe to conclude that the Germans have an RDF system'.[33] In a recent letter to the author, Professor R V Jones confirmed his statement and added,

> Despite Bainbridge-Bell's report, and other firm evidence that the Germans had radar, a general disbelief continued in Britain that such was the case. Only when we obtained the Auderville photograph and the associated transmissions was this disbelief dispelled. It was a classic case of authorities being unwilling to accept unwelcome evidence.

The British did not apply countermeasure jamming against *Seetakt* until mid-February 1941, while after the *Graf Spee* scuttling, knowing the British had gained access to *Seetakt*, the *Kriegsmarine* introduced their *Wismar* system of frequency-changing as a safety measure.

The Admiralty had approved the start of radar development for the Royal Navy in October 1935. The first operational set was a 7m wavelength air warning radar fitted to *Rodney* late in 1938.[34] In January 1938, development work was started on radar of 50cm wavelength which led to the installation of a production model main armament fire control radar in the cruiser *Birmingham* by December 1940 – allowing for trials, probably not an unreasonable time after Bainbridge-Bell's report. He was awarded an OBE for his work along with the sum of £2400 under the government's postwar 'Awards to Inventors' scheme. He gave it all away to charities and died at the age of sixty-three in 1956.

It is somewhat ironic the *Admiral Graf Spee* finally met her end about 1000 miles from where the German naval hero after whom she was named, met his. Admiral Maximilian Graf von Spee, whose battle honour 'Coronel' the pocket battleship displayed on her tower, was lost with his ship in December 1914 at the Battle of the Falkland Islands, a name to be in the news again sixty-eight years later.

Notes

1. R Watson-Watt, *Three Steps to Victory*, Odhams (London 1957), p351.
2. E F Sieche, 'German Naval Radar to 1945', *Warship 21* (London 1982), p3.
3. R V Jones, *Most Secret War*, Hodder & Stoughton (London 1978), p78 and 83.
4. *Ibid*, p42.
5. *Ibid*, p42.
6. *British Intelligence in the Second World War*, Vol 2, HMSO (London 1981), p245 footnote.
7. R V Jones, *op cit*, p106. *British Intelligence in the Second World War*, Vol 1 (London 1986), Appendix 5 'The Oslo Report'.
8. R Watson-Watt, *op cit*, p124.
9. Fritz Trenkle, *Die Deutschen Funkmessverfahren bis 1945* (Stuttgart 1986), pp116–117.
10. *Ibid*, p3. PRO file AIR 20/1678, ASI Report on German RDF in Air/Sea Warfare, 21 January 1943.
11. *British Intelligence in the Second World War*, Vol 2, p247. PRO file AIR 20/1629, DT (German RDF) 'Freya' and 'Wurzburg'.
12. *British Intelligence in the Second World War*, Vol 3, Part 1 p558 footnote.
13. PRO file ADM 223/69, River Plate, events subsequent to the action, Despatches from Naval Attaché.
14. PRO file ADM 1/9759, Report of Naval Attaché, Buenos Aires, 21 December 1939.
15. PRO file ADM 223/69, Despatches from Naval Attaché (sections 4, 9 and 11a).
16. PRO file ADM 223/68, *Graf Spee* 1939, the German story.
17. Pilot's Flying Log Book (Army Book 425) – L H Bainbridge-Bell.
18. R Watson-Watt *op cit*, p129.
19. *Ibid*, p223.
20. E G Bowen, *Radar Days*, Adam Hilger (London 1987), p8.
21. *Ibid*, p14.
22. R Watson-Watt, *op cit*, p167.
23. PRO file AVIA 7/216 Part 1, Aircraft for Experimental Work, RDF flights. (Ansons were K6260 and K8758).
24. Bainbridge-Bell's Flying Log Book. He used this to record his flights up to 1956.
25. PRO file FO 371/24265 (file A 1070/3/46). These Foreign Office files were closed until 1990.
26. PRO file FO 371/24265 (file A 952/3/46).
27. PRO file FO 371/24266.
28. PRO file FO 371/24265 (file A 1070/3/46).
29. PRO file ADM 116/4472 (case 6160), Wreck of the *Graf Spee*, Reports on Examination & Purchase.
30. PRO file ADM 1/19292, Battle with *Admiral Graf Spee*. NMM file 660178/1&2, Report on visit to *Graf Spee* wreck, April 1940, Parts I, II and III.
31. PRO file ADM 116/5475 (case 6579), Salvage of metal from *Graf Spee*.
32. PRO file ADM 1/10794, Use of RDF by *Graf Spee*, Sir Henry Tizard.
33. PRO file AIR 20/1624, Air Scientific Intelligence Report No 7, 17 July 1940.
34. R Watson-Watt, *op cit*, p349 and 350.

Acknowledgements

The author wishes to acknowledge and thank the following for help and assistance with information: R M Coppock and Alan Francis (MOD Directorate of Naval Staff Studies), Derek Howse, Professor R V Jones, F A Kingsley, Richard Lynes, CPO W Schultz (Asst Naval Attaché Office, Embassy of the Federal Republic of Germany), Erwin Sieche, John Stroud, David Topliss (National Maritime Museum).

JAPANESE MIDGET SUBMARINES
Kōhyōteki Types A to C

Originally produced under conditions of extreme secrecy, the first Japanese midget submarines were conceived as one of a variety of measures intended to reduce the numbers of the superior US fleet in the run-up to a decisive gunnery battle. War experience quickly revealed that this scenario was unlikely and the submarines were switched to attacks on shipping in harbours and anchorages. Jiro Itani, Hans Lengerer and Tomoko Rehm-Takahara describe the background to this unconventional weapon system and the main features of the craft themselves.

Midget submarines (*Kōhyōteki*) were introduced into the Imperial Japanese Navy (IJN) as a secret means to compensate for the quantitative discrepancy between the USA and Japan in terms of naval forces due to the arms limitation treaties of Washington and London. They were to be used en masse in the early stages of the planned 'decisive battle', the gunnery engagement between the US Navy's and IJN's main forces expected in the vicinity of Ogasawara Island, to change the ratio of capital ships in favour of the IJN. For this purpose they were to be transported aboard special carrier ships to the operational area and deployed as opportunity afforded. Their success largely depended upon surprise and therefore rigorous measures were taken to conceal their existence from foreign navies and even inside the IJN the number of people who knew about them was severely restricted.

The prototype obtained a world record in underwater speed (roughly 25kts) but improvements resulting from the tests slightly reduced the speed of the follow-on trials boats, but they still held the record. In later boats this was still further reduced by protective equipment like net cutters, propeller guards, ward-off lines, etc, but this equipment was found necessary to break through harbour defences and the like. When launching trials from the seaplane tender *Chiyoda* (converted to midget submarine carrier) proved successful, thirty-six *Kōhyōteki* were ordered in quick succession in 1940 to become the armament for three carrier ships.

In the months before the opening of the Pacific War the opinion grew that the decisive battle – a modified and extended repeat of the Battle of Tsushima in the Russo–Japanese War – would not take place at the beginning of the conflict. Therefore the midgets would lose their operational rationale, which lead to the proposal to use the midgets for attacking stationary ships in harbours and anchorages. Eventually the tacit agreement of the C-in-C Combined Fleet was obtained and five midgets attacked the US fleet in Pearl Harbor. Instead of the planned surface ships, large fleet type submarines were used as carriers. The operation ended with the loss of all boats and crews except one man, and the sinking of the single battleship for a long time credited to *Kōhyōteki* was in fact the result of the air attack. From the earliest design stage the Chief of the Navy General Staff had insisted that it was not to be a special attack weapon requiring the certain death of the crew but that they were to be given reasonable chance of escape to a carrier ship. This assumption was also stressed in order to obtain the C-in-C's agreement. In fact, the boats which attacked the enemy anchorages at Pearl Harbor, Sydney, Diego Suarez, and Guadalcanal (except in two cases) were all lost along with their crews and therefore judging from the results of real attacks the *Kōhyōteki* was indeed a special attack weapon irrespective of the theoretical chances of escape. Between the attack on Pearl Harbor and those on Sydney and Diego Suarez, several modifications were made relating either to the boat directly or to the transportation equipment of the carrier submarines but without any effect on the basic method of operation.

As a result of the steadily vanishing possibility of a decisive battle between surface forces, the enormous expansion of the Japanese sphere of influence, and the results of attacks on anchored ships, the potential of *Kōhyōteki* for the defence of local bases was recognised. This brought about the development of Types B (prototype) and C which had a 40hp/25kW diesel generator to charge the battery and used diesel propulsion for surface propulsion in order to increase the operational radius. Even though dimensions and displacement were slightly increased, Type C still very closely resembled Type A in appearance. With the construction of Types B and C the

Midget submarine No 69 *(a Type C boat) carrying out launching trials from the fast transport* No 5, *17 August 1944. (All uncredited photos from* Hans Lengerer Collection*)*

original offensive emphasis of *Kōhyōteki* changed to defensive operations and in fact except in a few cases when *Kōhyōteki* attacked ships underway, they operated in the Solomons, Aleutians, Rabaul, Truk, Halmahera, Philippines, Okinawa, etc to defend local bases and were either lost in action, destroyed by bombing or strafing, or scuttled by their crews immediately before the landing of enemy forces.

The total number of *Kōhyōteki* Types A to C built at Karasukojima and Ourazaki in secret factories is supposed to be 89 (58 boats of Type A, 1 boat of Type B, and 36 boats of Type C) but there is no firm evidence.

Historical Background

The Washington Naval Treaty (1922), which brought to an end the potentially destructive postwar arms race, imposed numerical and size restrictions on capital ships. This was followed in 1930 by the First London Treaty, and the combined effect of these was to limit Japan to 60 per cent of US tonnage of battleships, carriers and heavy cruisers, 70 per cent of light cruisers and destroyers, but parity in submarines.[1] By this time the USA had come to be regarded as Japan's most likely enemy, a position formally adopted on 28 February 1928 with a modification to the 'Principles for the Defence of Imperial Japan', a strategy originally drawn up on 4 April 1907 against Russia.

Not all Japanese military leaders were prepared to accept the inferior international status implied in these treaties, and in particular many believed that the London Treaty seriously undermined Japan's powers of defence. Once this view gained ground among the politicians the stage was set for a series of diplomatic incidents that were to change Japan's political position in the world. The outbreak of the Manchurian Incident (September 1931) and First Shanghai Incident (January 1932) intensified the impression in Japan that a military confrontation with the USA was not far off and so the impetus to military expansion was strengthened. The tendency was to build (sometimes secretly) fighting ships outside the treaty limitations, to concentrate maximum offensive power on minimum displacement and to develop air power which was outside any restrictions. At the same time the IJN strove to develop unique tactics and weapons, and intensified training.

The most serious problem was the aforesaid restriction on the number of warships. At that time the prevailing theory was that a decision in a sea battle was determined by the firepower of the capital ships, but in war the IJN's main fleet would face a superior US battle line. This

JAPANESE MIDGET SUBMARINES

A Type A midget washed ashore after the Pearl Harbor attack, 7 December 1941.

The Type A boat captured after the Pearl Harbor attack was exhibited in many American cities. Windows were cut along the centre section to allow spectators to view the interior.

problem oppressed every leader responsible for strategic and tactical concepts relating to the defence of the Japanese Empire. The best chance – allowing for the restrictions of the arms limitation treaties – was considered to be the construction of combat ships with superior offensive power and manoeuvrability, sacrificing habitability and to some extent defensive power, along with the successful execution of the so-called 'gradual reduction operation', intended through several phases to secure parity of power before the decisive artillery duel between the main forces.

IJN strategists were convinced that the US Navy would advance across the Pacific in order to annihilate the IJN in its home waters. As a counter the IJN planned a gradual attrition of the US fleet by submarines and surface torpedo attack forces, and then to intercept the reduced American battlefleet in an area westwards of the Ogasawara Islands and to destroy them in the main phase of the decisive battle.[2] In the night before the artillery duel, which was expected to begin at dawn, an attack by powerful torpedo forces above and below the sea was planned in order to shift the balance of strength in favour of the IJN and to disrupt the defensive formation of the US fleet.

From Piloted Torpedo to Midget Submarine

A strategy against the USA and the best tactics to achieve it were the subjects of much deliberation in the IJN and stretched even to circles outside the active service. For the development of the midget submarine the proposal of Captain (first reserve) Yokoo Takeyoshi was of special importance. Yokoo, who had participated in the Russo–Japanese War, was of the opinion that the use of piloted torpedoes would ensure that hits were obtained.[3] Captain Kishimoto Kaneji,[4] who at that time was the chief of the Second Section of the First Main Division of the Navy Technical Department (in charge of torpedoes), took up this idea and developed it into a practical weapon.

In December 1931 he explained his idea to Weapon Technical Commander Asama Toshihide, and ordered further research in extreme secrecy. This was to develop a small 'mother torpedo' intended to close with enemy ships very quickly and noiselessly to achieve certain hits from torpedoes launched at short range. Kishimoto's initial idea was a manned 'large torpedo' with the following characteristics:

1. It must be usable in numbers in the decisive battle. Therefore the underwater speed must be 30kts. This was 1.5 times as fast as the US battlefleet whose speed was supposed to be 20kts;
2. It must be equipped with two torpedo tubes (TT);
3. The cruising radius was set by the distance between the friendly and the enemy fleet, *ie* outside the range of the main guns, and must therefore be 60km (*c*35 nautical miles);
4. After launching torpedoes the 'mother torpedo' would remain in the battle zone to be recovered afterwards.

The research vehicle produced by Commander Asama was a manned torpedo-shaped midget submarine which was expected to reach 30kts in submerged condition by the use of small, light and high-powered batteries and an electric motor of similar characteristics, thus meeting the above requirements.

After this preliminary study which proved its feasibility, in the summer of 1932, Kishimoto explained his concept directly to the Chief of Navy General Staff (*Gunreibu sochō*) Admiral Prince Fushimi-no-miya Hiroyasu in order to maintain secrecy and to circumvent any obstacles in the hierarchy. Prince Fushimi asked if it was a special (suicide) attack weapon (*tokko heiki*)

General arrangement of prototype.

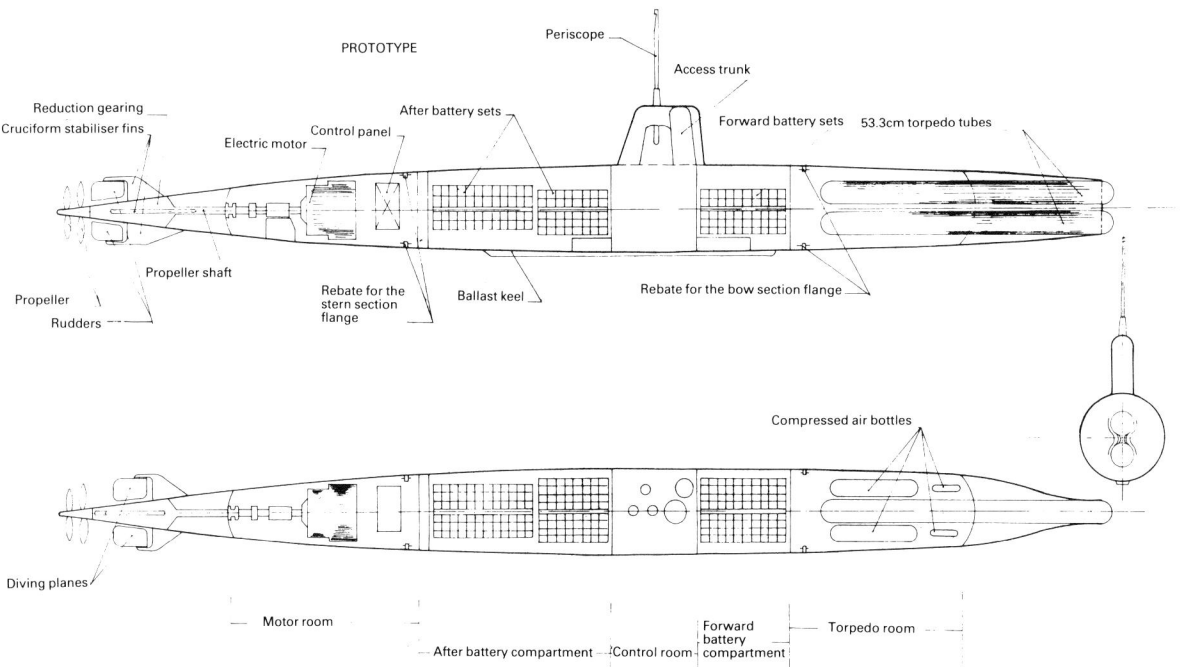

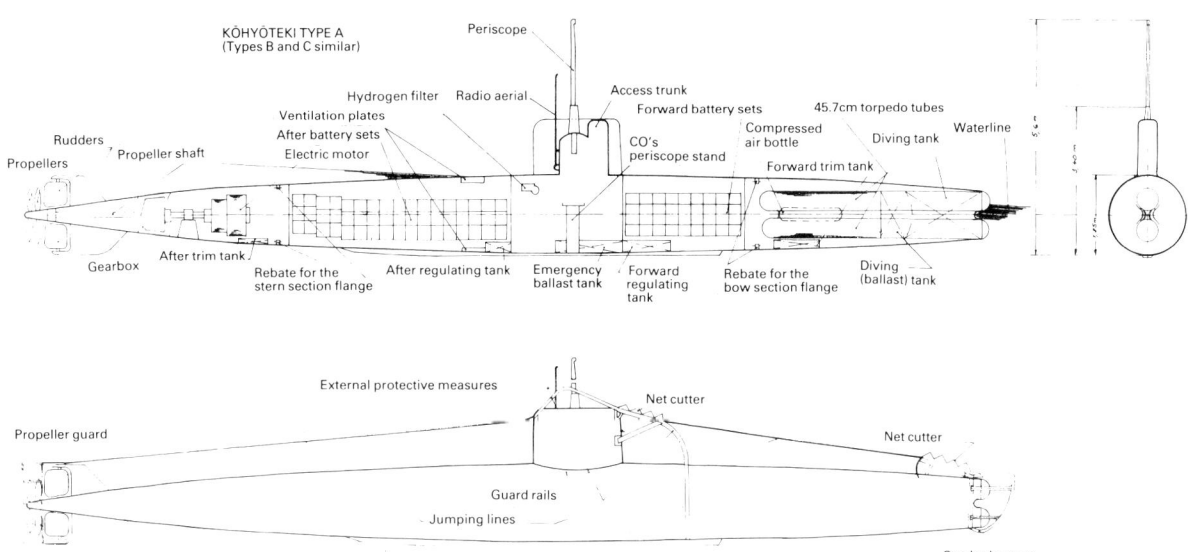

General arrangement of Type A.

intended to crash into the enemy ship (as was the original conception of Captain Yokoo), but Kishimoto explained that the operation needed death-defying courage but should never mean suicide because of the way rescue was contemplated.

With this assurance in mind Admiral Fushimi gave the go-ahead to the Navy Ministry with the proviso that the crew had to be given a possibility of escape.[5] The Navy Technical Department was ordered to begin research and Vice-Admiral Sugi, the chief of the IJN's highest technical organisation, assembled a committee which began design work in August 1932 under conditions of extreme secrecy (*saiko kimitsu gunki* = highest grade of military secret). Under the chairmanship of Kishimoto four officers, who were later also promoted to flag rank, worked out the main elements. Commander Katayama Ariki was responsible for the hull, Commander Asama for the torpedoes, Commander Nawa Takeshi for the battery and Commander Yamada Kiyoshi for the electric motor. They were assisted by navy engineers Ishiyama Saburo, Ishii Kinnozuku and assistant engineer Kusunoki Atsushi. The co-ordination was in the hands of Asama.

Because this was a technically unprecedented weapon some problems quickly came to the fore. How could the poisonous gas from the batteries be removed, the steering be simplified, the trim in submerged condition be stabilised? These were only a few of the difficulties, and many ideas were considered, such as cleaning (absorption) of the poison gas by palladium catalyser, speed control by changing the combinations of series and parallel battery connections, and automatic depth-keeping gear. By the constant efforts of all persons concerned other difficulties were also solved and the design was finished in the short period of roughly two months, as ordered.[6] A battery-driven[7] torpedo-shaped boat with underwater speed of 25kts[8] and radius of action of 60km was considered to be most suitable.

The Test Boat and Trials

In October 1932 the test production of one boat was ordered at Kure Navy Yard Torpedo Experimental Division (*Kure kaigun kosho gyōrai jikkenbu*)[9] to be carried out in strict secrecy. Construction was completed quickly and in August 1933 trials were ordered. They were begun that month at Iyo-nada in the Inland Sea (*Setonaikai* = the waters surrounded by Honshu, Kyushu and Shikoku) with unmanned tests at first using the automatic depth control system in order to test the general properties and especially the underwater speed. In runs between Yashirojima and Sada-misaki a maximum speed of 24.85kts was obtained.[10]

In October 1933, Lieutenant-Commander Kato Ryōnosuke and Sub-Lieutenant (Eng) Harada Shin carried out various tests in the Inland Sea and from the summer of 1934 ocean trials were conducted off Sukumo Bay (Kochi prefecture). These trials proved the potential of this manned torpedo in a decisive oceanic battle, but some problems remained. The periscope depth had to be increased in order to avoid exposure of the hull, the depth-keeping needed to be improved, and an increase in cruising range was demanded. The first problem was solved by the addition of a conning tower to the torpedo-shaped hull (this also enabled the crew to enter or leave the boat at sea), and by lengthening the periscope – necessary because of the greater wave heights in the open sea than inside the bay. The automatic depth-keeping gear was also improved but the cruising radius remained unchanged.

The fitting of the conning tower etc greatly increased resistance and in the trials following the modifications the underwater speed dropped to 22kts for 50 minutes; the boat could then run for another 8 hours at very low speed. In December 1934 the trials were finished and further alterations were made to the original concept. At that time the principle of re-testing was agreed and the secret design drawings and other papers relating to the boat were stored in the safe of the Naval Technical Department, while the prototype was stored under seal in a separate secret storehouse belonging to Kure's Torpedo Experimental Division.

Transport Ships

Even before the end of the ocean tests the Navy General Staff had worked out the operational concept of their massed employment in the decisive battle, where a large number of torpedoes would be fired from very short range to obtain the highest hitting rate. Because of their extremely short radius of action the boats would have to be transported by special carriers directly to the operational area. These carriers were to accompany the main fleet, deploy at the appropriate time and launch the boats in secrecy. Then the boats were to approach the enemy under their own power and attack the capital ships immediately before the start of the decisive gunnery engagement. Once the boats were launched the friendly main force was to lure the enemy into the area where the boats were waiting to fire their torpedoes.

This idea seemed practicable but the fundamental problems were how to deploy the carrier ships in order to obtain the most suitable launching point in the face of the enemy fleet, and how they could launch the midgets unnoticed by the enemy. The success of these boats would greatly depend on these factors. Therefore, the secrecy of their existence and that of the carriers was of decisive importance and had to be absolutely maintained. For this reason several code names such as 'Anti-Submarine Bombing Target' (*Taisen bakugeki hyōteki*),[11] 'TB Model' (*TB mokei*), 'Special Target' (*Tokushu hyōteki*) and 'A Target' (*A hyōteki*) were used, while the carriers were constructed as seaplane tenders (AV) and fast oil tankers (ships excluded from the limitations of Washington and London Treaties), but capable of conversion to midget submarine carriers within a few weeks should the situation require this measure.[12]

The budget for three ships (whose real purpose was known to only a few inside the IJN) passed parliament as part of the Second Fleet Replenishment Programme of 1934. They were called *Chitose*, *Chiyoda*, and *Mizuho* when completed.[13] Subsequently, one more was added in the Third Fleet Replenishment Programme of 1937, this time intended to be built as a minelayer but the plan changed later and this ship was also completed as the seaplane tender *Nisshin*. As designed, when operating as *kōhyōteki* carriers, twelve boats could be transported. Thus the Navy General Staff ultimately envisaged four carriers with twelve boats, each with two torpedoes, totalling 96 torpedoes, which should have been sufficient to cause enough destruction and confusion to improve significantly the chances of their own battle line.

Knowledge of the existence of the boat was limited to the personnel directly in charge of it. The Chief of Kure Navy Yard was given only very vague information by the Chief of the Naval Technical Department and did not learn any more from the Chief of the Torpedo Experimental Division, despite the fact that the latter was his subordinate. This division was remote from other departments and installations at the Navy Yard on Karasukojima, and only selected personnel were permitted to work on the boat. Others could enter the building only by special permission of the Navy Minister, but they were examined by the military police, finger-printed and issued with identity cards displaying their photograph. Documents relating to the boat were not sent by mail but by reliable couriers. Naturally, all personnel who came into contact with the boat were strictly enjoined not to say anything about it, any mention of the secret being a matter for prosecution. During the tests in the Inland Sea and off Sukumo Bay no officer below the rank of Commander took part, such was the desire to restrict knowledge of the boat.

With the carriers under design and trials of the prototype nearing an end, the Navy General Staff adopted the following outline plan for future development:

1. On the basis of the trials of the prototype, two more test boats were to be built as soon as possible at Kure Navy Yard to be tested with crew aboard under operational (war) conditions;
2. If these trials proved their unqualified suitability for the intended purpose, launching tests should be done from a carrier;
3. If this boat was then formally adopted after the tests, forty-eight of them should be built as soon as possible;[14]
4. The necessary base for storage and maintenance should be established near the Kure Navy Yard;

US netlayers Catclaw *and* Baretta *raising a* Kōhyōteki *lost in the Guadalcanal campaign. (Note that a similar view was reproduced in* Warship 1992, *p176, with the boat misidentified as a* Kōryū).

Another view of the Catclaw *and* Baretta *in action, showing the full length of the midget.*

5. At the same time crew induction and training should begin and three seaplane tenders should be converted to midget submarine carriers.[15]

The Second Test Boats

Despite the intention to build the next test boats as soon as possible, three years and eight months passed before any significant development was begun in August 1938. During this time the political situation had changed drastically, Japan resigning from the League of Nations, giving notice to abandon the London Treaty, carrying out anti-communism talks with Germany, and becoming embroiled in the Rokoryo Incident (7 July 1937) and Second Shanghai Incident (9 August 1937). Simultaneously, the IJN had been engaged in serious remedial measures to nearly all warships after the torpedo boat *Tomozuru* capsized (March 1934), the 'Fourth Fleet Incident' (September 1935), and the failure of the medium pressure turbine blades in the destroyer *Asashio* (December 1937).[16] Therefore, the construction of the carriers had to be delayed. However, preparations for series production were only begun after the termination of both arms limitation treaties (31 December 1936), the passing of the first naval expansion programme (the Third Fleet Replenishment Programme) in March 1937, the completion of *Chitose* (July 1938) and *Chiyoda* (December 1938), and the imminent completion of refits associated with the *Tomozuru* and Fourth Fleet Incidents.

Early in the summer of 1939 the modified design was accepted and in July the Navy Minister ordered the Torpedo Experimental Division to construct two boats and to carry out the tests mentioned in (1) and (2) above. In the order of the Navy Ministry the boats were officially called *Kōhyōteki* (A Target). The hull of the first boat was completed at the end of 1939 and launched in April of the following year. Two months later, at the end of June, the second boat was also 'commissioned'. However, Lieutenant Sekido Yoshimitsu and Sub-Lieutenant (Eng) Hori Toshio had already tested the first boat in the Inland Sea and off Sukumo Bay. This showed some difficulty when operating in open water and therefore its formal acceptance as a weapon was postponed until the launching trials from *Chiyoda* had been carried out.

The stern of *Chiyoda* was converted immediately after the boat's basic sea trials and the launching tests in the Inland Sea (Iyō-nada in Hiroshima Bay); attack trials in the somewhat rough sea south of Bungo Strait were carried out from the middle of July to 26 August. An unmanned *kōhyōteki* was launched from *Chiyoda* more than ten times at first, and then manned by Sekido and Hori in order to test the effect upon the crew and the boat at various combinations of force and direction of wind and waves and speed of the mother ship. At a speed of 12kts a comparatively flat angle in entering the water was observed and no abnormality was recognised even at 20kts. With *Chiyoda* at this speed the boat appeared

One of a pair of Type A boats raised after the attack on Sydney. This one is at the Australian War Memorial in Canberra. (Toshio Tamura)

about 1000m astern of the mother ship and began to move under its own power. Captain (Eng) Koyama Tei, a member of the Second Main Division of the Navy Technical Department and an observer at the tests, recalls that Vice-Admiral Toyoda Soemu (at that time Chief of the Navy Technical Department) who had been dissatisfied with the initial results of the ocean experiments, changed his opinion and seeemed to think they were usable for the decisive battle.

Lieutenant Sekido later stated that the aims of the second trials series were to test: 'launching from the mother ship; manoeuvrability', and 'practical capabilities in attack'. He complained about the sensitivity of various instruments and pointed to the rolling and pitching of the boat at periscope depth as especially disadvantageous, 'because it was very difficult to acquire the target by the periscope . . . and the conning tower was always exposed which cannot in fact be called submerging . . . for me, the impression did not disappear for a long time that attack on the ocean . . . was very, very difficult . . .'.[17] Despite some setbacks the second test series progressed relatively smoothly thanks to the adequate guidance and control of the Torpedo Experimental Division and the co-operation of the personnel concerned from other organisations.

As a response to the problems highlighted by the trials, improvements were carried out to the motor drive of the hydroplanes, to the torpedo tubes (to prevent the torpedo stalling in the tube), and to the storage and the battery containers in order to avoid local explosions. Changes of instruments or their improvement were also carried out[18] and on 15 November 1940 the weapon was adopted formally under the title *Kōhyōteki*. At that time the technicians argued that this boat was 'state of the art' and that any improvements could only be achieved by an increase in displacement, while the tacticians' opinion was that any larger boat would have trouble in relation to the mother ship (storage, launching). Furthermore the world situation would not allow further postponement so there was pressure to intensify training and formally adopt the boat. The decisive battle was still the sole thought of the tacticians; as yet there was no idea of attacking ships at anchor in a harbour.

Two weeks before the formal adoption, on 10 October, boats *Nos 3* to *12* were ordered. They were completed at the end of August in the following year and from this time one carrier could be completely equipped. Two months after the first order the Navy Ministry ordered the building of *Kōhyōteki Nos 13* to *36*; when they were completed three carriers could be armed. At the same time as the *Kōhyōteki* was formally adopted, training began for No 1 group (thirteen trainees in total). It was directed by Commander Kato Ryonosuke under the overall command of Captain Harada Kaku, who commanded the seaplane tender *Chiyoda*.[19] In April and October 1941 the training of No 2 and No 3 groups started and in November ten officers and petty officers of the first or second group left Japan aboard five carrier submarines to attack the American warships in Pearl Harbor.[20] The conversion of the large 'mother' submarines was a last-minute effort (in the case of *I 16* from 20 October to 10 November 1941), following the change in the tactical concept and the adoption of 'Hawai Sakusen'. At the same time, from 19 September to 5 November, launching experiments were repeatedly carried out with twelve boats from *Chiyoda* (whose launching mechanism had been operational since 10 January). The trials at Iyo-nada showed that with calm seas and trained crews the twelve midgets could be launched through the stern opening in about 20 minutes (best time was 17 minutes) or at the rate of about two boats every three minutes.

Technical Description of Kōhyōteki

The *Kōhyōteki* was a single hull midget submarine. The pressure hull was all welded, of 8mm cold rolled steel plates of MS 44 quality. By utilising all-welded construction Katayama intended to save weight and his policy of radical reduction of weight was successful in many respects. For non pressure resistant parts 2.6mm thick material of the same quality was used. The bulkheads which divided the boat into several compartments, were gas- but not water-tight because the panels between the stiffeners were only 1.2mm thick. This was one of the successful radical measures taken by Katayama to save weight. Construction was to fine tolerances and despite the necessity to reduce weight the collapse depth of the pressure hull was calculated at 200m but the safe diving depth remained unchanged at 100m, which was the same as the *A-Hyōteki*. Thus, the safety coefficient was roughly 1.4 times higher than that of a fleet submarine.

Compared to the *A-Hyōteki* the length was increased by 600mm (from 23.30m to 23.90m), the interior diameter in the control room section by 26mm (from 1.824m to 1.850m) and also the depth from the keel to the upper edge of the conning tower by 26mm (from 3.074m to 3.100m).[21] The shape of the cross-section changed only at the bow to become an oval by drawing in the sides. This shape was hydrodynamically better than the earlier *A-Hyōteki* but the hull still had the form of an enlarged torpedo with a conning tower. The fore and aft ends of the pressure hull were closed by convex shaped pressure bulkheads, beyond which was one external compartment. These formed the main ballast (diving) tank and the reserve ballast tank at the bow, but at the stern was a free flooding space.

Apart from the conning tower there were no superstructure and in order to keep water resistance to a minimum only small mooring cleats cut out of a 12mm thick steel plate were welded to the hull fore and aft. In addition, a 9.5m long ballast keel (50mm × 12mm) was welded to the bottom of the hull while the propellers and TT were surrounded by guards intended to protect them from damage. These guards were streamlined and did not extend beyond the extreme dimensions of the boat.

The hull was divided into three sections, in order to facilitate construction and especially to allow the pre-assembly fitting-out of the sections.

The forward section essentially contained:
1. Two 45.7cm (18in) TT mounted one over the other with about 250mm distance between;
2. Two high pressure air flasks and two impulse tanks;
3. The torpedo firing valves;
4. One diving tank (main ballast tank) and one reserve ballast tank;
5. 250 lead pigs weighing 1254kg.

Unlike the *A-Hyōteki*'s 53.3cm (21in) TT for the Type 89 torpedo, the calibre was reduced to 45.7cm due to the appearance of the Type 97 oxygen torpedo. The length of the TT was 5.4m and therefore they extended nearly the total extent of the forward section. Because there was no space behind the TT for a loading room and torpedo reloading hatch, the torpedoes were muzzle-loaded. The TT therefore had no breech but a spherical casting riveted and welded to the tube which contained an adjustable tail stop with a thick rubber buffer and a housing for the impulse air check valve. The TT were free flooding and venting so there was no need to flood them before firing and no TT doors were provided. The warhead of the torpedo projected out of the tube about 300mm, thus providing a streamlined closing.

The setting of the torpedo gyro and the depth-keeping gear was from the control room by means of mechanical linkages and shafts connected to the gyro and depth-

Net cutter and torpedo guard at bow of one of the Sydney boats.

setting spindles at the TT. The adjusted gyro angle and depth could be read on indicator dials in the control room. The torpedoes were fired by air pressure. Each tube had an impulse tank of 69.4 litres volume. Because the air was not lead to a tank a stream of bubbles appeared on the surface after firing and there was much complaint about this by the crews.

About 600mm behind the forward pressure bulkhead there was another transverse bulkhead, this space forming a reserve ballast tank which was filled with water when the torpedoes were fired, *ie* it functioned as a torpedo compensation tank. Even though this tank was filled through the sea valve comparatively quickly it was not sufficent to avoid the bow and the conning tower bobbing to the surface even if the speed was increased at the same time.

The forward external compartment, *ie* from the forward pressure bulkhead to near the muzzle of the TT, formed the main ballast (diving) tank of average 2.15m in length and a volume of 1336 litres covering the TT. The filling of this tank was usually carried out aboard the mother ship because the diameter of the vent valve and the opening of the pivoted flooding valve were somewhat small and it took time to flood this tank when the boat was in the water.

The centre section was separated into three compartments, namely the forward battery room, control room with conning tower and aft battery room. The *Kōhyōteki* had electric propulsion for surface as well as submerged operation. The main battery fed the main electric motor as well as the few auxiliaries and provided energy for the lighting and wireless equipment.[22] The main battery consisted of 192 trays[23] of two cells each and 4v per tray, connected in parallel-series combinations: 96 trays of the aft battery and 40 trays of the forward battery, totalling 136 trays (*ie* roughly 75 per cent of the total battery) were installed in the aft battery room, the remaining 56 (originally 88) trays of the forward battery in the forward battery room, arranged along the sides to obtain a centre passage for inspection and maintenance.

For the Pearl Harbor operation 32 battery trays were removed from the forward battery room (port side) in order to instal four HP air flasks and one oxygen flask to increase the operational endurance. Because the boat was not equipped with an air compressor the air flasks were filled by a charging connection through the pressure hull near the conning tower to the HP air manifold in the control room. Since the operation of the Kōhyōteki was limited to a few hours there was only one circulation ventilation system for the batteries. The hydrogen generated during the discharging of the battery had to be removed or at least reduced and this was done by suctioning air from the forward battery room, forcing it through a hydrogen absorbing cell and discharging it into the aft battery room after cooling and mixing with HP air.

Above the battery rooms an exhaust hole was cut into the pressure hull and closed by a ventilation plate of 150mm diameter around which several more small ventilation openings were arranged in rows. These openings could be closed from inside and ventilators were mounted beside it. When the battery was charged in the shipyard or aboard of the carrier vessel the hydrogen gas could be removed by means of this ventilation system. The necessity for this equipment had become obvious during the tests with the A-Hyōteki.

In the forward battery room there was the forward trimming tank with a volume of 357 litres; in the aft battery room the aft trimming tank with exactly 100 litres less volume. For flooding and pumping out, the bilge pump could be used but this had the disadvantage of often sucking up mud. Therefore a small pump with a capacity of half a ton per hour was used which could be driven by the periscope motor. The distance between the tanks was small and the trimming of the boat difficult. Therefore on the port side of the forward battery room 284 lead pigs weighing 1421kg were carried on trolleys which could be shifted manually from the control room by means of a handwheel, steel wire and cable drum.

As a result of the tests with the A-Hyōteki a very simple lightweight non-directional hydrophone was installed allowing the operator to recognise little more than right and left in his headphones. One microphone set was mounted on each side of the aft battery room (in the case of the boats which participated in the Pearl Harbor attack) or of the main ballast tank (for the Diego Suarez and Sydney attacks) in streamlined casings. In the boats carrying out the Hawaii operation a demolition charge was mounted in the aft battery room.

The control room was located between the forward and aft battery room directly below the conning tower with its single access hatch. It contained all control instruments for the steering of the boat, such as the automatic and manual depth control device for the hydroplanes on the starboard side forward, the manual steering device for the rudders along the starboard side, the master gyro compass on the port side forward, the directional gyro beside the latter, the TT controls, the raised platform for the commander when using the periscope on the centreline, the cable hoist drum to starboard, the motor for periscope hoist, a small trim pump, HP air manifold, small crystal radio, hydrogen detector, etc.

The Kōhyōteki had a depth-keeping system, driven by a pneumatic piston which was moved by the operation of a control valve combining automatic and manual control. The automatic element affected both trim and depth. A piston, exposed to sea pressure on the lower end, was

Table 1: MAIN TECHNICAL DATA

	Prototype	Kōhyōteki Type A	Kōhyōteki Types B and C
Design begun	1932	1938	1942
First boat completed	1933	1939	1943
Length overall (m)	23.30	23.90	24.90
Breadth, maximum (m)	1.824	1.850	1.880
Depth, conning tower to keel (m)	3.074	3.100	3.100
Displacement, submerged (tons)[1]	42	46	50
Main battery (type/no of cells)	?/928	Special D/224	Special D/224
Main motor (hp)	600	600	600
Maximum submerged speed (kts)	24.85	24	18.5
Radius of action, above water (kts/nm)			6/300
submerged	25/22.5	6/80	4/120
		19/50 min	18.5/55 min
Safe diving depth (m)	100	100	100
TT (mm)	2–53.3	2–45.7	2–45.7
No of torpedoes	2	2	2
HP air flasks (litres capacity × no)[2]	430 × 2	430 × 2	430 × 2
	24 × 1	15 × 1	15 × 1
Periscope, length (m)	2.75	3.05	3.05
Generator			40hp/25kW × 1
Fuel (tons)			0.6
Crew	2	2	3
Continuous activity (days)	–	–	1–2

Notes

[1] Round numbers are given, data varying according to source.

[2] Sometimes changed, see text.

attached to a manually set spring pressure on the top. It was connected to one side of a lever and a heavy pendulum was supported on the opposite side of the fulcrum. The purpose of the spring-loaded slotted arm was to reduce oscillation. The operational principle was as follows: if the depth of the boat was greater than set by spring tension on the piston, sea pressure moved the piston up which operated the control valves and moved the diving planes to the 'surface' direction. If the midget took an up or down angle, the pendulum, by remaining vertical operated the control valve in a direction to move the diving planes in the proper direction to bring the boat to an even keel.

This equipment was adapted from Captain (Eng) Tomonaga's automatic depth-keeping system for large submarines but could not incorporate its separate tanks due to restricted space.[24] This reduced control to the movement of the diving planes, which meant that it was necessary to obtain neutral buoyancy and good trim at the desired depth before the automatic depth-keeping system could possibly operate satisfactorily. For this reason the automatic system had a manual over-ride.

The *A-Hyōteki* was equipped with an electric steering motor only but it was replaced by a Japanese-made Type 97 gyro compass during the course of the trials. This was of the German Anschutz type, 1926 model, with some modifications. There were no repeaters connected to this master compass nor any provisions for same. The complete equipment was tested at the US Navy Submarine Base in Hawaii, the report finding that 'after several minor adjustments were made the compass operated satisfactorily and settled on the meridian . . . the operation is extremely simple, the only operation required is to turn on the 10 ampere snap switch . . . and the rest of the operation is fully automatic . . .' The directional gyro was installed as an auxiliary instrument for the main gyro compass in the control room forward of the periscope. The commanding officer could read the card through the medium of a reflecting mirror mounted on top of the outside housing. The equipment was of Japanese design and manufacture.

For the replenishment of oxygen and absorption of exhaled carbon dioxide an air regeneration system was installed in the control room. Air was circulated by the turbo ventilator, and carbon dioxide was chemically absorbed by sodium hydroxide. The installation of this air regeneration system was one of the results of the tests of the *A-Hyōteki*.

For surfacing the main ballast tank had to be blown and the bow of the boat broke surface comparatively quickly while the stern remained in nearly the same position. Due to this extreme angle it was some time before the conning tower was completely surfaced. To deal with this problem a so-called 'emergency tank' with a volume of 410 litres was fitted at the bottom of the control room. This not only reduced the extreme angle on surfacing but also brought the conning tower to the surface much faster. The fitting of this tank was also a result of the *A-Hyōteki* trials.

The pressure resistant conning tower consisted of two vertical cylinders arranged one behind the other. The forward cylinder was about 550mm in diameter and was the hatchway; the after one was for the periscope and had a diameter of about 700mm. At the after end of the conning tower was a 70mm diameter antenna for the ultra short-wave radio. It could be manually raised 815mm from the control room by means of worm gear.

The periscope was manufactured by the Japan Optical Manufacturing Co (*Nihon Kogaku Seiko KK*) to the design of the Naval Technical Laboratory and named 'Type 92'. In order to keep it secret it was referred to as 'special glasses' (*toku megane*) or 'special glass' (*tokuganhyō*). Compared to the *A-Hyōteki* the length was increased by 300mm from 2.75m to 3.05m. The diameter was 92mm. It had two magnifications of 1.5 and 6.0 and was almost a copy of the 9.14m long Zeiss periscope. This periscope could be raised 1.22m by an electric double reduction worm and gear drive to the cable drum. There

US drawing of the hydrogen filtration system based on examination of the boat captured after Pearl Harbor.

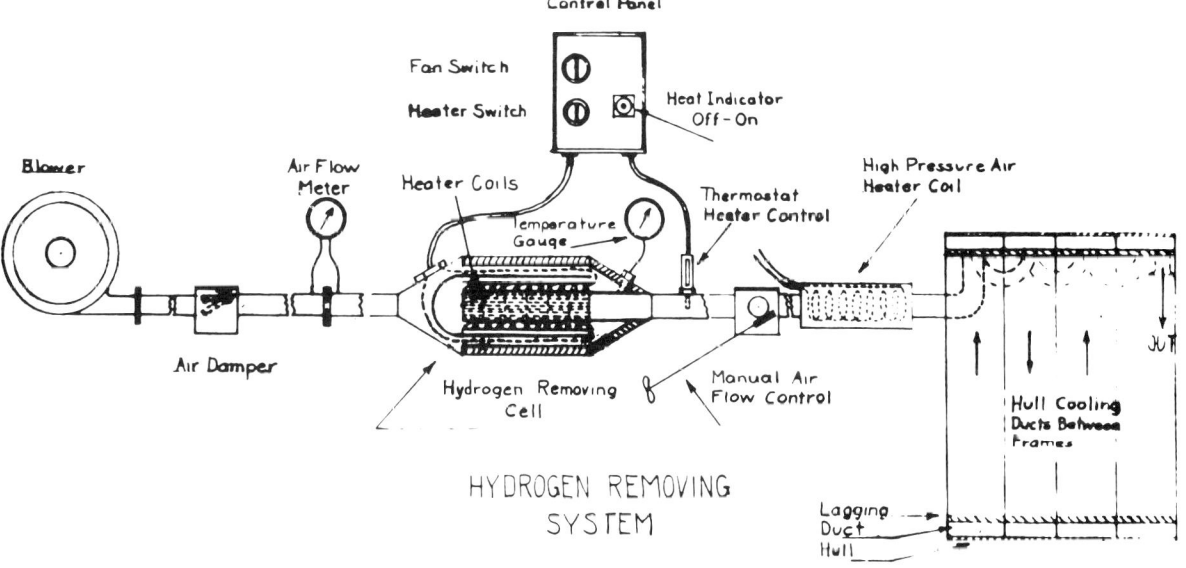

was a clutch arrangement which allowed a second worm gear with cable drum to be disengaged and the first worm gear to be connected to a shaft which had an eccentric on the lower end for actuating a small reciprocating drain pump.

The periscope well projected out of the conning tower for about 300mm, which was enclosed by a streamlined casing to reduce resistance, similar to the casing surrounding the conning tower itself. Between periscope cylinder and well and their casings there were openings for the battery ventilation exhaust.

The stern section consisted of the motor room and the tail assembly. The bulkhead which separated the motor room from the aft battery room was covered by insulation on the forward side, which also served for sound-deadening. It had a small access door and some breaks for shafts and wires etc.

The main motor for both submerged and surface operation was manufactured by Toshiba Co. Its output was 600hp at 1800rpm, and weighed about 1.5 tons. The control panel was arranged in front of the motor and operated from the control room by manual switches. Depending on battery combinations, the following speeds were available:

Full speed ahead	24kts
Half speed ahead	12kts
Slow speed (30 per cent revs)	8kts
Slow speed (20 per cent revs)	5kts
Astern	5kts

The rpm of the motor were reduced by a reduction gear in the ratio of 5.5:1 and then transmitted to the propeller shaft. This gearbox was behind the pressure bulkhead, ie

This and the following views depict a Kōhyōteki *exhibited at Etajima, Inland Sea. It was one of the boats lost at Pearl Harbor, returned by the US after salvage.* (Hayashi Yoshikazu)

outside of the motor room. The concentric propeller shafts drove contra-rotating tandem propellers: the forward propeller turned right-handed, the aft one left. Each propeller had four blades. The forward propeller was 1.35m in diameter, the aft one 1.25m. The pitch of the former was 2.52m, that of the latter 1.87m. The propellers were protected by hooped guards and skegs.

Vertical and horizontal stabiliser fins with vertical rudders and stern planes were installed just forward of the propellers which, of course, were on the axis of the pressure hull. The tail assembly was so arranged and faired as to offer no projections to catch a net. Before and partly below the main motor was the aft balance tank with a capacity of 180 litres. This tank was also one part of the trimming system, together with the forward balance tank (232 litres) which was located below the aft end of the lower TT in the forward section.

As already noted the submerged speed of the production version of the boats dropped to 24kts. A further decrease to 19kts followed the fitting of protective frames for TT and periscope, net cutters at the bow and conning tower, jumping lines from bow to stern and propeller guards. In the case of the boats for the attack on Pearl Harbor the submerged speed was reduced to 14kts since some of the batteries had been removed.

Bow view. The ends of the torpedo tubes and the guards are modern repairs following serious rust damage. (Hayashi Yoshikazu)

Training Kōhyōteki

The first crews (13 officers and petty officers) reported to *Chiyoda* in November 1940, and in the following January operational training began in Orako Bay (Kurahashi-jima). The intervening time had been used for theoretical education and classroom training. Later, training was shifted to Mitsukue Bay (Ehime prefecture) and was continued from 12 to 20 March.

In December 1940 *Kōhyōteki Nos 13* to *36* had been ordered and in March 1941 the training of the first group was completed. In April the second trainee group (22 officers and petty officers plus 12 mechanicians) was recruited. At the same time the base was shifted to Karasukojima. Theoretical education was carried out in the secret room of the Torpedo Experimental Division of Kure Navy Yard, chart manoeuvres and other exercises in the corresponding room of the Submarine School. Basic sea training then followed aboard the tug *Kure Maru*, transferred from the middle of May to the seaplane tenders *Chiyoda* and *Nisshin* (whose utilisation as a temporary *Kōhyōteki* carrier had been decided in April of the same year). On 20 August the training of the second group was completed.

Before the first trainee group had graduated, in February 1941, three training *Kōhyōteki* had been ordered. These were necessitated by the rapid increase in the number of boats and the continuous deterioration of relations with the USA, which required accelerated preparations.

In these training boats some of the batteries were removed from the aft battery room and this space was equipped as a second control room. Aft of the new, slightly shorter, control room was a stand-by room for six or seven trainees and two instructors. The forward half of these rooms was surrounded by a pressure resistant tank to compensate for the substantial reduction in weight due to the removal of the batteries. The size of the conning tower was doubled and two periscopes etc were fitted.

During training Sub-Lieutenant Iwasa Naoji investigated the tactical potential of *Kōhyōteki* and found that they might be used for attacking warships anchored in a harbour or base. After some argument the C-in-C Combined Fleet accepted the participation of *Kōhyōteki* in the attack on the US fleet at Pearl Harbor. Because the boats could not be carried by the *Chiyoda* or *Nisshin*, five 'C' type cruiser submarines were fitted for the transportation of the *Kōhyōteki* on the deck behind the conning tower. The control room of the midget (whose bow faced aft) was above the engine room of the carrier submarine and access between the two was possible through watertight cylinders and a hatch in the bottom of the midget. An arrangement of two close-fitting cylinders enabled access between carrier and midget submerged or surfaced and allowed maintenance during the transportation to the operational area.

The *Kōhyōteki* were carried on chocks and all fastenings could be controlled from inside the carrier submarine. A telephone cable was led to the connection in the conning tower of the midget.

Later Improvements

The First Special Attack Unit (*Tokubetsu kogekitai*) had been annihilated during the attack on Pearl Harbor and the capture of at least one *Kōhyōteki* by American forces became known in Japan very soon after. Therefore the Second Special Attack Unit tried to anticipate likely defence measures by corresponding countermeasures. Up to May 1942 the following improvements were carried out:

1. The control of rudders and diving planes was changed from HP air to oil pressure (hydraulic transmission). This was a big step forward since no battery trays had to be removed and the fitting of additional HP air and oxygen bottles was not necessary to increase the operational radius (in the case of the boats which participated in '*Hawai Sakusen*' the submerged operational time had been increased from 4 hours to 16 hours but underwater speed was reduced to 14kts). Submerged speed could be improved to 19kts again and another advantage was the noiseless power transmission to the steering elements.
2. The free-flooding TT were equipped with tube hatches. The hatches were fitted after the loading of the torpedoes and separated automatically when the torpedo was launched. The auxiliary ballast tank almost lost

Conning tower, eroded by its years underwater. (Hayashi Yoshikazu)

its function because the weight of the water which filled the TT after launching almost compensated for the torpedoes. The adjustment of the torpedoes aboard the carrier became easier but an air (oxygen) compressor on the after deck of the mother ship was still necessary in order to replenish the air (oxygen) in the chamber if necessary.[25]

3. Runners were mounted at the bow to enable the boat to creep along the sea-bed and to surmount defensive installations. But during operations the disadvantages became obvious, and one boat of the Second Special Attack Unit could not fire its torpedoes because the runners had been deformed during the penetration into Sydney harbour (31 May 1942).

4. Due to the fitting of the access cylinder (*Kōtsu to*) described above, the maintenance of the equipment and batteries of *Kōhyōteki* became possible and the health of the crew was decisively improved.

5. The habitability aboard the midget was quite naturally very limited (for example, the crew could not stand upright except in the control room). Apart from the aforementioned air regeneration system in the control room the habitability was improved by minor measures such as the modification of the seats etc.

After the Second Special Attack Unit all *Kōhyōteki* were prepared for the fitting of access cylinders. One boat of the Madagascar Attack Unit (which obtained the most notable success of all IJN midgets by damaging the battleship *Ramillies* and sinking the tanker *British Loyalty* in Diego Suarez) was ordered to attack a merchant ship in the Indian Ocean but lost its chance due to the long time needed to launch from the mother ship. The boat had to be abandoned (scuttled) because there was no equipment to re-embark the midget at sea. As a result of the evaluation of this experience the following modifications were carried out:

Detail of the eroded conning tower. (Kunio Kitamura)

1. The diameter of the access cylinder was increased.
2. A stabiliser was provided to shorten the time for running up the gyro compass.

These two measures shortened the time necessary for launching to 3.5 minutes.

3. It was also decided to equip the carrier submarine with a lifting apparatus for taking the *Kōhyōteki* aboard at sea and *I 18* was fitted with this gear. At the beginning of 1943 an experiment was conducted at Aki-nada (Setonaikai) in which *I 18* towed a *Kōhyōteki* and recovered the submerged boat while underway. Even though the experiment was judged a success this technique was never used operationally and the large carrier submarines were not equipped with this apparatus.

Kōhyōteki *Types B and C*

At the beginning of the Pacific War there was no chance for the decisive battle the IJN's strategists and tacticians had long planned for and for which secret weapons like the *Kōhyōteki* had been developed. Therefore *Kōhyōteki*, contrary to original thinking, were used to attack anchored ships or convoys if the opportunity arose. In view of the ever diminishing chance for the IJN's traditional decisive battle and the enormous widening of the Japanese sphere of influence, other tactical possibilities of the midget had to be investigated. It was quite natural that the defence of local bases came well to the fore, but it meant abandoning the original offensive character of the *Kōhyōteki*, which could not be achieved without improvement to the equipment. The biggest deficiencies were the shortage of range and the boats' inability to recharge the batteries themselves. Therefore, in June 1942, when the plan to station *Kōhyōteki* at Midway had been abandoned following the US Navy victory, the remaining *Kōhyōteki* were transported to the Aleutians by *Chiyoda*. The design of a boat suited to defensive operations was begun and the experimental *Kōhyōteki No 53* was built to new drawings. Naturally war lessons and general experience with the *Kōhyōteki* were incorporated in this design.

Details of stern, propellers and guards. (Hayashi Yoshikazu)

This boat was classified Type B (*Otsu*) to be distinguished from earlier *Kōhyōteki* which became Type A. Type B differed from its predecessor technically in several ways and with regard to the crew one more man was added and one of every three was a technical rank.

1. The hull was lengthened by 1.0m, and the diameter increased to 1.88m from 1.85m.
2. The displacement rose to 50 tons from 46 tons.[26]
3. In order to charge the battery and for surface operation a 40hp/25kW diesel-generator type B was fitted. Its capacity was minimal and the charging of the battery took 18 hours.
4. A fuel tank with 0.6 ton capacity was added.
5. The diesel increased the radius of action to 300–350 miles at 5–6.5kts (surface speed, of course). The speed depended on the extent and shape of the additional protective equipment added to the hull.
6. The submerged speed was reduced to 18.5kts by the increased dimensions and displacement. This speed could be maintained for 55 minutes. At 4kts the radius of action was increased to 120 miles.

Despite the minor changes in principal dimensions and displacement the character of the *Kōhyōteki* was changed completely by the battery charging facility and the use of a diesel for surface navigation brought it very near to a fully operable midget submarine.

About thirty-six Type C[27] boats (the production version of Type B) were built at Ourazaki before the autumn of 1944. Besides this specially built construction site, some civilian companies were engaged in the mass production of fittings and equipment parts. Even though the existence of the boat had become known to the allied forces on several occasions the IJN still insisted on secrecy and used terms like ①*sen* (*Maru ichi* ship, the term '*sen*' meaning a commercial ship in contrast to '*kan*' used for warships), *H-kanamono* (H metal item), *M-kanomono*, etc.

The stern section, even in its damaged state, reveals the design's origin, deriving from torpedo practice. (Kunio Kitamura)

The main electric motor removed from the boat and displayed separately. (Kunio Kitamura)

The contra-rotating propellers. (Kunio Kitamura)

Notes

1. Since the submarine played a major role in IJN strategy this was considered especially disadvantageous.
2. The artillery duel was the last main stage of the total operation for the destruction of the US battlefleet. The operation consisted of at least three stages (depending on circumstances), each of which had some subdivisions.
3. It is very interesting to note that the idea for this very special design for torpedo warfare, the piloted torpedo, did not originate in 1931. According to *Don Game Banashi (2) Sensuitei Shiwa* (*Don Game History (2) – History of the Submarine – Don Game* is a Japanese term for a turtle, used as a synonym for IJN's first submarines, the Holland boats *No 1* to *No 5*) by Fukuda Ichiro and published in *Umi to Sora* (*Sea and Sky*) September 1958, Yokoo had already proposed this idea in 1905 when he belonged to the First Submarine Division. At that time an attack on the harbour of Vladivostok was intended but the war ended before the submarines were ready for action. He proposed to be towed by a standard torpedo, guiding it during the attack, and was therefore quite unrealistic.
4. He was later promoted to Rear-Admiral and is famous as the inventor of the legendary oxygen-driven Type 93 torpedo.
5. Shigeru Fukudome, 'Hawaii Operation', in *The Japanese Navy in World War II*, US Naval Institute, Annapolis, Maryland, p14.
6. This was possible only because of preparatory work; for example, Commander Nawa designed the battery in July 1932.
7. There were also separate plans to use either a combined diesel and battery or diesel-only propulsion but the personnel responsible in the Navy Technical Department doubted the possibility of developing a lightweight fast-running diesel with high propulsive efficiency within a short period of time. Therefore, the decision turned out in favour of a solely battery-driven model because in this case the designers could fall back upon the lightweight high-powered battery (designed by Nawa) and the already proven electric motor of Tōkyō Shibaura Denki.
8. The initial 30kts called for had to be reduced to 25kts because many fittings and equipment for safety and recovery of the crew had to be added once the designers had been ordered not to construct a boat whose operation would mean certain death.
9. Because the first idea was a piloted torpedo (*ningen gyōrai*) the boat was considered to be a 'mother torpedo' (*bokan gyōrai*) or weapon (*heiki*) rather than a small submarine, and therefore the Torpedo Main Division (*Suirai Shumubu*) of the Navy Technical Department was responsible and ordered from the Torpedo Experimental Division.
10. According to *Kaigun Zosen Gijutsu Gaiyo* (Survey of Naval Shipbuilding Technology), Vol 3, p540, the boat ran at a top speed of 27.6kts during a test in Setonaikai but this seems unbelievable for a 41.5-ton boat powered by a 600hp motor and the speed given here may be considered more realistic. It was roughly the final planned figure and was still the world record for midget submarines with a battery-powered propulsion system.
11. When this term came to the notice of the Navy Air Force the 'Anti Submarine Bombing Target' was demanded for testing and the personnel in charge had much trouble in refusing the requests.
12. For a detailed description see: Hans Lengerer, Jiro Itani, Tomoko Rehm-Takahara, 'Verdeckte Schiffe der IJN: Die Chitose-Klasse – Kōhyōteki-Träger, Seeflugzeugträger, Schnelle Tanker, Hybrid-Flugzeugträger, Flugzeugträger', in

Marine-Rundschau 6/1989 (last published issue).

13 They were also considered for use as fast oilers, hybrid aircraft carriers or to be converted to full aircraft carriers (CV).

14 Twelve boats for each of the three carriers *Chitose*, *Chiyoda* and *Mizuho* and twelve units for reserve.

15 The term 'midget submarine carrier' was not officially used but referred to as 'second condition' which corresponded to this task. At that time the Third Fleet Replenishment Programme had not yet been worked out, so three ships only could be considered.

16 For a brief description see: Jiro Itani and Hans Lengerer, 'Die Grundplanung der japanischen Kriegsschiffe vor 1940 – Zwei Katastrophen wirkten sich auf den Schiffbau aus', in *Marine-Forum*, March and April 1992.

17 Lieutenant Sekido characterised the boat as follows: 'This was a high speed boat armed with high quality torpedoes manned by a smaller crew, which produced advantages, but I am afraid it also meant problems because of the contradictary requirements of superior capabilities and a small lightweight hull.' He complained that 'the Tokusen at that time relied on the mother ship with respect to charging, air supply and torpedo loading etc and some of the equipment called for high levels of skill and were easily deranged'. He believed 'if they [the midgets] were operated by mother ships with excellent tuning and repair capacities, these disadvantages would be greatly reduced. But the performance of the bases makes it difficult to get the most from the boats unless we control both maintenance and equipment . . .'. He also described the problems caused by a dummy torpedo which had come to a dead stop in the TT in half-launched condition.

18 Other problems related to alterations in the reserve buoyancy, maintenance of trim (which changed after about four hours submerged running), reduction of minimum submerged speed (mainly to avoid the periscope wake and this was closely related to the reserve buoyancy), lengthening of the periscope and wireless antenna, automatic control of the main motor, lifting of the bow when firing the torpedoes, etc. Even though some of them could not be entirely solved the improvements made underwater running more stable and observation easier and meant a change in designation from 'human torpedo' to 'special submarine' (*Tokushu senkotei* – this term was first used in a publication of the Imperial HQ after the attack on Pearl Harbor).

19 Harada was promoted to Vice-Admiral later and wrote a report '*Kōhyōteki* – Record and Impressions' between March 1943 and February 1944. A part (chapters 1 to 5) was published in *The Imperial Japanese Navy* (*Nihon no kaigun*), No 4 (25 May 1978), p25ff. His report contains a wealth of hitherto unknown details about the training and operational planning.

20 In August 1941 when the order was given for the conversion of *Chiyoda* to a midget submarine carrier, a study of the tactical use of the *Kōhyōteki* was made which pointed to the attack of enemy ships in port as the primary task (second was co-operation with the advanced submarine force and third the employment in the decisive battle) and this was agreed by a committee chaired by Vice-Admiral Tsunada, the Chief of Kure Navy Yard. According to Vice-Admiral Harada the originator of this idea was Sub-Lieutenant Iwasa Naoji, a member of the first trainee group and one of the nine 'war gods' of the attack on Pearl Harbor. On 5 October 1941 Iwasa, accompanied by Harada, reported to Admiral Yamamoto Isoroku, explaining the boat and the attack methods and pointed out his confidence in the attack on Pearl Harbor and his doubts with regard to an attack on Singapore or San Francisco. At that time he proposed to carry two *Kōhyōteki* on each large type submarine but this could not be achieved, nor could the enlarged submerged radius (the requirement was 26kts for 1.5 hours).

21 Measured without the casing of the periscope support. If measured to the top of the casing the depth was 3.40m. Along with the lengthened periscope this was an important factor for improving observation.

22 The navigational instruments, like the directional gyro, had an independent energy source (separate batteries).

23 This number refers to the boats which participated in the Pearl Harbor attack. In their case 32 trays had been removed in order to obtain space for four HP air flasks and one oxygen flask in the forward battery room. Originally the boats had 224 trays.

24 In the original system adopted by large submarines water inside the trim and adjustment tanks was shifted according to the change of depth or angle up or down, but due to insufficient space the *Kōhyōteki* had no such tanks.

25 As a rule the torpedoes were loaded the day before the planned attack.

26 According to data found in a captured *Kōhyōteki* the displacement was 44.7 tons.

27 Most sources state that about seventy-nine boats were built but according to available operational records no boat with a number higher than 85 was used but a photograph showing *No 89* exists. Therefore it is supposed that about thirty-six were built by the following calculation (52 boats Type A, 1 boat Type B, 36 boats Type C = 89 boats).

THE ADMIRAL SCHEER AT WAR

In general the employment of German heavy units as commerce raiders was only a limited success during the Second World War. However, the ship with the best record in terms of Allied vessels sunk was the *Admiral Scheer*. Her wartime career is chronicled by Pierre Hervieux.

The *Admiral Scheer* belonged to the well-known *Deutschland* class which comprised two other units, *Deutschland*, renamed *Lützow* in February 1940, and *Admiral Graf Spee*. These three ships were formally described as *'Panzerschiffe'* (armoured ships) but were commonly known as 'Pocket Battleships'. They were closer to heavy cruisers and were reclassified as such in 1940. The *Scheer* was built by the Wilhelmshaven *'Reichsmarinewerft'*, being laid down on 25 June 1931, launched on 1 April 1933 and commissioned on 12 November 1934. The particularity of the class was to be armed with six 280mm (11in) guns (2 × 3), which was a battlecruiser calibre and made them more powerful than usual heavy cruisers. They were also faster than battleships and in theory could only be threatened by five battlecruisers, the British *Hood*, *Renown* and *Repulse*, and the French *Dunkerque* and *Strasbourg*. Therefore they were a real menace to the commercial traffic of any potential enemy, and they successfully carried out their role of commerce raider during the Second World War.

The Outbreak of War

To accomplish those tasks, the *Scheer* was placed under the Commander-in-Chief, West, and was then under the command of *Kapitän zur See* Wurmbach and, on 4 September 1939, had the distinction of being among the four first German warships to be attacked by RAF Bomber Command aircraft. The *Scheer*, then lying in the Schillig Roads in the Heligoland Bight, was attacked from low altitude by five Bristol Blenheims of No 110 Squadron (Flight Lieutenant Doran) which made three hits with 250lb bombs. None of these exploded, probably because the height from which the bombs were dropped was insufficient to work off the safety device of the bomb fuses. The *Scheer* was out of action only until 10 October 1939, and she shot down one of the attackers with her AA guns. During the winter of 1939–40, she was in dock for a general refit and her tower mast was replaced by a tubular one, whilst a clipper stem was added and also a funnel cap. Because of this, the *Scheer* could not take part in the occupation of Norway operation in April 1940.

On the night of 19/20 July 1940, bombers of the RAF unsuccessfully attacked the *Scheer* and the battleship *Tirpitz* in the naval dockyard at Wilhelmshaven. The weight of attack which Bomber Command could then devote to this purpose on any one night was only some 25 to 40 bombers (Hampdens, Wellingtons and Whitleys) and this was insufficient to achieve any very favourable results. This explains the poor results obtained by 1042 bomber sorties which aimed 683 tons of bombs at the *Bismarck*, *Gneisenau*, *Lützow*, *Prinz Eugen*, *Scharnhorst*, *Scheer* and *Tirpitz*, between July and October 1940! The *Scheer* underwent trials in the Baltic from July to September 1940 and, on 23 October, under the command of *Kapitän zur See* Krancke, left Gotenhafen for commerce warfare operations in the Atlantic.

The First Atlantic Sortie

On 27 October the heavy cruiser, which had come through the Kiel Canal from the Baltic, received orders in Brunsbüttel to set out and reached Stavanger on 28 October. From 31 October to 1 November the ship passed through the Denmark Strait undetected, for no intelligence was received in London regarding these preliminary movements. The first news of the *Scheer*'s presence on the shipping routes was received when she attacked the homeward-bound Halifax convoy HX84, consisting of thirty-seven ships escorted only by the armed merchant cruiser HMS *Jervis Bay* (1922, 14,164 tons, 7–6in guns, commanded by Captain Fegan) at 52°45N/32°13W on the evening of 5 November.

Earlier on the same day, around 1430, the independently-routed British reefer ship *Mopan* (5389 tons) was in sight. The *Scheer* ordered her to stop, which she did, and after the crew took to the lifeboats, fired at the waterline with 105mm (4.1in) guns. The beautiful banana ship sank at 1605, at 52°59N/32°12W, and unfortunately failed to send a raider report. Had she sent a report, she might have saved the convoy, which the Germans knew as having left Halifax on 27 October through their Monitoring Service which intercepted the British radio traffic.

Armed with that valuable information, on the morning

A prewar view of Scheer *dressed overall. The shield on the bow contains the battle honour 'Skagerrak' (the German name for the Battle of Jutland), where Admiral Scheer led the German High Seas Fleet in the major battlefleet encounter of the First World War.* (CMP)

The Scheer *is still obviously in the fitting-out stage, with guns in different elevation positions. The three rangefinders are shipped but a searchlight and various other pieces of equipment are still missing.* (CMP)

The Scheer, *with the* Deutschland *behind, with markings for neutrality patrols painted on their main turrets, during the Spanish Civil War.* (Author's collection)

of 5 November, at 0940, the *Scheer* catapulted her Arado Ar196 floatplane, piloted by *Oberleutnant zur See* Pietsch, who was ordered to stay out of the convoy's sight and not to send any radio message. The Arado was back at exactly midday and Pietsch announced that he had discovered the convoy, which was sailing a few hours south from their position, without, he said, any escort – which was true in a way, for the *Jervis Bay* could be taken for a merchant ship. After that exciting announcement the *Scheer* turned at full speed towards Convoy HX84.

When approaching the convoy, at about 25km, the *Jervis Bay* asked her in morse 'What ship?'. The *Scheer* did not reply and closed the distance to 17km, to be able to use her 105mm guns too and, at about 1640, opened fire, firstly with her 280mm guns. The *Jervis Bay* sent up red flares and the convoy at once scattered and made good use of its smoke-making apparatus to cover its dispersal, then she unhesitatingly challenged her redoubtable adversary to a most unequal duel. The result was a foregone conclusion and, shortly after 1700, the burning *Jervis Bay* sank at 52°41N/32°17W. Her Captain, E S F Fegen, was awarded a posthumous Victoria Cross for his gallantry and self-sacrifice.

In the meantime, 150mm (5.9in) and 105mm (4.1in) guns from the *Scheer* opened fire on two ships and set them on fire, but a heavy smoke hid them and darkness came quickly too, enabling them to escape. They were the British troop transport *Rangitiki* (previously a passenger–cargo ship, 16,698 tons) and the British tanker *San Demetrio* (8073 tons), at 52°48N/32°15W. After the *Jervis Bay* went down, the *Scheer* sailed to the south, where the convoy was assembled less than an hour previously, but by then it had completely dispersed. As soon as a ship was seen German guns opened fire but, thanks to darkness, most of them escaped. The Germans thought they had sunk between six and eight ships, but they in fact attacked and damaged some of them twice. For instance, the *San Demetrio*'s crew had abandoned the ship, but later one of her boats resighted her, and a handful of the crew, under the second officer, boarded and got the fire under control. The engines were restarted and, in spite of the lack of almost all navigational aids, the ship was brought safely to port with the greater part of her valuable cargo intact. Nevertheless, the *Scheer* was able to sink five cargo ships from the convoy, all British, before *Kapitän zur See* Krancke ceased fire at about 2040, as about half her 150mm ammunition had been fired after four hours of a hide-and-seek fight, her victims being:

Maidan (7908 tons), 52°28N/32°08W
Trewellard (5201 tons), 52°27N/32°09W
Beaverford (10,042 tons), 52°05N/32°18W
Kenbane Head (5225 tons), 52°18N/32°28W
Fresno City (4955 tons), 51°47N/33°29W

A sixth British cargo ship, the *Andalusian* (3082 tons), was damaged.

On receipt of *Jervis Bay*'s distress signal, the Home Fleet deployed the battleships *Nelson* and *Rodney* with destroyer escorts to block the Iceland–Faroes passage, and the battlecruisers *Hood*, *Repulse* and *Renown* (recalled from Force H) with the light cruisers *Dido*, *Naiad* and *Phoebe* of the 15th Cruiser Squadron and six destroyers, to block the approaches to Brest and Lorient. But all these dispositions and searches were made in vain, because the *Scheer* was making a prolonged cruise and steamed immediately south into the central Atlantic.

Apart from sinking six merchant ships and one armed merchant cruiser, her sudden appearance on the Halifax route seriously disorganised the entire flow of shipping across the Atlantic. The next two homeward-bound Halifax convoys and also a Bermuda–Halifax convoy were recalled. Many ships were thus delayed, and the assembly ports became seriously congested. The normal convoy cycle was not resumed in the North Atlantic until HX89 sailed on 17 November. The loss of imports caused to the British by the *Scheer*'s sudden appearance on their principal convoy route was, therefore, far greater than the cargoes actually sunk by her, and that was exactly Grand Admiral Raeder's goal.

Had the British Force H been available to make a search to the west from Gibraltar, she might now have been intercepted, but Admiral Somerville was about to carry out Operation 'Coat' inside the Mediterranean and could not hunt for the *Scheer*, which was refuelled on 12 November from the tanker *Eurofeld* and, on 14 November, met her supply ship *Nordmark* at 22°..N/46°20W.

The Scheer *in heavy seas, before the war, with a Heinkel He60 floatplane on board which was replaced by an Arado Ar196 in August 1939.* (Author's collection)

The *Scheer* next moved towards the West Indies and, on 24 November, sank the British cargo ship *Port Hobart* (7448 tons), south-east of Bermuda, at 24°44N/58°21W. Her victim made a raider report, but did not say whether her assailant was a warship or a disguised merchant raider. Though the message caused the *Scheer* to move east towards the Cape Verde Islands, it did not help to clarify matters in the Admiralty or in the headquarters of the British Commanders-in-Chief overseas who were trying to catch the raider.

On 1 December, the *Scheer* sank the British cargo ship *Tribesman* (6242 tons), about 900 miles west of Bathurst, at approximately 15°N/35°W and then moved to the Pernambuco–Azores route, which she searched without result. After meeting the supply ship *Nordmark* again at a rendezvous just north of the Equator on 14 December, the *Scheer* steamed towards the route between Freetown and South American ports, crossing the Equator on 16 December. On 18 December, at 00°57N/22°42W, thanks to her Arado Ar196 floatplane which located the victim, she captured in broad daylight the British cargo ship *Duquesa* (8651 tons), which was loaded with foodstuffs, and deliberately allowed her to make a raider report in order to divert attention from the *Admiral Hipper* which, far away to the north, had just started to make her first foray into the Atlantic. In this purpose she succeeded, for Admiral Raikes sent the heavy cruiser *Dorsetshire* and the light cruiser *Neptune* westward from Freetown for 500 miles; the aircraft carrier *Hermes*, the light cruiser *Dragon* and the armed merchant cruiser *Pretoria Castle* met at St Helena and thence searched north-east; and Force K, which was on passage to Freetown from Britain, with the new aircraft carrier *Formidable* and the heavy cruiser *Norfolk*, passed west of the Azores. But the meshes of the net were far too big, and the *Scheer* escaped from them without difficulty.

To return to the refrigerated ship *Duquesa* which had

been captured on 18 December: on board there was a cargo of 14.5 million eggs and about 3000 tons of meat. The ship was taken as a prize, and used to supply other ships, including the armed merchant cruisers *Thor* and *Pinguin*, the supply ship *Nordmark*, prizes and blockade runners. On 26 December, at the 'Andalusia' point, at 15°S/18°W, the *Scheer* met the three above-mentioned ships and also the tanker *Eurofeld* and, of course, the very welcome *Duquesa*. Once completely unloaded, the ship was sunk by the *Nordmark* on 20 February 1941, in an unknown position (at the end of January 1941, the *Duquesa* was at 23°S/15°W).

When the New Year opened, the *Scheer* was still at the 'Andalusia' rendezvous with the other ships and, her repairs having been completed on 5 January 1941, after filling her fuel tanks, she moved on to the Cape Town–Freetown route on 8 January. Between 9 and 12 January, the *Scheer* tried without success to intercept the troop convoy (WS5A) which had already been attacked by the *Admiral Hipper* on 25 December 1940, also without success. On 18 January, at approximately 11°S/02°W on the Cape Town–Freetown route the Norwegian tanker *Sandefjord* (8038 tons) was sighted by *Scheer*'s Arado Ar196 floatplane, and once captured was sent as a prize to Bordeaux, where she arrived safely on 28 February.

The *Scheer*'s captain now changed his tactics and, instead of generally approaching his intended victims by night, adopted the ruse of making the approach in broad daylight whilst simulating the signals of a British warship. The first success obtained by this means was the capture of the Dutch cargo ship *Barneveld* (5597 tons), on the Freetown to Capetown route, on 20 January, at approximately 07°S/03°E; she was sunk later. A British cargo ship, the *Stanpark* (5103 tons), which happened to be passing through the same waters, in the opposite direction, was also seized a few hours after (at 09°00S/02°20E)

The Scheer *undergoing major refit during winter 1939–40.* (Author's collection)

and sunk later. The *Scheer*'s ruse was so successful that no distress messages were sent by either ship, and for many months the British Admiralty remained unaware of the cause of their disappearance. But the *Scheer* considered it advisable to leave the neighbourhood of these successes quickly. She therefore steamed back to a further meeting at 'Andalusia' with the raider *Thor* and the supply ship *Nordmark*. The *Scheer* refuelled, took on supplies, and passed prisoners over to the *Nordmark* between 24 and 28 January. Then she steamed off south-east, passed far south of the Cape of Good Hope on 3 February and entered the Indian Ocean. For nearly a week the *Scheer* searched the routes from Australia to the Cape, but did not sight a single ship.

From 14 to 17 February, at 13°S/64°E, she met the raider *Atlantis* and her two prizes taken in the Atlantic (*Speybank* and *Ketty Brövig*), and the German cargo ship *Tannenfels* which was en route from Massawa to Bordeaux as a blockade runner, for replenishment and exchange of information. At this time, the German supply and servicing organisation for their raiders was certainly working at a high pitch of efficiency. On the advice of the *Atlantis*' captain, the *Scheer* now moved to the northern exit from the Mozambique Channel and used her aircraft again for reconnaissance to good purpose. On the morning of 20 February, the Arado was catapulted and discovered the British tanker *British Advocate* (6994 tons), which was quickly captured by the *Scheer*, 07°10S/45°30E, and, with a prize crew, sent to the Gironde in France, where she arrived safely on 29 April.

On 20 February too, the Arado was catapulted a second time and located the Greek cargo ship *Grigorios C II* (2546 tons) during the night. She was sunk by the *Scheer* next morning. After the sinking of the Greek freighter, the floatplane was again catapulted and, flying back to the *Scheer* at 0915, reported having seen a third ship. She was the British cargo ship *Canadian Cruiser* (7178 tons), which was promptly located and sunk by the *Scheer* (07°36S/47°18E), but not before she made a raider report. Next day, 22 February, another victim, the Dutch cargo

Another view taken at Wilhelmshaven during the refit of 1939–40. (Author's collection)

ship *Rantaupandjang* (2542 tons) also sent a report before being sunk by the *Scheer* (08°24S/51°35E). It was therefore clear that a longer stay in this fruitful area might be dangerous. This, and recently received orders recalling her to Germany by the end of March, made an early return to the South Atlantic necessary.

However, before the *Scheer* had left the region of her recent successes, British countermeasures, taken in response to the raider reports already mentioned, nearly succeeded in intercepting her. Four-and-a-half hours after the Dutch ship made her raider report on the morning of 22 February, the light cruiser *Glasgow*'s aircraft sighted the *Scheer* at 08°30S/51°35E. The cruiser at once signalled an enemy report. Her intention was 'to attack by night and shadow by day'. The British aircraft shadowed the *Scheer* for about 30 minutes, and *Oberleutnant zur See* Pietsch asked for permission to attack and shoot down the dangerous intruder. Captain Krancke refused, for the Arado's wings had been damaged by heavy seas and, only repaired superficially, so it could have been risky to fire the two wing 20mm guns. By the afternoon, the British Commander-in-Chief, East Indies, Vice-Admiral Leatham, deployed from Mombasa the aircraft carrier *Hermes* and the light cruiser *Capetown*, the light cruisers *Emerald* and *Hawkins* (which were relieved by the light cruiser *Enterprise*) from the convoy WS5B, and the heavy cruiser *Shropshire* from the Somali coast. The Australian heavy cruiser *Canberra*, relieved on 20 February off Colombo by the New Zealand light cruiser *Leander* whilst with the convoy US9, was sent from the Maldives. Until 26 February all these ships searched in vain for the *Scheer*, which turned away 100 miles to the south-east to shake off the pursuit, and then resumed a south-westerly course. The only British hope of catching the *Scheer* lay with the *Glasgow* keeping in touch, but her aircraft lost contact when visibility became reduced. The *panzerschiffe* passed 400 miles to the east of Mauritius and then steamed well to the south of the Cape of Good Hope on 3 March and was back into the Atlantic without further incident.

On 9 and 10 March, there was a third meeting at 'Andalusia' between the *Scheer*, the supply ship *Nordmark* and the armed merchant raiders *Pinguin* and *Kormoran*, exchanging stores and prisoners. She also rendezvoused with the *U-124* (*Kapitänleutnant* Schulz) which had brought some vital spare parts for *Scheer*'s radio. On 11 March, having completed refitting her engines and cleaning her boot-topping, the *Scheer* steamed away to the north on her homeward passage, meeting the German freighter *Portland* on the same day. She crossed the Equator on 15 March and the Halifax convoy routes on the night of 22/23 March. The *Scheer* met the German freighter *Alsterufer* on 24 March and obtained supplies from her. After waiting two days, she found favourable weather for the break-through on the 27th and entered the Denmark Strait. Unnoticed by the enemy, the *Scheer* evaded the light cruisers *Fiji* and *Nigeria*.

It is certain that the operations of the battlecruisers *Gneisenau* and *Scharnhorst* diverted British attention from the *Scheer*'s return, and that was planned by the

The after 280mm (11in) turret and shells aboard Scheer's *sister-ship* Lützow *at Gotenhafen (January 1940). The guns could fire these 661lb shells at a maximum range of 39,890yds with a rate of fire of 3 per minute.* (ECPA)

German Naval Staff, as most of the British battleships and cruisers were looking for the two more powerful and dangerous raiders. On 28 and 29 March, the Admiralty warned Admiral Tovey that wireless intelligence indicated a possible break-through by a warship from the south of Iceland on the latter date. The battleship *King George V* and four cruisers were ordered to the eastern end of the Iceland–Faroes mine barrier. Air searches were requested, but bad weather prevented any being flown. It was, in any case, too late by 48 hours, for the *Scheer* had already passed through those waters and was now off the Norwegian coast.

On 30 March, the *Scheer* reached the area off Bergen, anchored for a day for the first time in five months, in Grimstadfjord and, escorted by the destroyers *Z-23, Z-24* and the torpedo boat *Iltis*, reached Kiel on 1 April. She had steamed 46,419 sea miles and had demonstrated the excellent qualities of her class for commerce raiding. During her cruise, the *Scheer* had sunk fourteen merchant ships and one armed merchant cruiser, and captured two merchant ships which were sent to France with prize crews (see table for details), making a grand total of 113,223 tons.

Baltic Interlude

British photographic reconnaissance on 4 September showed that the *Scheer* (now under the command of *Kapitän zur See* Meendsen-Bohlken), had left Kiel after a complete overhaul, her funnel cap having been raised and been made more pointed, similar to *Lützow*'s. She was sighted shortly afterwards, by British aircraft, passing north through the Great Belt and, later, at the entrance of Oslofjord. She was seen in Oslo the same day and on 5 and 8 September unsuccessful bombing attacks were made by a small number of No 2 Group RAF B-17s. The *Scheer* returned to Swinemünde and, on 10 September, British air patrols were organised to deal with a possible break-out. Five days later the *Scheer* could not, of course, be found in Oslo, nor did a search of all likely ports succeed in finding her at once. But on 18 September photographs of Swinemünde showed that she was back in her original base, and tension therefore relaxed.

In fact, at that time, the *Scheer* became part of the so-called 'Baltic Fleet' which was transferred to the Åland Sea to prevent a possible break-out by the Soviet Fleet into the Baltic. As the military situation was much in favour of German troops on land, it was also eventually expected that the Soviet Fleet would make a dash to be interned in Sweden. The German 'Baltic Fleet' comprised two groups: the Northern group with the battleship *Tirpitz*, the heavy cruiser *Admiral Scheer*, the light cruisers *Köln* and *Nürnberg*, the destroyers *Z-25, Z-26, Z-27* and the torpedo boats *T-2, T-5, T-7, T-8, T-11* and some motor torpedo boats: and the Southern group with the light cruisers *Emden, Leipzig* and motor torpedo boats. The German ships were keen and enthusiastic at the idea of confronting and fighting the Red Fleet, but the Soviets made no attempt to oblige them.

THE ADMIRAL SCHEER AT WAR

The Scheer *at sea after her refit. Noteworthy are the new tubular mast and bridge structure, the clipper stern, the enlarged funnel cap and the Arado Ar196 floatplane on its catapult.* (ECPA)

On 23 September, the fleet left Swinemünde. On the next day, after German air attacks on the Soviet warships in Kronstadt and Leningrad from 21 to 24 September, the *Tirpitz*, the *Scheer* and two torpedo boats were recalled, and three torpedo boats joined the Southern group. In October, the *Scheer* moved to Hamburg and again caused anxieties to the Admiralty, this time aggravated by extremely bad weather, which made air reconnaissance difficult and increased the likelihood of the enemy choosing such a time for a break-out into the Atlantic. Not until 28 November was she again found back in Swinemünde.

Transfer to Norway

On 21 February 1942, the *Scheer* and the heavy cruiser *Prinz Eugen* were transferred from Brunsbüttelkoog to

The Scheer *seen from the* Prinz Eugen *as they were sailing for Norway on 21 February 1942.* (Author's collection)

Norway, escorted by the destroyers *Richard Beitzen, Paul Jacobi, Hermann Schoemann, Friedrich Ihn* and *Z-25*. At 1110, British reconnaissance aircraft located the ships, but an aircraft maintaining contact was shot down by German fighters. Of the British aircraft which were then deployed, only one bomber found the force. It dropped its bombs near the *Prinz Eugen* and was shot down by the fleet's AA guns. Whereupon, the German squadron reversed course, a ruse that was successful in avoiding the expected air attacks. During the evening, they resumed their passage north as the weather worsened. Early next morning, 22 February, two British bombers, employed on other missions, located and attacked the ships as they were entering the inner leads. They were unsuccessful and, after halting for fuelling in Grimstadfjord, just south of Bergen, the ships sailed north again on the same evening, bound for Trondheim.

The bad weather continued, with the result that the destroyers were unable to keep up with the heavy cruisers. The destroyers *Beitzen*, *Jacobi* and *Ihn* lost touch and returned to Bergen. Now only escorted by two destroyers, on 23 February the force was attacked by the submarine *Trident*, which torpedoed and damaged the heavy cruiser *Prinz Eugen*. The *Scheer* went on to Aasfjord and took up a berth not far from the *Tirpitz*. The British Home Fleet (Admiral Tovey) made an attempt, with the aircraft carrier *Victorious*, the heavy cruiser *Berwick*, four

The Soviet icebreaker Aleksander Sibiryakov *on fire after having been shelled by the* Scheer *on 25 August 1942.* (Author's collection)

The sinking Aleksander Sibiryakov *seen from one of the* Scheer's *motor boats with Soviet survivors on board.* (Author's collection)

destroyers and the battleship *King George V* as cover, to intercept the German ships on their way north, but it failed.

On 9/10 May, the *Scheer* was transferred from Trondheim to Narvik with the fleet tanker *Dithmarschen* and the torpedo boats *T-5* and *T-7*. From there, on 3 July, the *Scheer*, her sister-ship *Lützow* and six destroyers sailed for Altafjord to join another group comprising the battleship *Tirpitz*, the heavy cruiser *Hipper*, four destroyers and two torpedo boats. The purpose was the attack on convoy PQ17, and they all joined up, being reinforced by two destroyers, but on 3 July the *Lützow* and three destroyers went aground near Narvik and took no further part in the operation. On 5 July, German air reconnaissance and U-boats reported the breaking up of the convoy and the departure of the cruisers westwards. Then the *Tirpitz, Scheer, Hipper*, seven destroyers and two torpedo boats put to sea from Altafjord between 1100 and 1130. When it became clear that the scattered convoy was already suffering heavily at the hands of the U-boats and aircraft, Admiral Raeder cancelled the operation. At 2130 that evening, Admiral Schniewind reversed his course. Though preparations were made to attack him with carrier aircraft and submarines, the *Scheer* and her consorts reached Narvik safely on 6 July.

Because of the threat they had represented, the German surface ships were undoubtedly responsible for the departure of the escorting cruisers and destroyers, leaving the merchant ships almost defenceless against the German U-boats and aircraft. So, indirectly, they played their part in the destruction of PQ17, twenty-three transports being sunk out of thirty-four. Among the ships was the British fleet tanker *Aldersdale* (1937, 8402 tons) and a twenty-fourth vessel, the British rescue ship *Zaafaran* (1921, 1559 tons) was also lost. It was a real disaster, for, in addition to the ships themselves, with them was also lost a cargo of 3350 vehicles, 430 tanks, 210 aircraft and 99,316 tons of other war equipment. A whole army could have been equipped from those supplies. The total German losses amounted to only five aircraft lost!

The *Scheer*, with the *Tirpitz* and *Hipper* were now using Narvik as their main base, instead of Trondheim. On 6 August they were joined by the light cruiser *Köln*. With the deployment of such big ships in northern Norway, the *Kriegsmarine* began to search for a new use for them. On 16 August, the *Scheer* was sent on Operation 'Wunderland', a commerce raid into the Kara Sea, east of Novaya Zemlya. It was a bold foray, into previously uncontested waters, designed to disrupt or even shut down the Soviet Arctic shipping route that sent convoys from the Pacific to Murmansk during the brief summer. The *Scheer*, however, found little traffic, even going as far as about 78°N/100°E. She had been escorted by the three destroyers remaining in Northern Norway, *Richard Beitzen*, *Friedrich Eckoldt* and *Erich Steinbrinck*, but only as far as Bear Island.

Following ice reconnaissance by *U-251* and *U-601*, the *Scheer* passed Cape Zhelania on 19 August and proceeded eastwards through the Kara Sea. The ship's Arado Ar196 floatplane reported parts of three-ship convoys with the icebreakers *Krassin* and *Lenin* on 20 August, near the island of Krakovka, and on the 23 August in the Vilkitski Strait, but mist and ice prevented an approach. On 25 August, 180 miles south-west of Cape Čeliuskin, the *Scheer* encountered the Soviet icebreaker *Aleksander Sibiryakov* (Captain Kacharev). The action took place near Belucha Island and the icebreaker was sunk by gunfire, after a courageous defence. The *Sibiryakov* (ex-*Belleadventure*) had been built in 1909 and had a displacement of 1384 tons. Her armament comprised two 76mm (3in) and two 45mm AA guns, and some machine-guns. Out of her complement of 128, 21 were saved by the Germans.

On 27 August, the *Scheer* shelled Port Dickson, the major Soviet Arctic port. Conditions of visibility were not very good, being restricted to 6/7km, and it was therefore difficult to know how many ships were in the harbour but, according to the Soviets, there were three 'transports' and the patrol boat *Dezhnev*. The three following Soviet ships were damaged: patrol boat *SKR-19 Semyon Dezhnev* (1937, 3578 tons, former icebreaker), coaster *Revolutsioner* (1909, 433 tons) and icebreaker *Taymir* (1909, 1358 tons). The tanker *Valerian Kuibishev* (1914, 4629 tons) was also damaged and exploded later. In addition, extensive damage was done to the docking facilities. This is the *Scheer*'s gunnery log of the time, firing at ships which are probably trying to escape:

0037/0045 with 280mm guns at a ship which was identified as being probably the *Taymir*
0038/0042 with port 150mm guns on a different ship, thought to be a tanker [probably the *Valerian Kuibishev*]
0043/0046 with 105mm AA guns on a ship who probably was the same as the first one [*Taymir*]
0044/0046 with starboard 150mm guns on the first target [*Taymir*].

Firing ceased when both ships disappeared behind the head of a peninsula. As the *Scheer* fired too at the shore

The Scheer's *main forward turret shelling Port Dickson on 27 August 1942. Note the quadruple 20mm AA mounting on the turret.* (Author's collection)

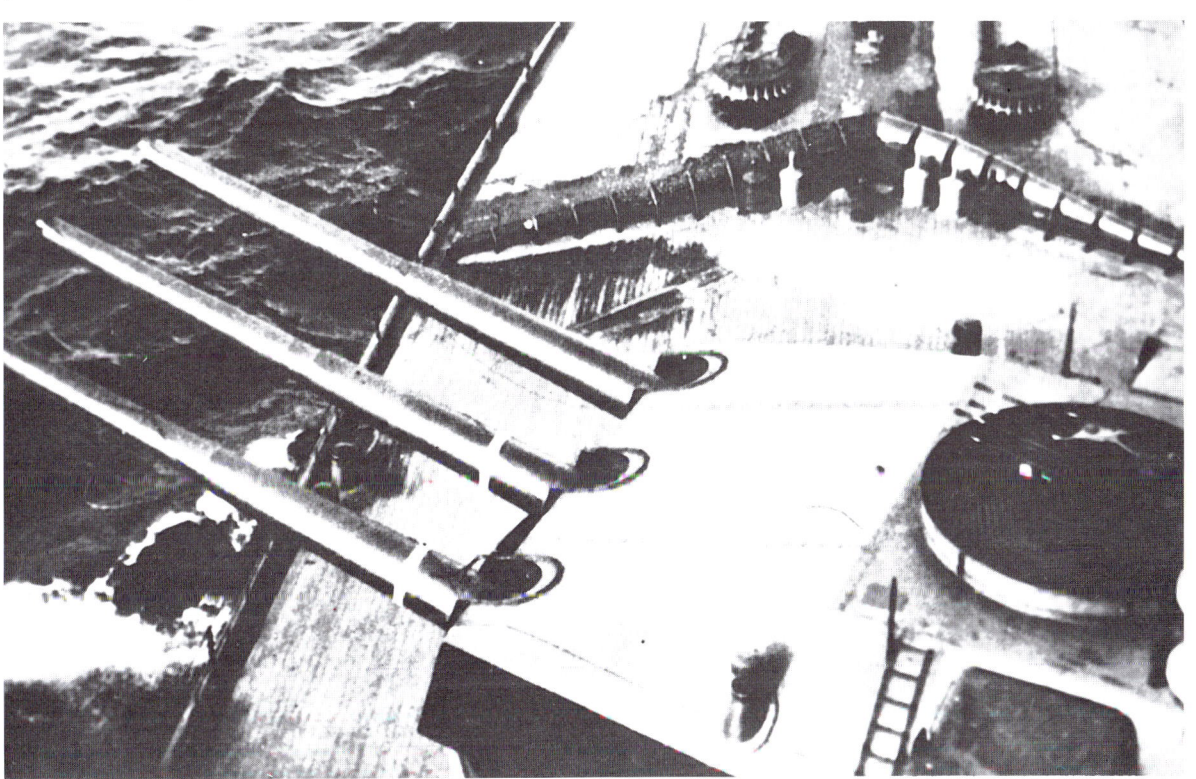

installations, the other two ships must have been damaged in the harbour. There were three Soviet coastal batteries, with 152mm (6in), 130mm (5.1in) and 45mm guns, and they fired at the *Scheer* which shelled Port Dickson from a distance of about 5500m, but the *Scheer* was not hit. On 30 August, the *Scheer* was back in Narvik again and there were no Soviet attempts to intercept the ship by surface forces or to attack her with bombers.

On 10 September, the *Scheer*, with the *Hipper*, *Köln* and some destroyers moved from Narvik to Altafjord. They were sighted by British submarine patrols, but only the *Tigris* got in an attack, but she missed. On the 13th, the German Group Command North wished to sail the Altafjord squadron to attack convoy QP14, but Hitler warned Raeder that, because the ships were so important to the defence of Norway, he must not take undue risks. Raeder thereupon cancelled the operation, and the surface ships remained idle throughout the convoy's passage and also for convoy PQ18.

The Eastern Front

Early in November, the *Scheer* was sent to Germany to refit, arrived in Kiel on 11 November and, passing through the Kiel Canal, was transferred to Wilhelmshaven and went into dock. After the refit, the ship came under the command of *Fregattenkäpitan* Gruber, who had been the Communications Officer on the heavy cruiser *Blücher*. In February 1943, he was succeeded by *Kapitän zur See* Rothe-Roth, under whom the *Scheer* served as a training ship in the eastern Baltic, her main base being Gotenhafen.

From October 1944, under the command of *Kapitän zur See* Thienemann, the *Scheer* joined the Special Combat Units 'Rogge' and 'Thiele', created very hurriedly to help fight Soviet land forces in the East. These two units comprised the heavy cruisers *Scheer*, *Lützow* and *Prinz Eugen*, the light cruiser *Leipzig* and also the old battleship *Schlesien*, escorted by destroyers and torpedo boats. The results of their gunnery on Soviet tank concentrations was devastating, and their 280mm, 203mm, 150mm, 127mm and 105mm shells were responsible for halting numerous Soviet attacks, helping German troops to hold on and thousands of refugees and soldiers to escape from the Soviets.

During the Soviet offensive against the Sworbe Peninsula, the *Scheer* intervened on 22, 23 and 24 November with the destroyers *Z-25*, *Z-35* and the torpedo boats *T-3*, *T-5*, *T-9*, *T-12*, *T-13* and *T-16*, particularly on the night of 23/24 November, when all ships opened fire to keep the Soviet attackers' heads down; nineteen ferry barges of the 24th Landing Flotilla, escorted by M- and R-boats, evacuated the last 4694 German soldiers, with some guns and equipment, leaving the Soviets to find a deserted peninsula the following morning. The Soviets replied with shore batteries, but the *Scheer* increased the range, firing from a distance of 35km (22 miles) and the Soviet shells were useless, whilst the German fire, even from that long distance, was still accurate thanks to German Army forward observers reporting the fall of shots.

Frustrated, the Soviets decided to attack the *Scheer* with aircraft and at dawn on 23 November, when the *Scheer*, escorted by four torpedo boats, was shelling target areas designated from Sworbe, the first fighter-bomber attacks on the battle group began. From 0820 on there were continuous air attacks by small units, which were partly turned back by German fighter protection, but this was only available in flight strength. By 1215 the German fighters were absent and an enemy bomber formation attacked the battle group. From the end of 1944, the *Scheer*'s light anti-aircraft armament had been increased to a total of 8–37mm, 6–40mm and 24–20mm guns, which ensured a warm welcome for the Red planes. When at 1335 the *Scheer* and the torpedo boats were attacked by about thirty aircraft, the German ships put up a strong defensive fire and managed to outmanoeuvre all the torpedoes, only a small bomb causing minor damage on the *Scheer*'s foredeck. The torpedo boats also suffered only minor damage, caused by the mine-like effect of the bombs with time fuses.

This attack was carried out by one group of Soviet planes flying low with torpedoes, one at medium altitude with bombs, and one flying very high with bombs. Each group comprised four to six aircraft, accompanied by four fighters. The low-flying planes fired their machine-guns and the bombs of the group at medium altitude had delay fuses which gave them the effect of mines. The *Scheer* suffered the effects of shock, but received no lasting damage. The high-flying planes dropped heavier bombs (at least 250kg, 550lb) with short delay fuses. The torpedo planes came very

The crew of one of Scheer's *105mm AA turrets watches one of her victims slipping beneath the waves, after having been pounded by 150mm and 105mm shells.* (Author's collection)

THE ADMIRAL SCHEER AT WAR

Appearance changes to Admiral Scheer *1935–1945 (from* Gröner's German Warships 1815–1945*)*

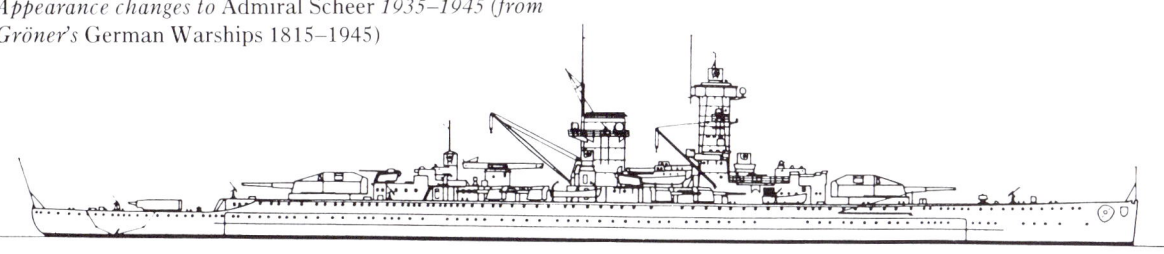

1935   Admiral Scheer (1935). *Mrva*

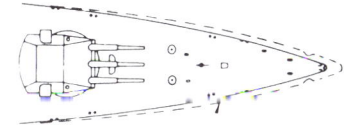

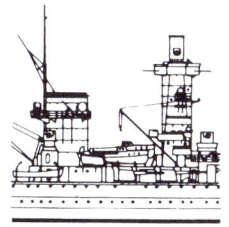

1939   Admiral Scheer (1939). *Mrva*

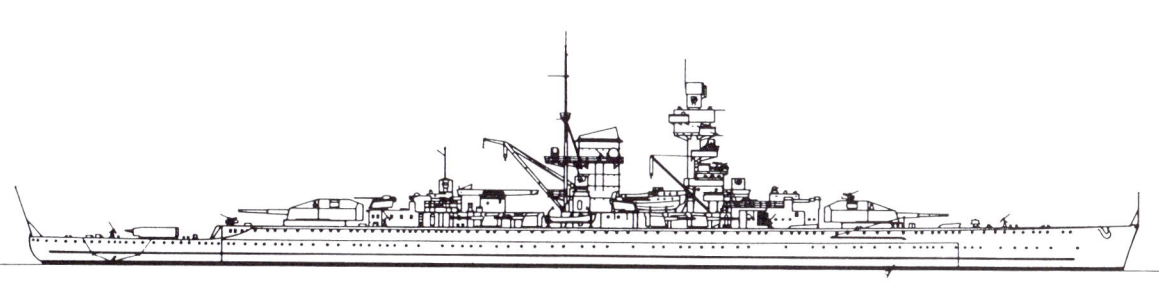

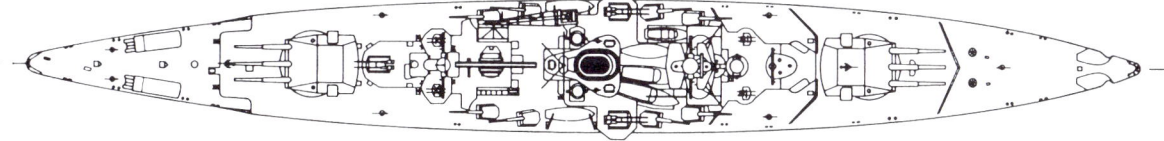

1940   Admiral Scheer (1940), second form. *Mrva*

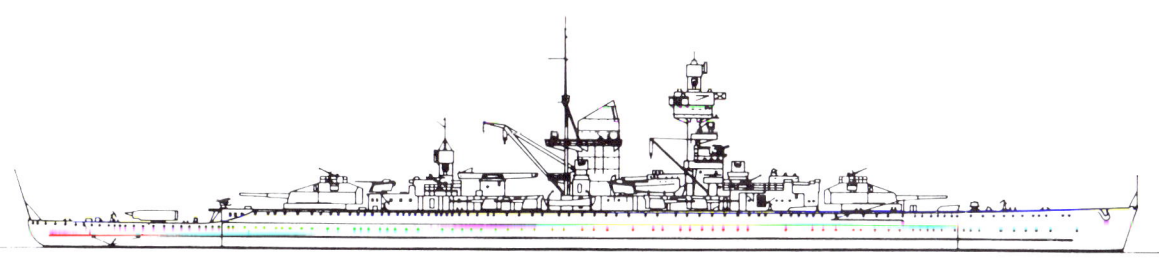

1945   Admiral Scheer (1945). *Mrva*

close to the ship, without regard for the anti-aircraft fire. After it had exhausted nearly all its ammunition, the battle group was recalled, and *Lützow*, with three destroyers, was held in readiness, but was not needed because the evacuation of the Sworbe Peninsula had been completed.

The Soviet fleet, as yet untried, was still in the easternmost part of the Gulf of Finland, despite the fact that a safe channel for large ships through that gulf had been swept in the late autumn of 1944. Large Soviet ships could have left the Gulf of Finland at any time, between the Finnish coast and the island of Dagö. The Soviet excuse for keeping the larger ships in the Kronstadt–Leningrad area was that the Germans had laid many mines, but the road from the Gulf of Finland into the open Baltic was evidently free from mines not later than December 1944 or January 1945. However, the only surface ships of the Red fleet encountered during the fighting all along the coast from the Courland beachhead to the mouth of the Oder river were MTBs – and their successes were minimal. In reality, the Red fleet's battleship *Oktyabrskaya Revolutsiya*, the heavy cruisers *Kirov* and *Maksim Gorky* and the twelve destroyers did not have the confidence to face big German units, destroyers and torpedo boats.

Soviet submarines began operating in the shipping lanes from the Courland coast down to Pomerania on 5 October 1944 and that is why it is, as the German Admiral Eastern Baltic reported, 'almost incomprehensible that the two battle groups (*Scheer* and *Prinz Eugen*) could stay in the same area five days consecutively without sighting a single (Soviet) submarine'.

Between 2 and 5 February 1945, the *Scheer* stood by at sea, escorted by the torpedo boats *T-23*, *T-35* and *T-36*, in support of a German counter-attack in Western Samland. On 4 February, the Soviet submarine *L-3* (Captain Konovalov) attacked and missed the *T-36*. In support of German forces against advancing Soviet armies, the *Scheer* with the destroyer *Z-34* and the torpedo boats *T-23*, *T-28* and *T-36*, intervened with their 280mm, 150mm and 105mm guns, near Frauenburg on 9 and 10 February and in returning to Gotenhafen put in at Kahlberg to take refugees on board. To support a German counter-attack which took place between 18 and 24 February, the *Scheer*, the destroyers *Z-38*, *Z-43*, the torpedo boats *T-28* and *T-35*, shelled concentrations of Soviet forces near Peyse and Gross-Heydekrug, on the south coast of Samland, on 18 and 19 February. On 3 March, the *Scheer*, the destroyers *Jacobi*, *Z-31*, *Z-38* and the torpedo boat *T-36* were employed to cover the bridgehead opposite Wallin.

As her gun barrels were worn out, *Scheer* was ordered to Kiel for repairs, en route taking on about 200 wounded and 800 refugees at Pillau. During the night of 9/10 April, RAF Bomber Command dropped 2634 tons of bombs on the harbour installations at Kiel. Among the sunken and damaged ships was the *Scheer* who suffered hits from 5.4 ton 'Tallboy' bombs and capsized. The old 'soldier' received an unusual funeral, being partially scrapped from July 1946, and buried later under rubble when the basin was filled in for building a new quay.

Ships Sunk, Captured and Destroyed by the Admiral Scheer

Date	Vessel
5 November 1940	British cargo ship *Mopan* (5389 tons)
5 November 1940	British armed merchant cruiser HMS *Jervis Bay* (14,164 tons)
5 November 1940	British cargo ship *Maidan* (7908 tons)
5 November 1940	British cargo ship *Trewellard* (5201 tons)
5 November 1940	British cargo ship *Beaverford* (10,042 tons)
5 November 1940	British cargo ship *Kenbane Head* (5225 tons)
5 November 1940	British cargo ship *Fresno City* (4955 tons)
24 November 1940	British cargo ship *Port Hobart* (7448 tons)
1 December 1940	British cargo ship *Tribesman* (6242 tons)
18 December 1940	British cargo ship *Duquesa* (8651 tons), captured and sunk on 20 February 1941 by the *Nordmark*
15 January 1941	Norwegian tanker *Sandefjord* (8038 tons), captured and sent as a prize to France.
20 January 1941	Dutch cargo ship *Barneveld* (5597 tons), captured and sunk later
20 January 1941	British cargo ship *Stanpark* (5103 tons), captured and sunk later
20 February 1941	British tanker *British Advocate* (6994 tons), captured and sent as a prize to France
21 February 1941	Greek cargo ship *Grigorios C II* (2546 tons)
21 February 1941	British cargo ship *Canadian Cruiser* (7178 tons)
22 February 1941	Dutch cargo ship *Rantaupandjang* (2542 tons)
25 August 1942	Soviet icebreaker *Aleksander Sibiryakov* (1384 tons)
27 August 1942	Soviet tanker *Valerian Kuibishev* (4629 tons), damaged and blew up later

Total = 19 ships (2 warships, 17 merchant ships) representing 119,236 tons. The *Scheer* was, in number of enemy ships sunk, the most successful heavy German unit.

THE AKIZUKI CLASS

With an armament principally dedicated to anti-aircraft and anti-submarine defence – the torpedoes were an after-thought – the Japanese navy's *Akizuki* class destroyers foreshadowed the postwar fleet escort. Designed for the protection of carrier task forces, they were also noteworthy for extremely long range, which with their unique armament are the main emphases of this study by Hans Lengerer, Jiro Itani and Tomoko Rehm-Takahara.

The *Akizuki* class destroyers are among the most notable warships built immediately before the outbreak of the war in the Pacific and unique among the destroyer classes of the Imperial Japanese Navy (IJN). Until their appearance Japanese tactical doctrine always required destroyers to be designed with a view to employment in the decisive fleet engagement between the IJN and its assumed enemy (which from 1907 was the US Navy), laying emphasis on heavy torpedo armament, speed and defence against counter-attacking small surface ships by low angle guns. In this conception the potential danger of aircraft was not sufficiently appreciated and the IJN's attempt to adopt dual-purpose guns in prewar destroyers was not successful, mainly due to the absence of a suitable fire control system.

The introduction of submarines and aircraft as naval weapons changed the character of sea operations, adding a second and third dimension to naval warfare in quick succession. Even the largest ships were threatened by aircraft but the attempts of naval architects to avoid or minimise damage from aerial weapons followed rather than anticipated the threat. In truth, the development of defensive measures always limped behind the rapid advances in aircraft design and weaponry. Considering the fundamental differences between warships, with their long design and building times, high costs and comparatively long life, and the small, cheap, mass produced and relatively low value early aircraft, this was inevitable. One of the first responses to the danger of air attack was the Royal Navy's conversion of several old light cruisers to AA ships in the 1930s and the US Navy and IJN followed this example later.

The IJN emphasised the naval air force, and especially its carrier-based arm, altering the tactical concept of the decisive battle as their importance grew, to widen the part the carrier-based air force would play. Accordingly, the importance of protecting the vulnerable carriers increased and finally resulted in the requirement of the Navy General Staff for a specialist AA escort ship to protect the carrier task forces against attacks from the air. This meant that speed and radius would have to match those of the IJN's modern fleet carriers. This ship would also take over the duties of the old fleet destroyers which had escorted the carrier division since 1928.

Due to the quantitative and qualitative restrictions to practically all kinds of warships imposed by the Washington and London Treaties (which limited the IJN's strength to roughly 60 per cent of the US and Royal Navies), IJN tacticians always specified multi-purpose ships and considered specialised (single-purpose) design a luxury the

Akizuki, *taken a few days after commissioning.* (All illustrations from author's collection)

WARSHIP 1993

Akizuki *under attack by US carrier aircraft, 29 September 1942. The ship was undamaged.* (US Naval Historical Center)

Another view of Akizuki *under attack in Empress Augusta Bay on 29 September 1942. This was the first photograph of the new class obtained by the US Navy and was analysed with much care.* (US National Archives)

THE AKIZUKI CLASS

'poor' IJN could not afford. Therefore, when the rough calculation of the Navy Technical Department produced a ship displacing more than 4000 tons the tactical authority reduced the demand for speed and radius but at the same time requested a torpedo armament, thus adding an element of offence to the hitherto entirely defensive nature of the concept. Thereby the classification was changed to B Type destroyer (*Otsu gata*) from AA escort ship (to distinguish this class, unofficially called AA destroyers – *Boku Kuchikukan* – from the A Type fleet destroyer); this produced the *Akizuki* class, for which there was no precedent in the IJN.

The main features of this class were:
1. Armament of four twin 65cal type 98 10cm (3.9in) HAG guns mounted in closed gunhouses, superimposed fore and aft;
2. Separate type 94 fire control systems for fore and aft batteries;
3. Main engines and fuel stowage for large radius and corresponding extended operations;
4. One funnel combining three uptakes;
5. High but well shaped and equipped bridge structure;
6. Excellent seaworthiness, good handling and good habitability;
7. Fine lines and smart appearance due to clear and functional arrangement of superstructure.

The IJN wanted to build thirty-nine ships of this class in total (in three building programmes), but only twelve were completed between 1942 and 1945 and twenty-six were cancelled (and one suspended after laying down). The twelve completed ships belonged to three sub-types, namely *Akizuki* (seven), *Fuyutsuki* (four), and *Michitsuki* (one), but only the first two had different basic design numbers (F 51 and F 53). The later types were gradually simplified to shorten the building period as the war situation deteriorated but the measures taken were insufficient; among them was the unbelievable decision to dispense with the fire control system for the after gun battery (even though this was primarily a result of inadequate production capacity). Six ships were sunk during the war, two were used as a breakwater after the war and one each was turned over to the USA, Britain, China and the Soviet Union. Of these, *Natsuzuki*, delivered to Britain, was returned to Japan and broken up at Uraga.

Design Background

In 1922 the Washington Treaty not only laid down the ratio of capital ships to be allowed the signatory powers but also placed an upper limit on the total tonnage of aircraft carriers, thus giving evidence of their great importance. This limitation was maintained in the London Treaty of 1930 which restricted the IJN's overall fighting strength to a ratio of 70 per of the US and Royal Navies. At the same time a number of fundamentally difficult problems hindered plans to increase the number of such vessels. Apart from the great expenditure required to maintain aircraft carriers, it was constantly necessary to modify and even reconstruct this special kind of warship to accommodate the frequent and rapid changes that were taking place in aircraft design and performance. Therefore it was not before the building of *Sōryū* and *Hiryū* that the fundamental basis for the IJN's naval air force was established; the construction of *Shōkaku* and *Zuikaku* in the Third Naval Replenishment Programme of 1937, following on the Second of 1934, was the improved continuation of their design philosophy.

Akizuki *in 1944. This photo was captured by US forces on Kwajalein and shows the ship after a major repair at Mitsubishi's Nagasaki yard, 6 July – 31 October 1943. The bow of the incomplete* Shimotsuki *was transferred to her to speed up the repair.* (US National Archives)

Unidentified Akizuki *class destroyer photographed on 15 September 1945 at Moji harbour, Kyushu. She still has all her main armament but some of the light MG mountings have been removed. The ship is probably* Natsuzuki. (US National Archives)

The rapid technical progress of aircraft suggested broader tactical utility, especially since air-launched torpedoes and dive bombing offered better chances of hits than horizontal bombing and greater destructive power, especially in the case of the torpedo. This fundamental development stage came to an end around 1935 but thereafter still more rapid progress was manifested in the Type 0 Kansen ('Zero' fighter), Type 97 Kanko ('Kate' torpedo bomber), Type 99 Kanbaku ('Val' dive bomber), and in the carriers *Shōkaku* and *Zuikaku* and the heavily armoured *Taiho*.

At the same time the tactical concept of the naval air force altered initially from the original roles of reconnaissance, ASW and gunnery spotting, to the attack on enemy capital ships in support of the main fleet gunnery duel, and eventually to concentrate on the destruction of enemy carriers and their aircraft in order to gain air supremacy for the surface gunnery battle. Thus the decisive air battle was to precede the decisive gunnery engagement and the outcome of the latter much depended on the result of the former. This change in tactical thinking was prompted by the increase in the size of the enemy's naval air force, the progress in aircraft design and the improvement in bombing and torpedo attack techniques. It also reflected an enormous growth in the perceived importance of the carrier, culminating in the belief among a minority of the IJN that the days of the battleship were numbered and that it should be replaced as the capital ship by the aircraft carrier itself.

Battleships could be protected by extra AA guns, horizontal armour and torpedo bulges, making them resistant to bomb and/or torpedo hits. In contrast the carrier was a most vulnerable warship, having large ammunition magazines, gasoline tanks with a widely distributed supply pipe network to refuelling points in the hangars and on the flight-deck, whose structure – from stability and other considerations – was relatively light; even one bomb hit could destroy the function of the ship in a moment.

In the first two tactical scenarios mentioned earlier, the carrier force would operate near the main battle force to be under the protection of these ships, but as the numbers and capabilities of the carrier and its aircraft improved, the tactical concept was radically revised so that the operational area became separated from that of the capital ship group.[1] Therefore, the IJN felt the necessity to give its very important carriers adequate protection against enemy air attacks by using small ships capable of operating alongside the carriers. The HA guns of these ships would form an AA screen, to reinforce the carriers' AA armament and escort fighters (interceptors).

Appearance of the AA Cruiser in RN and USN and Development of New Type HA gun in IJN

The idea of providing protection against air attacks by so-called AA ships was first conceived by the Royal Navy, which converted the old light cruisers *Coventry* and *Curlew* after analysing the invasion of Abyssinia by Italy in 1935 in which air power for the first time had seemed a menace to the fleet. Larger, purpose-built AA cruisers followed in the Royal Navy (*Dido* class) and US Navy (*Atlanta* class). The characteristic feature of this ship type was powerful main armament of AA guns. For the IJN the capabilities of the 12.7cm (5in) twin HA gun were satisfactory but in view of the future development of aircraft and their attack technique a better HA gun became necessary. This problem was solved by the development of the long (65cal) 10cm (3.9in) HA gun in twin mountings, and the decision to adopt this new type HA gun in several ship classes in the Fourth Naval Supplementary Programme of 1939, notably the carrier

Taiho, *Yamato* class battleships and the light cruisers of the *Ōyodo* class. This decision allowed the new AA ship to mount a new type HA gun of excellent characteristics.[2]

In April 1928 the IJN's First Carrier Division (*Akagi*, *Hosho*) was constituted for the first time and after May 1933 the Carrier Division acted as one element of the fleet in the peacetime organisation (participating in naval manoeuvres etc). When operating at sea the Carrier Division was accompanied by two old type destroyers (one destroyer to each carrier) to assist in take-off and landing operations by rescuing the pilot and crew in case of forced landing and, in rare cases, to recover the plane; in wartime they were to form the ASW escort. Because of this first duty these destroyers were nicknamed 'dragonfly lifters'. With the completion of the new carriers the cruising speed rose from 16kts to 19kts and the radius increased to more than 8000 miles. The old destroyers had a cruising speed of only 14kts and a comparatively small radius of action. Moreover, when they steamed at 16–18kts the consumption of fuel increased and they needed to refuel every few days. In the early 1930s refuelling the escorts from the carrier had not yet become an established technique, so it was necessary to be accompanied by oilers, which were also slow. Furthermore, refuelling could only be carried out with the ships stopped, making them vulnerable to attack.

Apart from their small radius of action there were other shortcomings which clearly showed the very limited tactical value of the old type destroyer, namely their small size and insufficient seaworthiness to steam at the same speed as the carrier in rough seas, and lack of AA fire power (since they were old ships the main guns were low angle weapons only and the small number of machineguns were of no practical value). Therefore, it became evident that the potential of the Carrier Division was circumscribed by the inadequacies of the escorting destroyers. However, the IJN was stressing air power more than other navies as one factor that might offset the quantitative inferiority of its naval strength so the effective use of the carrier force required escort ships to perform the following tasks:

1. to act together with the carrier even in rough seas (*ie* to have a large radius and possess excellent seaworthiness);
2. to be able to rescue the pilots and crews and also the planes in case of forced landings, etc;
3. to form a powerful AA screen; and
4. to provide the ASW escort to the task force.

In fact, defence against air attacks became the primary factor in the design, in combination with a large radius and excellent seaworthiness.

Naval General Staff Initial Requirements and Modifications

From the viewpoint of the tactical employment of the ships the requirements of the Navy General Staff set out in Table 1 were quite natural and also corresponded to the prevailing wisdom, but from the technical point of view the achievement of this extreme range[3] and very high speed in such a small warship was practically impossible.

Table 1: INITIAL REQUIREMENT OF THE NAVY GENERAL STAFF AND MODIFICATIONS

Date	July 1938	Sept[1] 1938	April[2] 1939
Displacement (standard) (tons)	2200	2350	2700
Speed (kts)	35	33	
Radius (kts/miles)	18/10,000	18/8000	
Armament (rounds)			
10cm long HA twin (400)	4	4	
25mm MG twin (6000)	2	2	
61cm TT quad (8)	–	1	
DC thrower (type 94)	–	–	2
distant	4	4	nil
near	2	2	nil
DC dropping rails	–	6	
DC type 95	–	–	54
distant	40	40	nil
near	36	36	nil
Other characteristics			
Good seaworthiness	Yes	Yes	
Rescue derrick for planes	Yes	Yes	
Smoke-screening equipment	Yes	Yes	

Notes
1. Change following rough calculations by Navy Technical Department.
2. Changes from September 1938.

When the Fundamental Design Section in the Fourth Main Division of the Navy Technical Department made the rough calculations it became evident that this ship would displace more than 4000 tons in trial condition (full load exceeding 4500 tons) in order to meet these requirements. According to these calculations more than 1200 tons of fuel had to be carried and apart from the increase in displacement, it was anticipated that the consumption of this extraordinarily large amount would produce serious stability problems (especially in light load condition). This would be exacerbated by the arrangement of the four twin HA gun mounts[4] whose most effective positioning – superfiring – would mean a rise in the centre of gravity (CG). These were very serious considerations for a navy which had experienced the capsizing of the torpedo boat *Tomozuru* only four years previously.

Ignoring the unrealistic displacement specified, another difficulty was that no existing main engine type could be used, and to satisfy the more than 35kts required, the design of new high powered main engines and boilers using high pressure, high temperature steam was essential (although this would bring about the very welcome by-product of reduced fuel consumption). However, this would be contrary to the decision to use the newly designed high powered (HPHT) turbine only in the experimental destroyer *Shimakaze* and to await the results of the trials of this ship.

Akizuki class ships were to escort the advancing carrier task forces and their damage and loss together with the carriers was predicted. Therefore, they had to be considered as expendable, which necessitated the construction of many ships of relatively low unit cost.

The surrendered Hanazuki *at Kure on 16 October 1945. Much of the armament has been removed but the director and radar antennas are visible. The ship was turned over to the US Navy as DD-934 and was broken up in 1948 following a period of trials.* (US National Archives)

Such considerations automatically prohibited the construction of escort vessels with the displacement of a small cruiser. Thus, in order to solve these problems there was no other way than to reduce the displacement. After much discussion between the responsible tacticians and technicians, consent was obtained to reduce the maximum speed to 33kts from more than 35kts and the radius to 8000 miles from the required 10,000; also to allow any absolutely essential increase in the displacement, but not to cut the HA armament which was the *raison d'être* of this class.

On the other hand no torpedo armament had hitherto been demanded for these AA escort ships because during exercises the old destroyers accompanying the carrier force had had almost no opportunity to fire their torpedoes. However, during the conferences the opinion gained ground that this high speed ship lacked any offensive power[5] and the tactical authority wanted to compensate by adding one quadruple 61cm (24in) TT and eight torpedoes, *ie* half the torpedo armament of a fleet destroyer. One more logical reason for adding the torpedo armament was the reduction in the number of fleet type destroyers to be built in the Fourth Naval Armament Fulfilment Programme in order to allow the construction of the *Akizuki* class. By widening the capabilities of the *Akizuki* class to take in the fleet destroyer role, the reduced numbers would be compensated for to some degree and, in addition, it would make it easier to obtain the agreement of the Finance Ministry.

Even though the British *Dido* class and the American *Atlantas* were armed with torpedoes, these cruisers cannot be compared meaningfully with the *Akizuki* class, where the torpedo was a heavy, useless weapon[6] which was mounted against the strenuous opposition of the Shipbuilding Division of the Naval Technical Department in the mistaken belief that concentrating as many weapons as possible on a ship gave it maximum flexibility of employment. With the adoption of torpedoes the classification was changed from 'close escort ship' (*Chokueikan*) to destroyer (*Kuchikukan*) because the IJN classified this kind of warship by the existence of the torpedo equipment.

Table 2: BUILDING PROGRAMMES

Programme	No of ships	Temporary names	Standard disp	Main requirements of Navy General Staff[1]	Budget Per ship	Total	Remarks
Maru Yon Keikaku	6	104–109	2600	8–10cm HA 4 TT 33kts 18kts/8000 miles	12,090,000	72,540,000	All completed
Maru Kyu Keikaku	10	360–369	2700	8–10cm HA 4 TT 33kts	17,820,400	178,204,000	Six ships completed
Maru Go Keikaku	16	770–785 (V 7)	2980	AA DD Type *Akizuki* (10cm HA)	18,387,000	294,192,000	Cancelled
Kai Maru Go Keikaku	23	5061–5083	2701	33kts	19,194,000	441,462,000	Cancelled

1. Data as given in the budget proposal submitted to parliament.

THE AKIZUKI CLASS

Another surrendered member of the class, Yoizuki at Kure on 16 October 1945; in much the same condition as Hanazuki. (US National Archives)

Building Programmes and Design Changes

The preparatory work on the (Fourth) Naval Armament Fulfillment Programme of 1939 (*Showa 14 nendo kaigun gunbi jūjitsu* or abbreviated *Maru Yon Keikaku*, ie circle 4 programme) came to the discussion stage in spring 1938, and its content was discussed by the Vice Chief of Navy General Staff and Vice Navy Minister in June of that year. This was roughly one year sooner than originally intended but the passing of the Second Vinson-Trammell Act on 17 May 1938 to continue the expansion of the US Navy changed the situation radically.

The draft of the Navy General Staff became the object of several revisions (a normal procedure) and in order to be able to build the new type close escort ship it was decided to reduce the scheduled twenty-four fleet type destroyers of 2000 tons standard (A Type, temporary designation W 106) to sixteen units and use the budget for these eight deleted ships to construct six of the larger escort ships. These newly adopted ships were classified as close escort ships as mentioned previously on 11 July 1938, and were temporarily designated W 115.

In September 1938 the Navy Minister proposed the budget for six B Type destroyers to the Finance Ministry, calculating the building cost at 12,580,000 yen per ship, outlining the main characteristics as 2600 tons standard displacement, speed 34kts, eight long 10cm HA guns, four torpedo tubes, four 25mm MG; this was accepted.[7] The final draft was approved by the Navy General Staff, Navy Ministry and Finance Ministry as well as the Premier in the 74th session of the Diet (opened on 26 December 1938) and passed on 6 March 1939. Notwithstanding the reduction of the building costs of the total programme to 1,205,780,000 from 1,263,000,000 yen the Type B destroyer programme was passed unchanged, being given the temporary names from No 104 to No 109.

In the budget application of the Navy Ministry to the Finance Ministry the necessity of building this kind of ship was explained by:

1. the requirement to operate the carriers as the main body of the naval air force;
2. the very important carrier division required protection at all times against each kind of threat, especially aircraft and submarines;
3. the capabilities of the carrier division were restricted by the poor characteristics of their existing escorts;
4. the absolute necessity to construct special ships in order to make the most of the fighting potential of the carrier unit;
5. the characteristics of these special ships emphasised their main duties – AA and ASW defence, rescue of aircraft crew and planes, but also to meet subsidiary demands (such as torpedo warfare).[8]

The background to the requirement for these ships and the tactical conception are summed up perfectly in this statement.

The escort force for the carriers required four Destroyer Divisions of four ships each to form one Destroyer Squadron, so ten more ships of the *Akizuki* class were demanded in the Emergency Wartime Warship Building Programme (*Senji kyuzo kansen kenzo keikaku*, abbreviated *Maru Kyu Keikaku* = the 'fast' character inside a circle) which was actually begun in September 1941, before its formal approval. In fact, the budget for three ships was passed in the 79th session (opened on 26 December 1941) and for seven ships in the 81th session (opened on 26 December 1942). These ships were given temporary names from No 360 to No 369 and the building costs for one ship were calculated at 17,820,400 yen.

However, in the USA the Third Vinson-Trammell Act (11 per cent Naval Expansion Act), followed within about a month by the Two Ocean Fleet Act (the so-called Stark proposal), had already been passed to increase the power of the US Navy to dimensions never seen before. The former programme was the direct cause of the premature start of the *Maru Kyu Keikaku* and also the Naval Armament Replenishment Programme of 1942 (*Showa 17 nendo kaigun gumbi hoju keikaku*; or abbreviated *Maru Go Keikaku* = 5 inside a circle) which was the IJN's direct response to the 11 per cent Naval Expansion Act. The

Two Ocean Fleet Act stimulated the drafting of the next building programme (abbreviated *Maru Roku Keikaku* = 6 inside a circle), but this remained a purely paper programme and the author has no definite information on how many of the thirty-four scheduled destroyers in this programme were to be of the *Akizuki* or modified *Akizuki* classes.

In *Maru Go Keikaku* the building of sixteen B Type destroyers of the *Akzuki* class were scheduled to form the second Destroyer Squadron of this kind of ship. They were temporarily called V 7 and included in this programme as No 770 to No 785.[9] Their standard displacement was given as 2980 tons in contrast to the 2700 tons of the former programme and the building costs per ship rose to 18,387,000 yen.

The commencement of this programme (the schedule specified 1 April 1942) was postponed because of insufficient building capacity, the lessons of the war which revised prewar strategic and tactical concepts, and the effects of the disastrous defeat at Midway. The content of this programme was totally changed (30 June 1942) to become *Kai* (Modified) *Maru Go Keikaku* (5th Programme), in which the sixteen ships of the original 5th programme were taken in but the number of fleet destroyers was reduced from sixteen ships to eight and their budget allocation was used to increase the Type B by seven ships to make twenty-three in total. This, undoubtedly, was a reflection of the immense increase in the importance of air defence and the belated realisation of the practical impossibility of a decisive fleet engagement. In addition, the former sixteen ships were changed to *Maru Yon Keikaku* type (ie *Akizuki* class of 2701 tons standard displacement), but the additional seven ships were to belong to a new type. On 3 August 1942 these twenty-three ships were given temporary names from No 5061 to No 5083 at building costs of 19,194,000 yen per ship; they were approved by the 81th session. The names of these ships were unofficially decided on 27 October 1943.

Much the same reasons had caused the postponement of *Maru Go Keikaku* and its radical change to *Modified Maru Go Keikaku* altered the shipbuilding programme several times in order to respond to the worsening war situation. When the *Maru 19 Sempyo* (the scheduled building table containing shipyards, laying down and completion dates, material data, etc) was drawn up, eighteen ships (from No 5066 to No 5083) were cancelled in 9 June 1944. The orders for the remaining five ships were placed (the last being *Harugumo* – No 5065 – on 14 August 1944 at Maizuru Arsenal), but the orders were suspended in April 1945 when new building was limited to special attack weapons (air, surface and submersible). Thus, none of these twenty-three ships was actually built and plans were abandoned for a high speed *Akizuki* class (with the same main engines and boilers as the *Shimakaze*, rearranged engine and boiler rooms, 37kts, altered main dimensions, one sextuple torpedo tube mounting, etc).[10]

When the *Akizuki* class was designed in 1938–39 (the fundamental design, number F 51, was decided on 4 December 1939) the IJN's shipbuilding stressed the quality of the individual ship rather than quantity. In contrast to the view that this kind of ship was an expendable weapon, too many requirements were incorporated, making the design very complex to produce and in wartime this very strong but lightweight ship required too many man-days for construction. This policy was changed within two years, and in order to shorten the building period various simplifications were introduced. This became the new fundamental design number F 53 (F 52 was the destroyer *Shimakaze*) decided on 21 October 1942, and this was applied from *Fuyutsuki* as the modified *Akizuki* class.[11]

As in the design of Type D (*Matsu* class) destroyers and the escort ships (*Kaibokan* = coast defense ship) Types C and D (*No 1* and *No 2* classes), priority was given to shortening the building time, and the hull lines became

The surrendered Natsuzuki *at Kure in October 1945.*
(US National Archives)

Table 3: NAMES OF AKIZUKI CLASS (ACTUAL AND SCHEDULED)

1. Maru Yon Keikaku
- 104 Akizuki
- 105 Teruzuki
- 106 Suzutsuki
- 107 Hatsuzuki
- 108 Niizuki
- 109 Wakatsuki

2. Maru Kyu Keikaku (Names decided on 19 June 1942)
- 360 Shimotsuki
- 361 Fuyutsuki
- 362 Harutsuki
- 363 Yoizuki
- 364 Natsuzuki
- 365 Michitsuki
- 366 Hanazuki
- 367 Kiyotsuki
- 368 Ōtsuki
- 369 Hazuki

3. Maru Go Keikaku
(Programme changed to Kai Maru Go Keikaku)

4. Kai Maru Go Keikaku (Names decided on 27 October 1942 but no ship was begun – see Table 4)
- 5061 Yamazuki
- 5062 Urazuki
- 5063 Aogumo
- 5064 Benigumo
- 5065 Harugumo
- 5066 Amagumo
- 5067 Yaegumo
- 5068 Fuyugumo
- 5069 Yukigumo
- 5070 Okitsukaze
- 5071 Shimokaze
- 5072 Asagochi
- 5073 Ōkaze
- 5074 Kochi
- 5075 Nishikaze
- 5076 Hae
- 5077 Kitakaze
- 5078 Hayakaze
- 5079 Natsukaze
- 5080 Fuyukaze
- 5081 Hatsunatsu
- 5082 Hatsuaki
- 5083 Hayaharu

more straight-lined with knuckles instead of curves and the material changed from special steel (DS and HT) to common mild steel (MS).[12] The reduction of building times from 12–13 months to 8–9 months was intended but given the actual situation of the shipbuilding industry in Japan in 1944 this was only an ideal. This simplification was to be introduced from *Michitsuki* (No 365). Her keel was laid down on 3 January 1945 at Sasebo but construction was halted in April 1945 when 16 per cent complete. Therefore *Hanazuki* (No 366) became the only completed ship of this simplified type.

General Arrangements

The hull had the greatest length of IJN destroyers but was also of the raised forecastle type like all IJN destroyers except the early 3rd class type. This class was a small cruiser rather than a destroyer in displacement and the main dimensions were settled accordingly. The excellent hull form offered better seaworthiness than the old 5500 ton light cruisers and with improved habitability the *Akizuki* class were very popular with their crews. The shape of the bow was different in each sub-type, ranging from the typical IJN rounded forefoot to the simplified straight-lined type and the last type with a cutaway foot. The stern was rounded with the knuckle directly below the waterline and drawn in above this line. The degaussing coil was positioned very near to the upper hull edge to give the impression that the radiused deck edge of the early destroyers had been revived, but this was an optical illusion.

As stated earlier, the superstructures were very logically arranged and consisted of, from bow to stern:

1. the fore HA turret group of which No 2 was mounted on an extended substructure to maximise the arc of fire;
2. the bridge structure surmounted by the type 94 director at the rear[13] and the air defence station in front and below it. The distance to No 2 gun turret allowed the turret to revolve through 360° and the height of the bridge was sufficient to overlook this superfiring No 2 gun;
3. the tripod foremast standing over the forward part of the funnel and connected to the rear side of the bridge by a platform;
4. the combined funnel (ie the casing which covered the three uptakes from the three boilers; the fore and aft tubes curved to touch the middle one in the upper part);
5. the ship's boats (two on each side, placed abreast the fore and aft edge of the funnel casing);
6. a small deckhouse with sided MG platforms and a 2m HA rangefinder and direction-finder antenna between;
7. the quadruple type 92 torpedo tubes (TT) with shield (this was the standard TT adopted in 1932);
8. the quick reloading system for torpedoes (this was placed behind the TT and therefore the deckhouse described next had a rectangular cut-out on the port side);
9. a large deckhouse used as the substructure for the searchlight position, aft tripod mast, aft type 94 HA director and No 3 gun (superfiring);
10. No 4 gun turret (mounted on the upper deck);
11. two type 94 depth charge (DC) throwers each with one type 3 loading platform;
12. three DC dropping rails to starboard and to port.

In addition, on the upper deck one pair of rails ran from starboard at the level of the TT shield around to port in front of the TT and then running aft near to the DC thrower position at first parallel with the upper deck edge but slightly to starboard from the aft edge of the deckhouse. These rails were used for the transportation of torpedoes and one davit for loading torpedoes was mounted at each end. This arrangement allowed the moving of torpedoes from both sides of the hull.

Hanazuki *in North Harbor, Manila Bay, on 29 October 1945. She is on repatriation duties, with reduced armament and some temporary structures on deck.* (US National Archives)

Main Armament

Among several unusual features of the *Akizuki*, the HA main armament and extended radius may be considered the most significant and therefore only these will be dealt with in detail due to limited space. The 65cal type 98 10cm HA gun and the type 94 HA fire control system are described in *Warship 1991* and readers who want the essential data are referred to this source.

As stated earlier the main armament consisted of eight HA guns mounted in four twin type A closed gunhouses (pseudo-turrets) arranged in two groups fore and aft in superfiring style. In order to direct the fire of each group against independent targets two type 94 fire control systems were mounted, one above the bridge structure and one aft forward of the rear gun mountings. The guns and turret mounts were the best developed by the IJN (a comparison with similar HA guns of the US, British and

Table 4: BUILDING DATA

Temporary name	Name	Builder	Laid down	Launched	Completed	Fate	Removed from list
104	*Akizuki*	Maizuru NYd	30.7.40	2.7.41	11.6.42	25.10.44 sunk	10.12.44
105	*Teruzuki*	Mitsubishi, Nagasaki	13.11.40	21.11.41	31.8.42	12.12.42 sunk	20.1.43
106	*Suzutsuki*	Mitsubishi, Nagasaki	15.3.41	4.3.42	29.12.42	[1]	20.11.45
107	*Hatsuzuki*	Maizuru, NYd	25.7.41	3.4.42	29.12.42	25.10.44 sunk	10.12.44
108	*Niizuki*	Mitsubishi, Nagasaki	8.12.41	29.6.42	31.3.43	5.7.43 sunk	10.9.43
109	*Wakatsuki*	Mitsubishi, Nagasaki	9.3.42	24.11.42	31.5.43	11.11.44 sunk	10.1.45
360	*Shimotsuki*	Mitsubishi, Nagasaki	6.7.42	7.4.43	31.3.44	24.11.44 sunk	10.1.45
361	*Fuyutsuki*	Maizuru, NYd	8.5.43	20.1.44	25.5.44	–	20.11.45
362	*Harutsuki*	Sasebo, NYd	23.12.43	3.8.44	28.12.44	–	5.10.45
363	*Yoizuki*	Uraga Dock	25.8.43	25.9.44	31.1.45	–	5.10.45
364	*Natsuzuki*	Sasebo NYd	1.5.44	2.12.44	8.4.45	–	5.10.45
365	*Michitsuki*	Sasebo NYd	3.1.45	–	–	17.4.45[2]	–
366	*Hanazuki*	Maizuru NYd	10.2.44	10.10.44	26.12.44	–	5.10.45
367	*Kiyotsuki*	Maizuru NYd	–	–	–	14.12.44[3]	–
368	*Ōtsuki*	Sasebo NYd	–	–	–	14.12.44[3][4]	–
369	*Hazuki*	Maizuru NYd	–	–	–	14.12.44[3]	–

Notes
1. Unserviceable at the end of the war.
2. Destroyed on the building slip.
3. Construction halted.
4. Estimated halt of construction.

THE AKIZUKI CLASS

APPEARANCE DIFFERENCES

Akizuki *as completed. Note absence of radar, weak close range armament, but two main armament directors.*

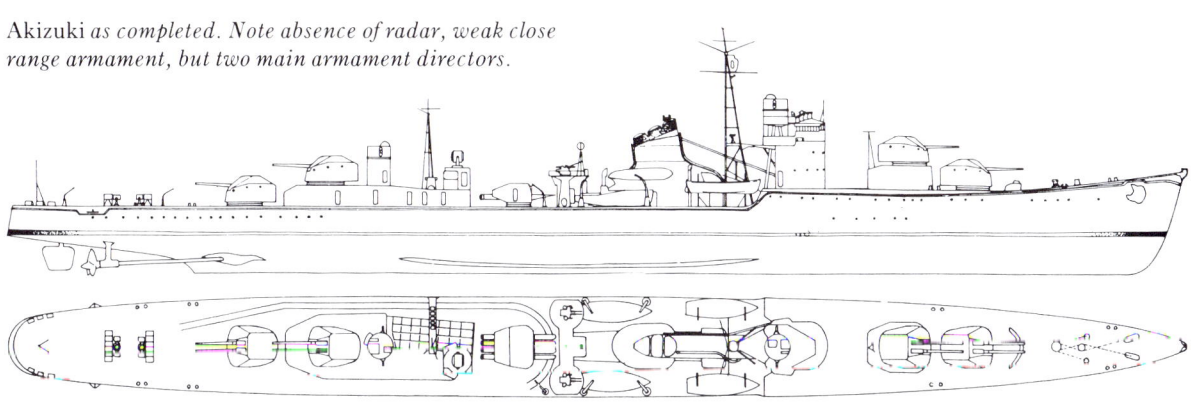

Hatsuzuki *in 1943. Type 21 radar antenna on fore mast platform.*

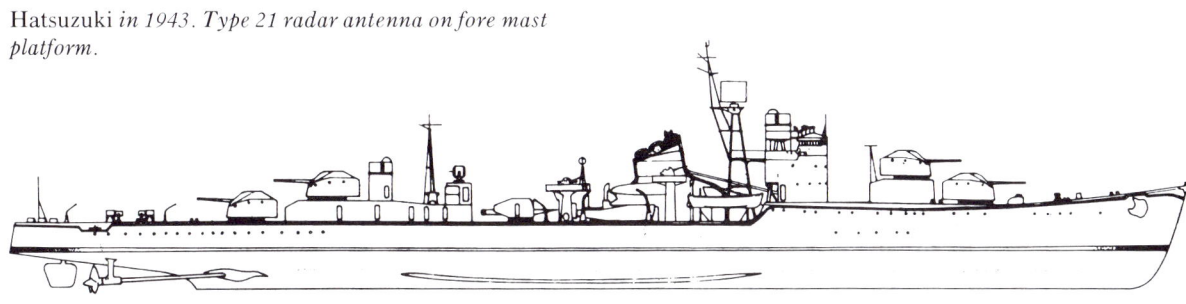

Fuyutsuki *in August 1944. Note Type 13 radar antenna on mainmast and enhanced AA armament.*

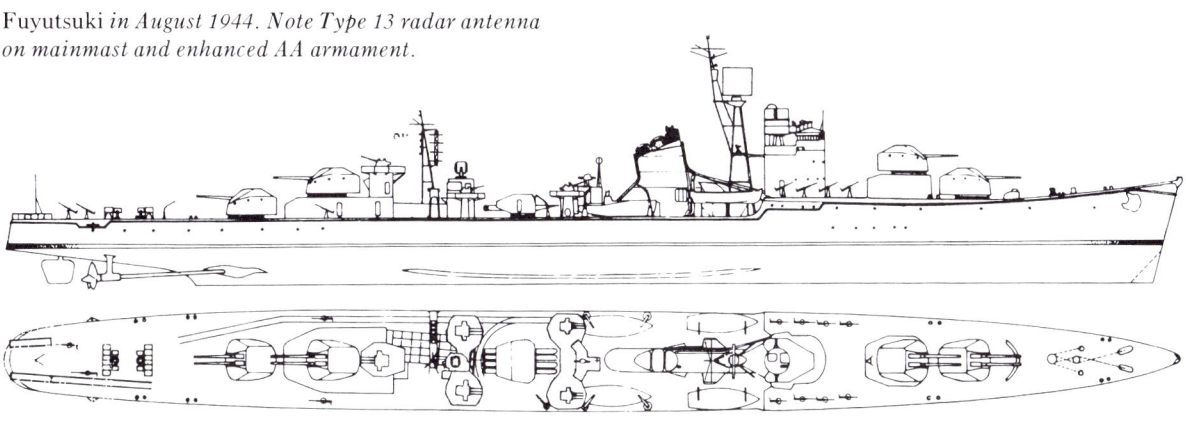

Suzutsuki *in 1945.*

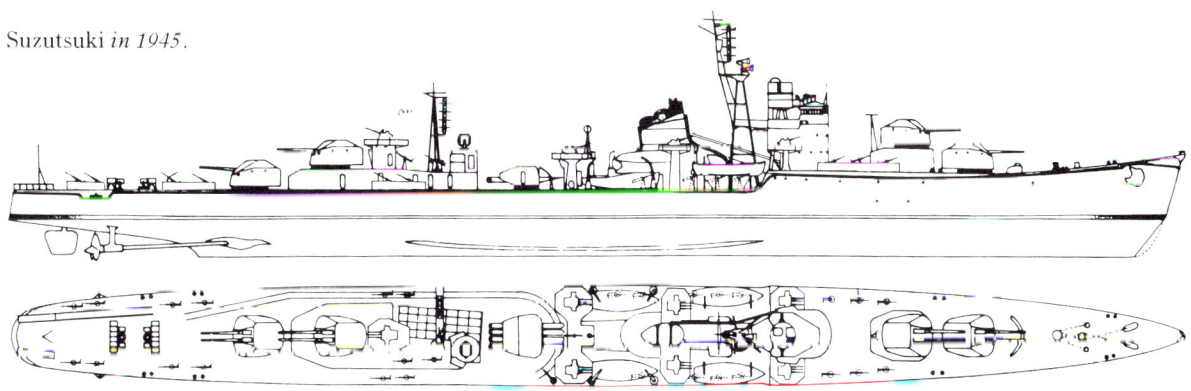

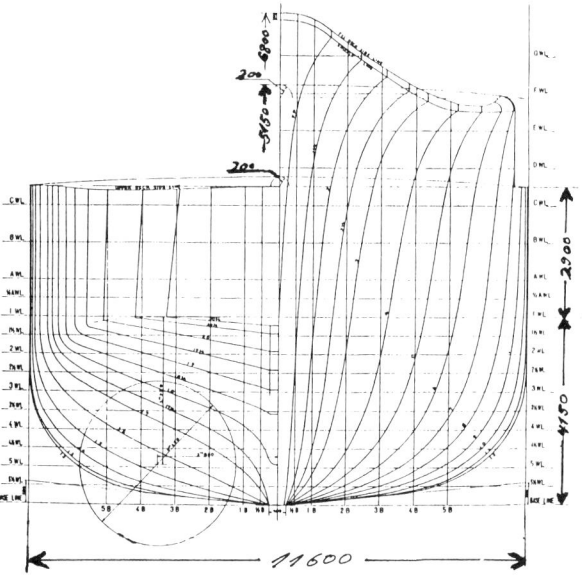

Body plan. Note the shape of the forecastle deck line and knuckle.

Table 6: ARMAMENT DATA

Main guns	65cal type 98 10cm HA (4 twin)
Machine-guns	type 96 25mm (2 twin)
Shell hoists	vertical type (3)
	horizontal type (1)
Rounds per gun (10cm)	12/300
Rounds (25cm)	100/3000
Fire control director	type 94 4.5m triaxial HA
	rangefinder (2)
Computer	2 units
Transmitter	2 units
Smoke-making system	type 91 (2 tons usage) (1)
Light MG	Type 96 (4)
Pistols	type 14 (14)
Rifles	type 38 (55)
Gas masks	types 93 and 97 (336)
Ready-use ammunition lockers	4 units
Torpedo tubes	type 92 model 4 (1 quad)
Director	type 97 model 2 (2)
Computer	two units
Air compressor	type Yu (3)
Air pump	type Kampon (1)
Second air compressor	(oxygen generator) (1)
DC thrower	type 94 (2)
DC rails	hydraulic type 3 (2)
	manual handling type 1 (4)
Small paravane	type 1 mod 1 (2)
Torpedo	type 93 model 1 mod 2 (8)
Head for training	torpedo 5 units (?)
DC	type 95 (54)
DC for training	type 88 (3)
Hydrophone	type 92 (1)
Sonar	type 93 model 3 (fitted later)
Echo-sounder	type 99 model 3

Table 5: MAIN TECHNICAL DATA

Length, overall	134.20m			
Length, waterline	132.00m (trial)			
Length, perpendicular	126.00m			
Width, maximum	11.60m			
Width, waterline (trial)	11.60m			
Width, max below waterline	11.60m			
Depth (at maximum width)	7.05m			
Freeboard, fore	6.80m			
aft	2.70m			
amidships	2.70m			
Displacement, trial (metric tons)	3470	(3452)	[3485]	
Draft, fore	4.15m	(4.12m)		
aft	4.15m	(4.12m)		
mean	4.15m	(4.12m)		
Displacement, full (metric tons)	3888	(3878)	[3899]	
Draft, average	4.51m	(4.47m)	[4.50m]	
Displacement, standard (long tons)	2701		[2750]	
Fuel, full (tons)	1080	(1097)	[1066]	
Radius of action (kts/miles)	18/8000			
Speed, full (designed)	33kts			
shp	52,000			

Notes

(Data for *Akizuki* taken during inclining trial, 16 May 1942.)
[Data for fundamental design of *Fuyutsuki* on 21 October 1942.]

Notes

Armament as completed; MG later increased considerably. Radar, sonar and hydrophone fitted or improved or increased later. Electrical equipment and wireless which the IJN intended with 'armament' omitted. Official data for *Akizuki* design dated 4 December 1939.

Plan of midships area showing torpedo tubes and reloading arrangements. Further forward the boiler room ventilation trunking is annotated.

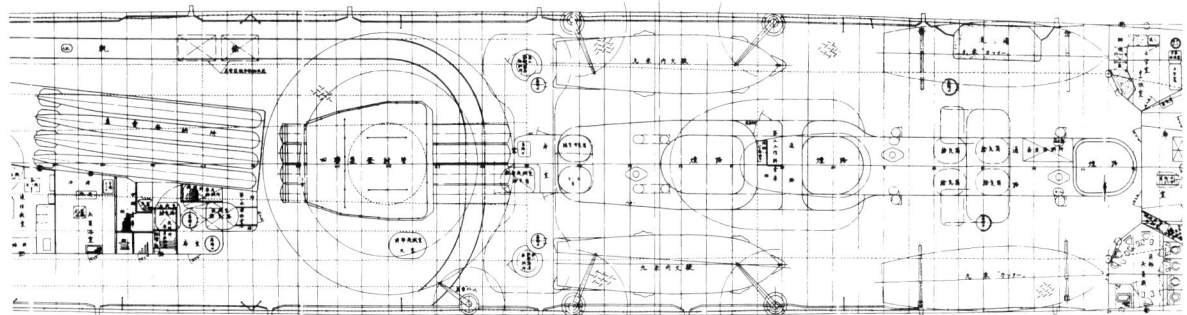

THE AKIZUKI CLASS

Fuyutsuki *and the battleship* Yamato *during the Okinawa Special Attack Operation. The destroyer is firing her after 10cm guns against the attacking aircraft.* Yamato *is down by the bow, so presumably already heavily damaged, but* Fuyutsuki *survived the near-suicidal operation.* (US National Archives)

German navies designed in the mid 1930s show that most of the ballistic and other essential data were superior; the drawbacks were the inability to adapt the fire control system from optical to radar direction, and the failure to develop the proximity fuse). Some Japanese sources state that when B-29 bombers attacked Kure near the end of the war at an altitude of more than 9000m (30,000ft), defence by shipborne HA gun could be provided only by *Akizuki* class ships since the 40cal type 89 12.7cm (5in) HA gun mounted on other ships in the port was effectively outranged. Even for the long 10cm the flight time of the shell was roughly 20 seconds and combined with the aforementioned shortcomings the success rate was very low. In contrast to this, authors like Hori Motoyoshi and Azuma Seiichi point to the shooting down of two low flying B-17s by *Akizuki* when she was attacked by three bombers on her way from Truk to Shortland in September 1942 as proof of the excellent performance of this HA gun. On that occasion *Akizuki* used the fire control systems for the fore and aft group independently and shot down two aircraft with the first 3-4 rounds.

In view of such results it is regrettable that the IJN used these special ships for supply operations in the Solomon area, which produced inevitable losses, so the number of operational ships never exceeded one Destroyer Division (four ships). It was not before 1944 that these ships could operate as a unit, in the Marianas and Philippines (Cape Engano) but by that time enemy air superiority was overwhelming and the results of their AA fire is uncertain. Furthermore, in December 1943, it had been decided to replace the fire control system for the after guns on newly built ships with one 25mm triple MG. On the official drawing of the *Fuyutsuki* type there is a note that the aft

Close-up of the mainmast of an Akizuki *class destroyer in 1945, showing the type 13 radar antenna. The searchlight remains but MG mountings and after rangefinder have been removed. The torpedo transportation rails are visible on the deck abaft the searchlight platform.* (US Naval Historical Center)

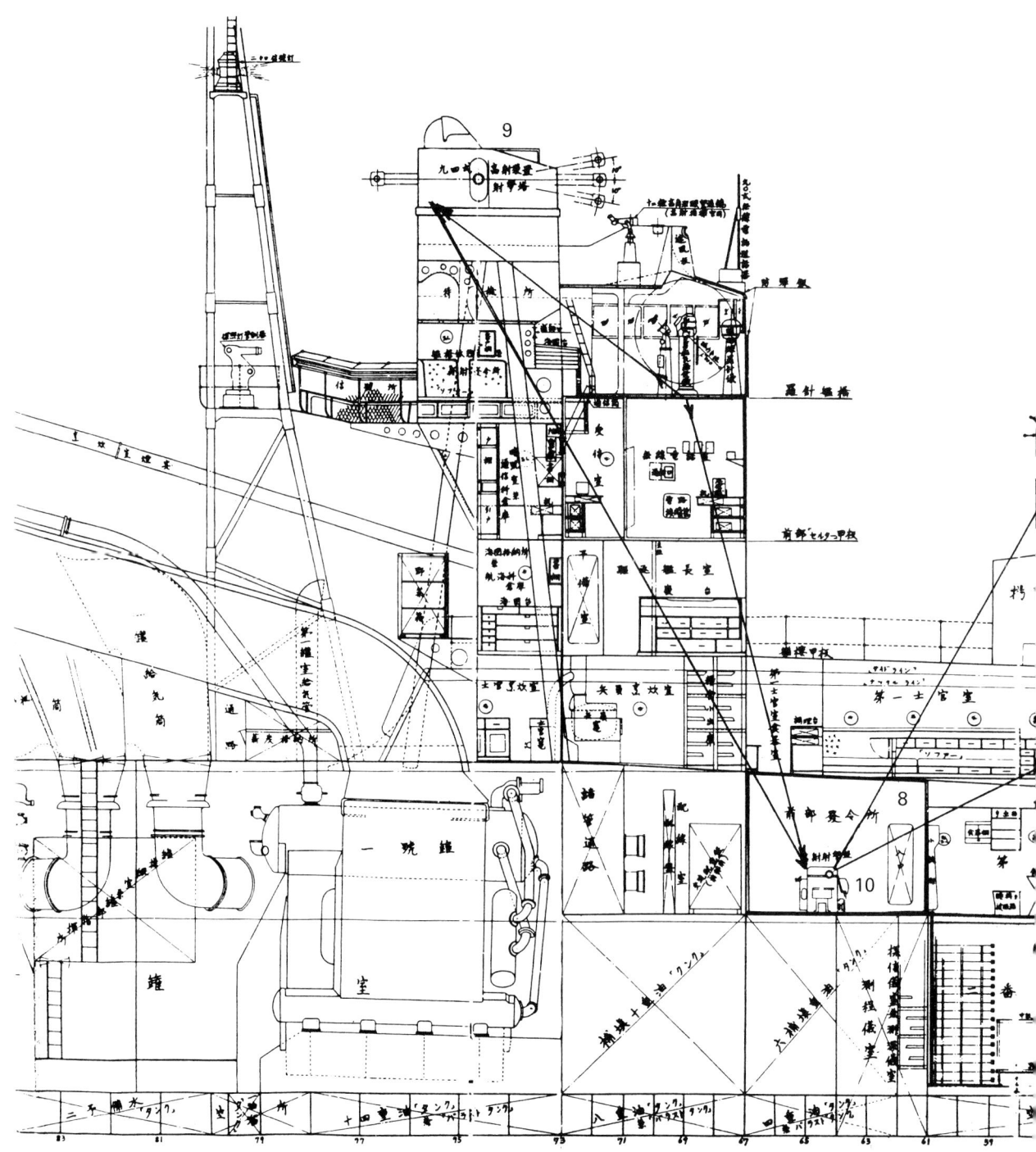

director must be arranged so as not to obstruct No 5 MG platform (at that time an increase in MG armament had already been achieved by changing twin to triple mounts and by the addition of two more 25mm triple MGs on platforms projecting out from the searchlight positions; these platforms were numbered 1 to 4). In fact no after director was mounted in *Fuyutsuki* and the following ships, so the effective firepower in the later ships was reduced, and it is no wonder that only the firing of *Wakatsuki* was particularly noteworthy in the two big battles of 1944. Even if the deletion of the fire control system for the after guns was mainly the result of inadequate production capacity, the notation on the *Fuyutsuki* drawing gives an impression of its reduced priority by that time as far as the highest technical authority was concerned.

In order to arrange the gun mounts in superimposed fashion several technical problems had to be solved. Before designing the *Akizuki* class the IJN had tried a superfiring arrangement of four 50cal *3nen shiki* 12.7cm guns in two twin mounts on the forecastle of the destroyers of the *Hatsuharu* class, but this proved to be a

Ammunition supply and fire control.

1. Shell hoist for 'A' mounting.
2. Shell hoist for 'B' mounting.
3. Shell supply position, 'A' mounting.
4. Shell supply position, 'B' mounting.
5. 'A' (No 1) magazine.
6. 'B' (No 2) magazine.
7. Fuse magazine.
8. AA fire control computer station.
9. Type 94 director providing data to computer.
10. Computer calculates firing data and supplies to gunhouses (11 and 12).

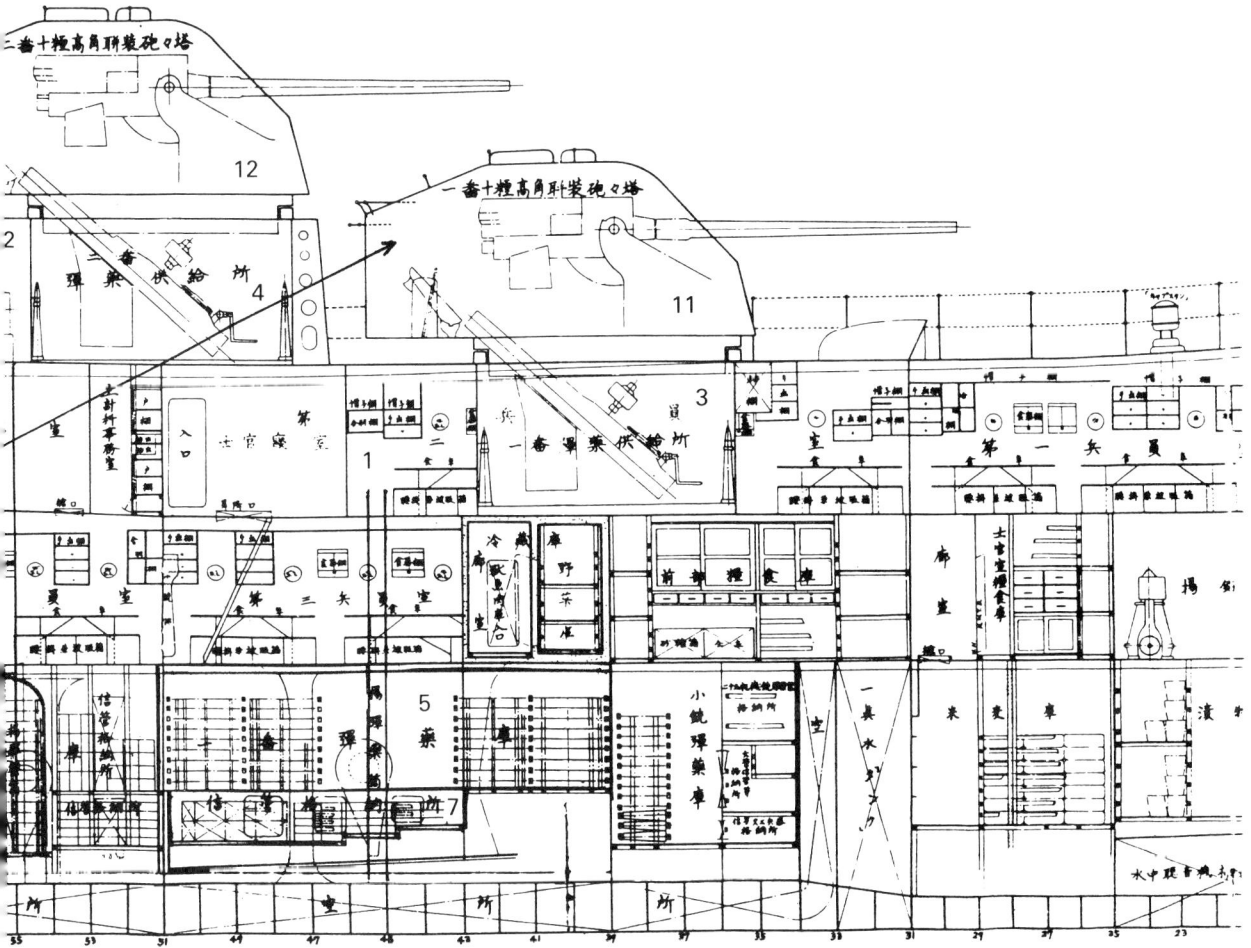

failure when the superfiring mount had to be removed due to stability problems. Therefore this kind of disposition was not popular with naval architects, because it raised the centre of gravity (CG) and reduced stability. There were many arguments against the superfiring arrangement which had to be seriously considered alongside the associated problems: the extended radius (1096 tons of heavy oil had to be carried, producing a continuously rising CG as it was consumed); the heightening of the bridge to provide adequate vision over the upper mounting (also raising CG); and the unification of the boiler exhausts into one funnel casing (caused indirectly by this disposition of the gun turrets which determined the position of the bridge which in turn greatly influenced the position of the funnel). In each case the increase of the side wind pressure area had also to be taken into consideration. On the other hand, air defence was the main duty of this ship and the most effective positioning of the HA gun mounts had to be given priority. This, without doubt, was the superfiring disposition, so adequate countermeasures particularly for maintaining stability in light load condition became necessary, and a solution to the stability and related problems was given next priority after the arrangement of HA gun turrets. This was a sensitive issue in a navy which faced a most serious crisis in ship design a few years previously with the *Tomozuru* capsize and Fourth Fleet incidents.

These stability countermeasures covered a wide range, such as fitting a ballast keel, partly reducing the deck height in the forecastle, lowering the forecastle deck around the position of No 1 gun turret (this was to reduce the height of the superimposed mount and was a feature also seen in *Yamato* class battleships), installation of the air defence station without a roof, and the provision of ballast tanks. After completion the stability of this class was good as can be seen from the figures in table 7.

For each gun 300 shells were stowed in the magazine and the supply to the gun was in two stages. The first step was to bring up the shell vertically from the magazine by pusher hoist to the shell supply room in the barbette-type support of the gunhouse. The second step was to transport them to the waiting position inside the gunhouse by the inclined hoist. From the waiting position the shell was transported manually to the loading tray of the gun and then continuously loaded by the automatic breech action. The manual handling (from the top of the lower hoist to the lower end of the upper hoist and from waiting position to loading tray) was hard work for the relevant personnel concerned due to the high firing speed that the mounting could maintain.[14]

The principles of ammunition supply and fire control are shown in an accompanying diagram for the forward HA guns. The after group was supplied and controlled in the same way, with the firing order station and its computer always located on the lower deck level; magazines were near by but in the hold. When the chief gunnery officer tracked the target with his binoculars in the director, the pointer and trainer of the director had the bearing and elevation figures in their instruments and turned the wheels to meet the target at which the chief gunnery officer aimed. Their movements were fed to the indicators inside the gunhouses and the barrels were elevated and trained by the traditional 'follow the pointer' system. From the rangefinder above the director the measured range was fed to the computer in the order station which was also supplied with many other data including compass course (bearing of the ship) in order to calculate the fuse time (as a function of future range) and to transmit it to the HA gun fuse-setters. These firing elements were continuously transmitted and according to Endo Akira's *High Angle Guns and Anti-Aircraft Ships* it took four seconds to open fire when the chief gunnery officer of *Suzutsuki* pushed the button for the first time.

As pointed out earlier the weight of *Akizuki*'s twin mounts was greater than the twin mount of a fleet type destroyer, and in addition she carried one more mounting. This may give the impression that *Akizuki* was an over-armed ship, but this is not actually the case. The total weight of armament amounted to 447 tons in the *Akizuki* class, much exceeding the light cruiser *Yubari*'s 324.3 tons and also more than the destroyer *Kagero*'s 345.7 tons and therefore the *Akizuki* class must be called a heavily armed ship, but as a percentage of trial displacement the *Akizuki* class had 12.7 per cent, and the *Kagero* class 14.2, so the armament was not an abnormal proportion of the design.

The demilitarised Harutsuki *in October 1945 being refuelled for repatriation duties (bringing back prisoners from Saigon, in this case).* (US National Archives)

THE AKIZUKI CLASS

Table 7: TRIALS DATA

	Trial condition Design	Trial condition Actual	Full load condition Design	Full load condition Actual	Light load condition Design	Light load condition Actual	Supplementary light load condition Design	Supplementary light load condition Actual
Displacement (t)	3458	3452	3899	3878	2451	2407	2810	2747
Average draught (m)	4.150	4.120	4.500	4.360	3.200	2.580	3.500	2.810
Trim (m)	–	–	0.07	0.21	1.07	1.130	1.370	1.260
Height of CG above keel line (base) (m)	4.410	4.380	4.250	4.220	5.000	4.980	4.750	4.740
Metacentric height from CG (m) GM	1.070	1.110	1.130	1.160	0.7	0.73	0.86	0.88
Maximum angle of stability (°)	44	–	45	–	42	–	44	–
Maximum righting lever (m)	0.67	2.421	0.73	3.194	0.35	0.727	0.5	1.236
Range (°)	92.6	93.9	99.4	101.1	71.5	73.4	80.5	81.4
Height of CG above WL (m) OG	0.26	0.26	−0.25	−0.25	1.80	1.830	1.25	1.300
Ratio of wind pressure $^{above}/_{below}$ WL	1.67	–	1.44	–	2.25	–	2.23	–
Supplementary ballast tank (capacity in tons)	–	–	–	–	–	–	351.1	339.37
Reserve buoyancy (t)	4265	–	3851	–	3299	–	4940	–
Assumed draught (m)	3.65	–	–	–	–	–	–	–
Rudder (design data)	One balanced, underslung							
	area (m²)		10.48					
	ratio of rudder area/ immersed hull side area (A/Am)		1/45					
Turning Cycle	Advance (ratio of maximum longitudinal direction) (DA)		4.5					
	Tactical Diameter (DT) (ratio of maximum lateral direction)		5.0					
	Maximum heel angle (°)		14					
Rolling period (°)	(trial)		9					
Displacement (tons)			3485					
8/10 trial speed (kts)			31					
Rudder angle (°)			35					

Notes

Akizuki inclining trials, 16 May 1942.

Trim is additional draught aft in each case.

Machinery and Fuel Stowage

The long radius and concomitant stowage of an extraordinary amount of fuel was as important a feature as the main armament. The heavy oil capacity occupied about 24 per cent of trial displacement and this was a big contrast to about 15 per cent for a fleet destroyer. The oil was carried in the lowest position as much as possible but tanks were also fitted from the double bottom to the upper deck just below the bridge structure. This would appear to have a negative influence upon stability but was in fact one method of maintaining stability by using the oil in the higher tanks first (and thus initially lowering the CG). Certainly the maintenance of stability caused the biggest difficulties in the design of this class and the arrangement of the oil tanks around the machinery spaces and the layout of the latter were principally dictated by this consideration.

The general arrangement of the machinery spaces was different from earlier destroyers, with two boiler rooms (BR) one abaft the other and the placement of two boilers in the forward BR, one in the aft BR, abandoning the past rule of using one boiler in one BR. The layout of the main engines was also a first for the IJN in placing the turbine set connected with the port shaft in the forward engine room (ER) and the other geared to the starboard shaft in the aft ER. This arrangement was mainly brought about by the necessity of providing tanks for the huge amount of fuel but at the same time also aimed to reduce the vulnerability of the machinery to battle damage. Therefore oil tanks were located on both sides of the BR and starboard and port of the fore and aft ER respectively. This was a transitional arrangement and it was not until the design of the Type D destroyer (*Matsu* class) that the IJN adopted the more rational system for limiting damage, namely the BR-ER-BR-ER system. Despite this criticism the arrangement of two BR and two ER instead of three BR and one ER and the deletion of the previous electrical generator space abaft the ER in favour of placing the electrical generators in the forward ER can be regarded as typical of the latest IJN destroyer design.

The total length of the engineering spaces was 51.5m; the BR were 26.2m (the forward 17m, aft 9.2m) and the ER 25.3m (the forward 13m, aft 12.3m). Compared to the previous *Kagero* class (BR 28.5m, ER 22.0m; total length 50.5m), this meant a slight increase in length but was actually a reduction when the fuel oil tanks and the hitherto separate generator room are considered.

The frame spaces used in the ER are a good example of detailed design and the deep consideration given to hull strength after the Fourth Fleet Incident. The forward ER was located between F 99 and F 113; the after one covered the length between F 113 and F 126. Within the total length of 25.3m, frame spaces of 1000mm (F 103 to F 107, F 110 to F 117, F 123 to F 126), 925mm (F 117 to F 121), 900mm (F 99 to F 103) and 800mm (F 107 to F 110, F 121 to F 123) were utilised. This complicated structure provided sufficient hull strength with comparatively little weight but on the other hand contributed also to the rather long building period.[15]

Main Engines

The reduction of the required speed to 33kts from 35kts made it possible to employ the same main engines as the *Kagero* class, developing 26,000shp per unit, 52,000shp in total. Actually, some modifications were incorporated to facilitate mass production and the engine specifications for Class 360 ships (*Shimotsuki* and other ships of *Maru Kyu Keikaku*) called for extended use of substitute materials and more modifications in the fittings, such as nuts and bolts to better respond to war demands. The biggest difference was the new design of the reduction gears for a propeller shaft speed of 340rpm instead of 380rpm in the *Kagero* class. The engine specifications for Class 104 ships (*Akizuki* and other ships of *Maru Yon Keikaku*) were concluded on 19 April 1939 and approved on 15 September. On 20 June of the following year, the rpm and shp of the cruising turbines were changed. The specifications for Class 360 ships were initiated in October 1941. The design called for the development of the aforementioned specified full power of 52,000shp under test conditions at a trial displacement of 3470 tons and using ordinary steam conditions. At standard speed (18kts) the fuel consumption was to be within 0.46kg/hr/shp.

General arrangement of machinery. Note that the exhaust from the forward engine room was led up both sides of the searchlight platform.

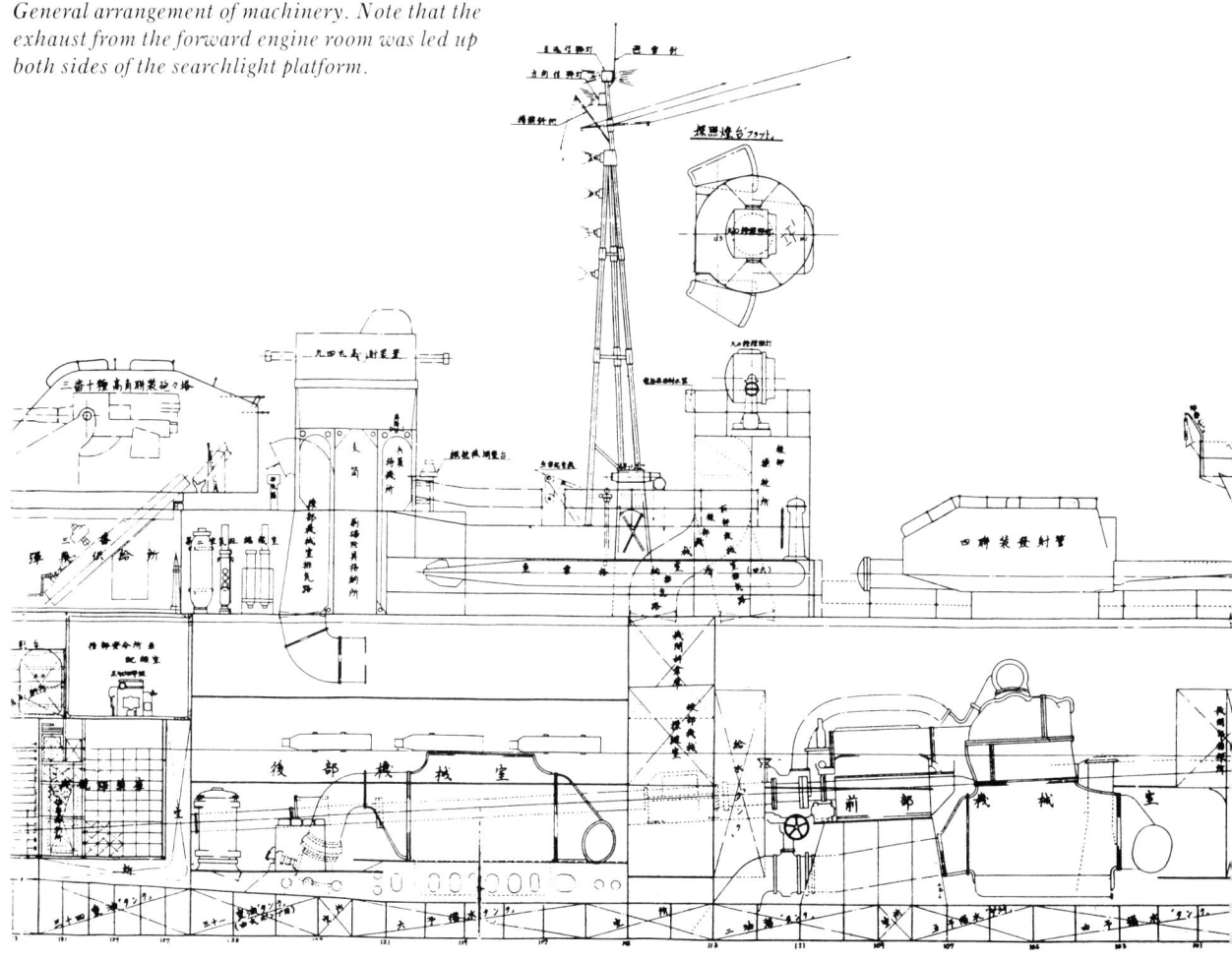

Each main engine consisted of high (HP), intermediate (IP), and low pressure (LP) turbines geared to a single shaft by the main reduction gearing. It was further geared to two cruising turbines (HP and LP) through the cruising reduction gear and a claw coupling which rendered possible the connection to the IP turbine. The exhaust steam of the cruising turbines was led directly to the steam chamber of the LP turbine. The cruising turbines developed 2700hp in each shaft, totalling 5400shp, when turning the shafts at 163rpm. Under overload condition they developed 7800shp, ie 3900shp in each shaft at 185rpm. However, when used in conjunction with the main turbines, they developed a total of 18,500shp, with up to 244rpm in order to achieve 26kts.

The astern turbine was in the LP casing and could develop 5000hp for each shaft, total 10,000hp, when turning the shaft at 198rpm. The steam was fed from the IP turbine in the usual way.

For each shaft there was one condenser of the single flow surface system type. Due to the installation of the electric generators in the forward ER it was arranged below the LP turbine in that room but placed beside the LP turbine in the after ER.

Boilers

There were three Kampon *shiki Ro Go*[16] Mark 3 B model 17 oil-fired boilers equipped with superheaters and air preheaters. The superheater was located in the middle of each bank, the air preheater over each bank. The fire room was not under air pressure since double casings with air pressure between were used. In contrast to the usual arrangement, the superheater steam tubes did not pass through the steam drums.

In the boiler feed system cross connections were provided so that in an emergency any pump could supply any boiler. No 1 and No 2 boilers were fitted in the forward BR (No 1 BR), No 3 boiler in the aft one (No 2 BR). The BR were of the open type and No 1 was located between F 73 and F 90 having a uniform frame space of 1000mm; No 2 one between F 90 and F 99 with either 1020mm (F 90 to F 95) or 1025mm frame space.

Funnel and Uptakes

One of the most noticeable features of the *Akizuki* class was the utilisation of one combined funnel to which the exhaust tubes (smoke outlets) from each boiler were

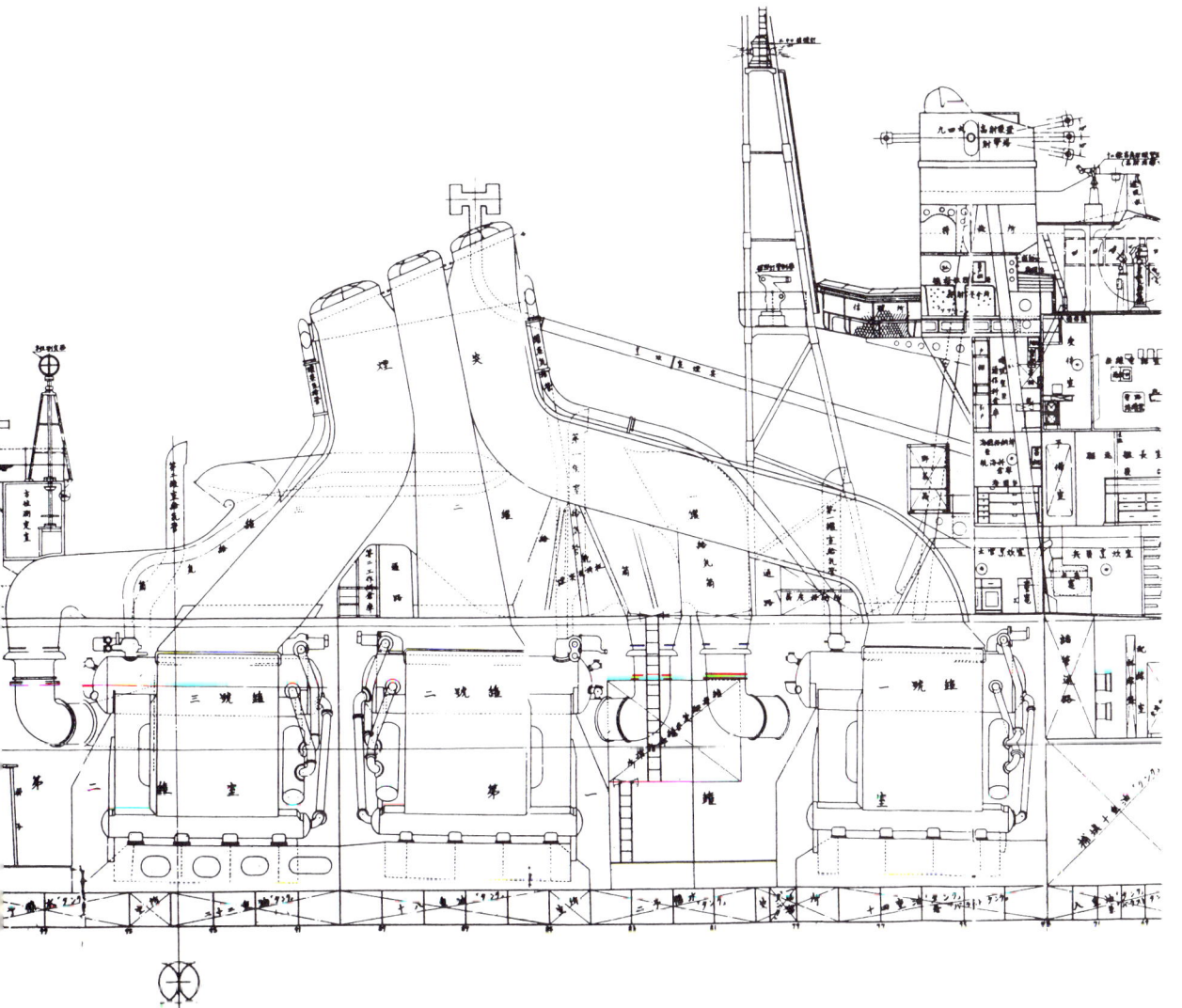

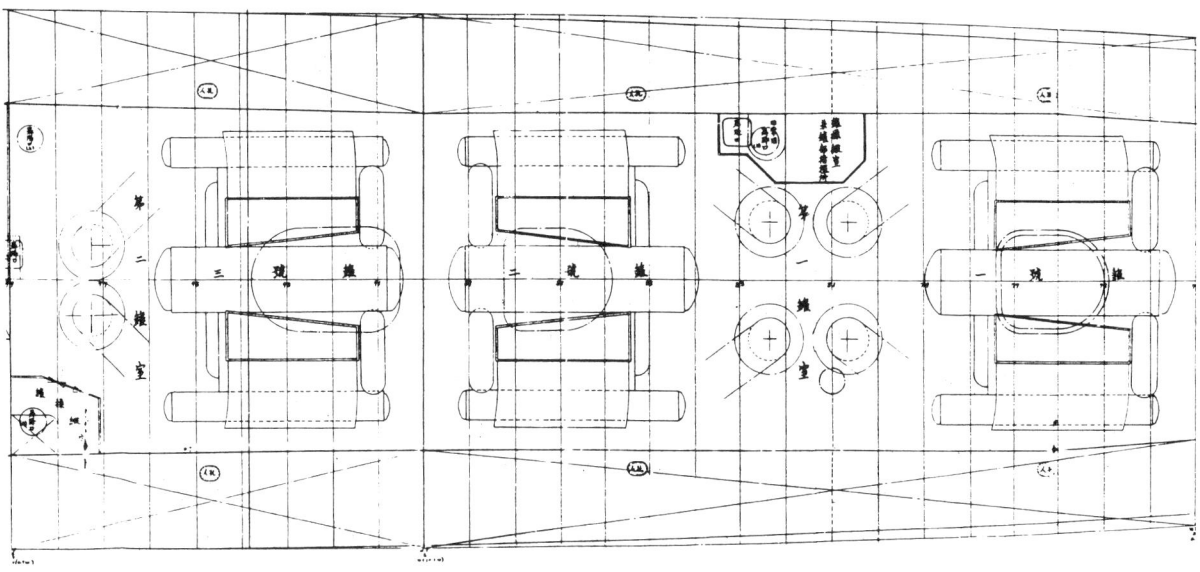

Plan of boiler rooms.

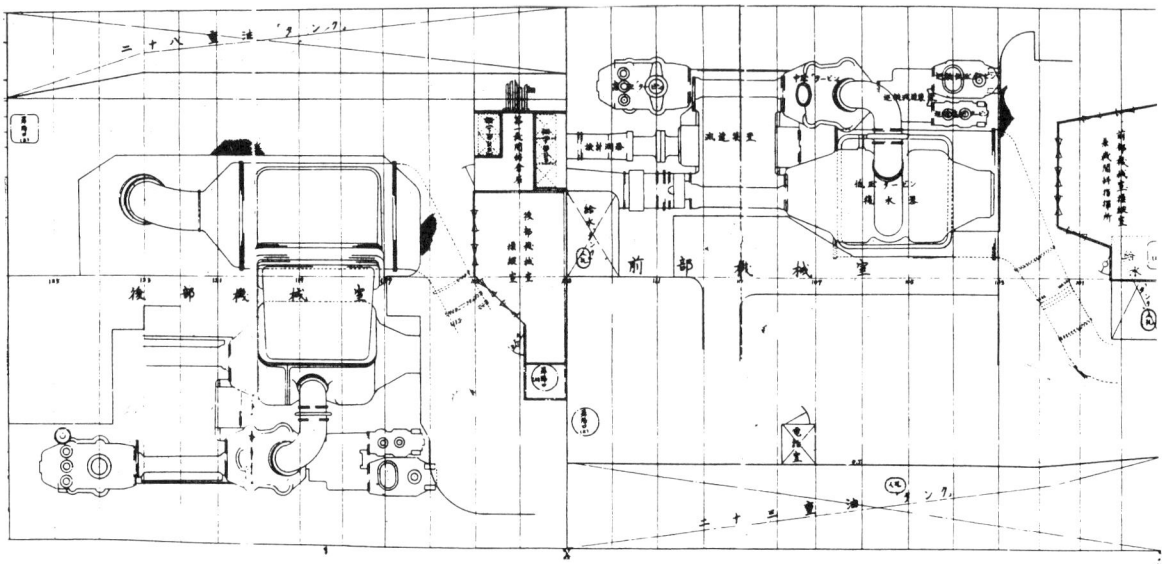

Plan of engine rooms.

connected. Like the arrangement of two BR this reduction to one funnel was also adopted in IJN destroyers for the first time. However, the interior of the funnel was divided into separate uptakes for each boiler. The forward division came out directly behind the bridge structure, taking the shape of a horizontal S to obtain sufficient distance from the bridge and fore tripod mast. The second division was inclined aft 10 degrees, while the lower part of the third was directed forward and then the upper part was adjusted to the inclination of the second uptake. These divisions were covered by one casing giving the appearance of a single funnel and this peculiar arrangement was also practised in other ships of that period, such as the light cruisers of the *Agano* and *Oyodo* classes.

The height of the funnel was about 8m (measured at 90 degrees to the upper deck to below the funnel cap of the second uptake) and the wind pressure area was the biggest of any IJN destroyer. Therefore it was supported by three steel wires of 8mm diameter on each side, guyed between the casing and the hull sides. Along the funnel casing several steam pipes and air supply and discharging tubes were assembled, together with the chimney from the galley with its characteristic H-shaped end projecting above the funnel.

The space between the uptakes and inside the casing was used for the arrangement of the air supply ducts to the BR and athwartship passage (two in case of *Fuyutsuki* and later ships, one in earlier ships).

THE AKIZUKI CLASS

Table 8: *Weight of Machinery*

Main engines	214
Shafts and propellers	92.5
Auxiliary machinery	55.5
Boilers	195.5
Smoke stacks and flues	13
Pipes, valves, and cocks	97
Miscellaneous	56
Water	78
Oil	5.5
Total	724.5 (without water and oil)
	808 (with water and oil)

Notes
Unit: Metric tonnes
Condition: Design (estimated) weight under official test conditions. Actual weight of *Akizuki* was 835 tons in trial condition and 840 tons in full load condition. In light load condition and supplementary light load condition the weight was 735 tons in each case.

Ventilation

Perfect fuel combustion requires the supply of sufficient air to obtain the greatest efficiency of burning and least visible smoke column from the boiler. The amount of air is very great and can be supplied only by forced ventilation, principally using centrifugal type fans and air passages or ducts from outside the weather deck to the boilers. In the case of the *Akizuki* class two different shapes were used. For No 1 BR an independent structure was employed as the ventilation duct (a combination of round and square shape), positioned alongside the curved No 1 funnel division with the square-shaped opening directed aft and downwards (it had some similarity with a standing pipe). No 2 and No 3 boilers were supplied by the characteristic IJN air supply ducts utilising the lower part of the funnel to make a ventilation passage within the space of the outer casing. The upper end took the form of a cap facing downwards. This arrangement was designed to heat the air before it was suctioned into the double casings of the boilers.

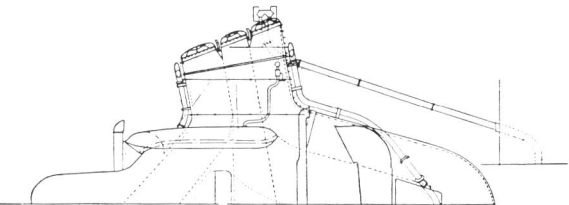
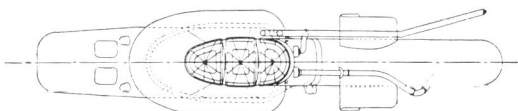
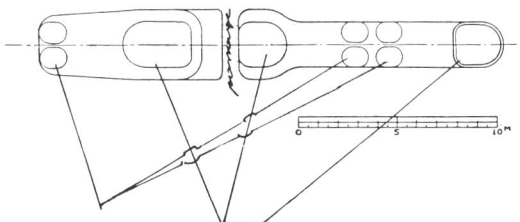

Funnel and ventilation arrangements. (Drawn by Kozo Izumi)

The cap type air supply duct was deleted in *Fuyutsuki* and later ships and was replaced by a ventilation duct for No 2 boiler between No 1 and No 2 funnel uptakes while No 3 boiler was supplied by a ventilation duct integrated into the midships MG platform to obtain a completely independent air supply system for each boiler.

The ER were also supplied and exhausted by forced ventilation and in their case the ventilation ducts were widely integrated into the midship MG platform (back to back with the duct for No 3 boiler) in the case of No 1 ER

Ventilation system for boiler rooms. (Drawn by Kozo Izumi)

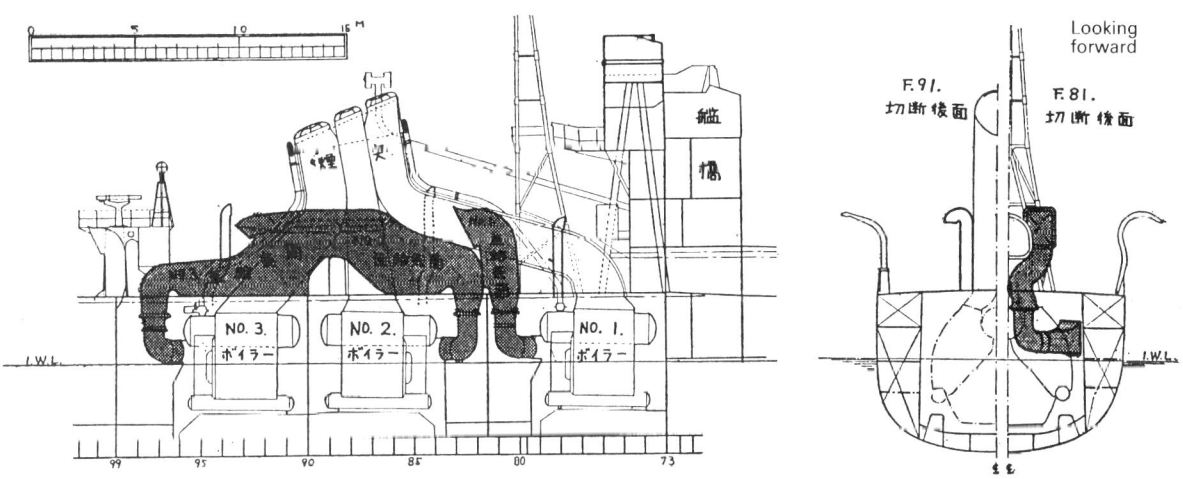

and the searchlight platform with regard to No 2 ER. Exhaust (or discharging) passages were in the searchlight platform for No 1 ER and aft director platform (No 2 ER). Due to the large amount of air needed in these rooms the dimensions of the ventilation ducts and exhaust passages were substantial. Ventilation duct openings above the weather deck were closed by steel shutters like Venetian blinds.

Auxiliary Machinery

The pumps and their turbines were relatively small and accessible for overhaul. When members of the US Naval Technical Mission to Japan inspected several IJN warships after the Pacific War they judged them to be of good quality. This progress had been obtained at the beginning of the 1930s by the necessity to reduce weight as much as possible in order to be able to keep down displacement or use the saved weight for increasing fighting power. This principle had already been adopted after the Washington Treaty but intensified after the further limitations of the London Treaty.

The steering engine was an electrically driven hydraulic plunger type capable of turning the rudder from hard port to hard starboard (or the opposite), a change of rudder angle of 70 degrees, in 30 seconds against a maximum torque of 45.6 metre-tons when going full speed astern.[17] In order to correspond to this requirement the hydraulic cylinder, which was the four-plunger type, was equipped with two electrically driven hydraulic pumps having this necessary capacity. For emergency use, *ie* when these electric pumps were inoperative, the hydraulic cylinder was equipped with a small manually driven hydraulic pump which could operate the rudder at speeds up to 30kts.[18]

Rudder and propeller arrangement. Right aft on the quarterdeck is a depth charge rail, with two type 94 DC throwers and associated type 3 loading platforms further forward on the centreline.

Official Speed Trials

Akizuki's speed at full power trial on the measured mile was 33.39kts at 52,193shp and 343rpm of the propeller shafts. The displacement was 3472 tons. At overload full power shp rose to 54,763, speed to 33.77kts, and rpm to 348.2. The standard speed of 18.26kts was obtained with 4839shp and 161.2rpm of propeller shafts. In this case the fuel consumption was 0.43kg/hr/shp as against 0.332kg/hr/shp at full power but cruising distance (radius) was 9062 miles in the case of the former speed (due to the low HP value) and only 1877 miles in the case of the latter.

The speed trials complied with the rules worked out by the Fifth Main Division (Machinery) of the Navy Technical Department and during that time distilled water with a salt content of three parts per million or less was used as feed water. At standard speed, whenever it was possible from the standpoint of safety, economy of fuel was attempted. It was also a standard procedure to use the auxiliary exhaust steam to heat the feed water. It was led off to the main water condenser when under full power and full cruising power, since there was an excess, and used in the main turbine under other circumstances.

Notes

[1] At the end of 1939 the tactical concept of the Navy General Staff was as follows:
 1. Organisation of a task force for the main purpose of attacking the enemy carriers (*Kido Kokubutai* composed of the fleet carriers *Shōkaku, Zuikaku, Taiho, Hiryū, Sōryū*);
 2. Formation of a close combat group to attack the enemy battleships at the beginning of the gunnery engagement (*Kessen Yasubutai* composed of *Akagi, Kaga, Junyō, Hiyō*, etc)
 3. Drawing up of a close protection force for its own capital ships to give AA protection and carry out ASW (*Chokuei Butai* composed of *Hosho, Taigei* and *Ryūjō*)

[2] For the properties of this HA gun see *Warship 1991*, pp81ff.

[3] This radius was the largest among IJN warships of this period and the only equivalent ships were the battleship *Haruna* (after conversion) and carrier *Taiho* (designed at that time)

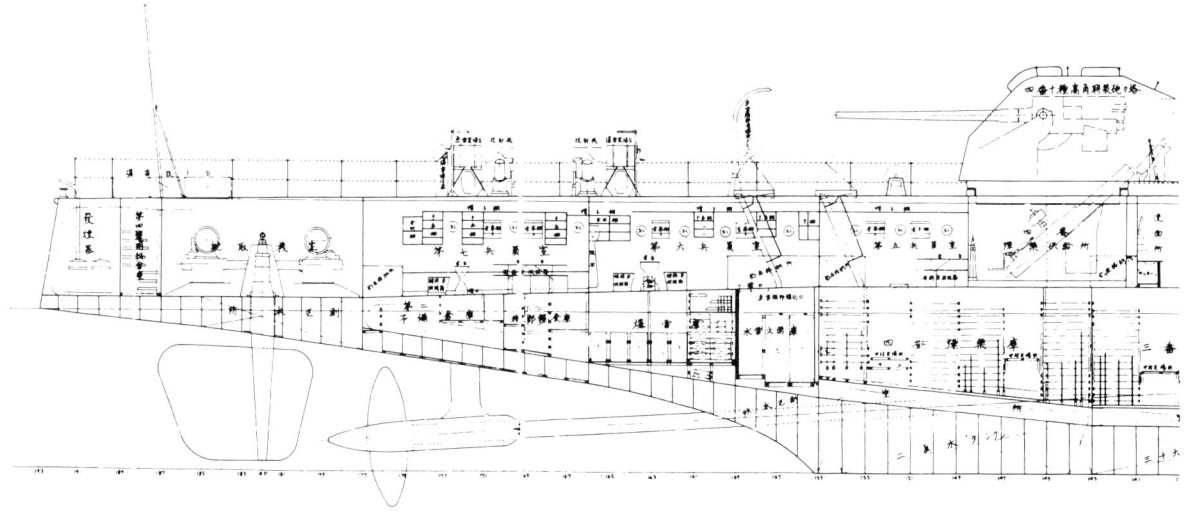

Ventilation system for engine rooms. (Drawn by Kozo Izumi)

Japanese flotilla craft being broken up at Sasebo in 1948. The substantially larger hull of the Fuyutsuki *stands out from the old destroyers* Yanagi *and* Hinoki *(to port) and the escorts to starboard.* (US Naval Historical Center)

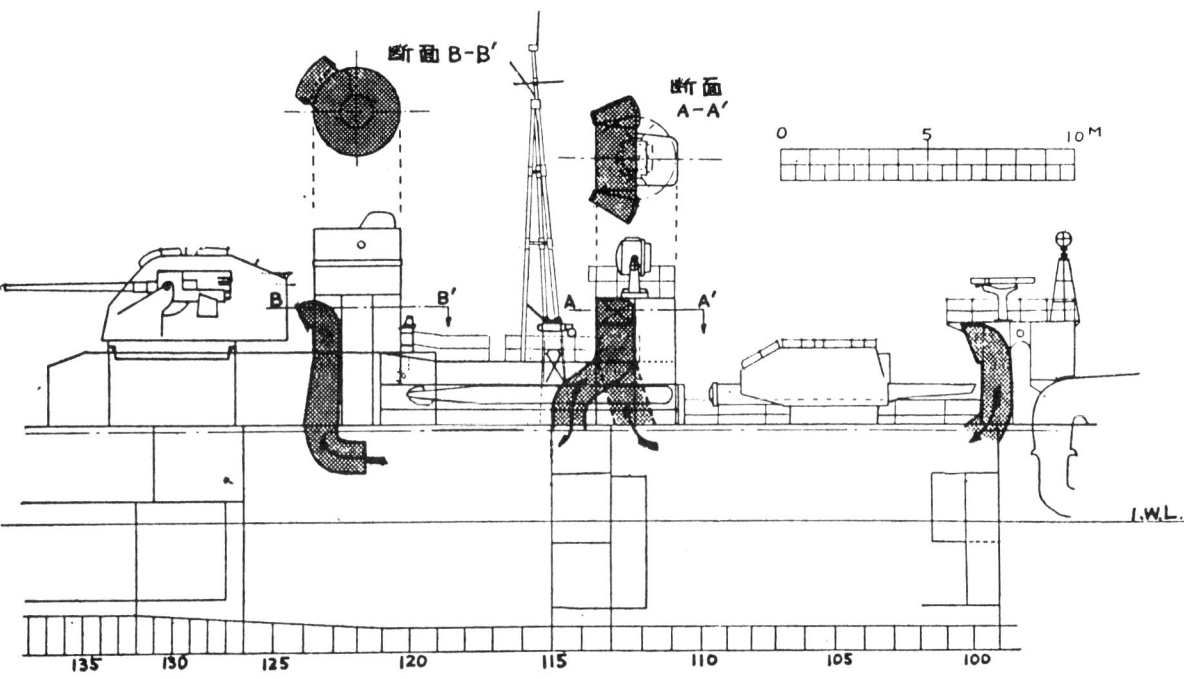

4. followed by *Tone* class cruisers with 18kts/8000 miles (the maximum for cruisers) and the most modern destroyer classes generally achieved 18kts/4000 miles (*Yamato* had 16kts/7200 miles). Thus, the radius of the *Akizuki* class was to be 2.5 times greater than other destroyers.
4. Compared to destroyers of the *Fubuki* to *Yugumo* classes this meant one more gun mount of heavier weight (34.5 tons as against 32 tons and a total of 138 tons against 96 tons) but the superimposed arrangement would exert an even greater influence on stability.
5. This opinion was expressed not only by technicians of the Second Main Division ('Torpedo Division') of the Navy Technical Department ('it is quite regrettable that such an excellent ship does not have the power of the destroyer at all by limiting the duty to AA defence only') but also by tacticians ('the chance of carrying out a torpedo attack can result from the condition of any sea battle, therefore a minimum torpedo armament is requested').
6. The author is well aware of the action of *Hatsuzuki* in the Battle of Cape Engano but the brave fight by this ship only delayed its sinking and does not invalidate this fundamental judgement.
7. This description makes it clear that the addition of the torpedo armament must have been decided between July and September 1938, but this author lacks a conclusive explanation for giving the speed as 34kts. In order to distinguish this destroyer type from the fleet type (A Type) the AA destroyer became B Type (C Type was *Shimakaze*, D Type *Matsu/Tachibana* classes).
8. This is abstracted from the original explanation.
9. By this programme the construction of an AA light cruiser (temporarily called V 18) was also scheduled. The author intends to deal with this project in a separate article together with other ones relating to air defence.
10. Due to space limitations the author intends to cover this design in a separate article.
11. This brought about the reduction of roughly 5500 man-days in the shipbuilding work.
12. See note 10.
13. The foundation of the HA director was supported by two posts on each side to prevent vibrations degrading the precision of the instrument.
14. For more details see *Warship 1991*, p93. It must be added that the decision not to fully automate the ammunition transportation was one of the reasons why the IJN selected the calibre of 10cm for HA, *ie* to reduce the weight of the round (shell and cartridge combined) for ease of manhandling.
15. As more extreme examples, the frame spaces of 1000mm, 1020mm, 1025mm can be mentioned to strengthen the impression of the very detailed calculation undertaken.
16. Navy Technical Department type No B.
17. When going ahead the maximum torque was 25.2 metre-tons.
18. This pump proved its value when *Suzutsuki* was heavily damaged while escorting *Yamato* on her last sortie and managed to return to Sasebo by going astern and using this hand-operated pump for steering throughout this voyage. During this night four sailors operated this pump by hand and were changed every 20 minutes.

USS TRITON
The Ultimate Submersible

Although best remembered for the first submerged circumnavigation of the globe, the exact purpose of the *Triton* was unclear at the time, and has remained so to this day. Robert P Largess and Harvey S Horwitz investigate the mysterious genesis of this huge nuclear submarine, using the testimony of many of those involved with her design and operation.

On 19 August 1958 the United States Navy launched the USS *Triton* (SSRN-586). At her officially released figures of 5940 tons standard displacement and 6670 submerged, and 447½ft in length, she was the largest submarine in the world, eclipsing the Japanese Navy's *I-400* class of huge submersible aircraft carriers designed to attack the Panama Canal. According to these figures she retained this title only a few months, till the completion of the Navy's first ballistic missile submarine *George Washington*, at 6020 tons standard, 6700 tons submerged displacement. Reportedly, however, *Triton*'s submerged displacement was much higher, 7900 tons. Much of the additional submerged tonnage was represented by the extensive ballast tankage necessary to give *Triton* the reserve buoyancy required for her to behave respectably at high speed on the surface. Thus, *George Washington*, designed like most modern submarines entirely to operate submerged, was actually much smaller in overall volume, and *Triton* may not have been exceeded in size until the *Ethan Allen* class several years later. Most subsequent ballistic missile submarines match or exceed her in size, but very few submarines of other types have ever done so. The only attack submarine to exceed her standard displacement is the *Los Angeles* class; and the only submarine to exceed her in both length and displacement other than the SSBN's is the huge Russian *Oscar* class cruise missile submarine.

And she remains a ship of superlatives. She has the highest surface speed of any submarine ever built,

USS Triton *(SSRN-586) on trials in 1959 doing what no other submarine could do – imitating a destroyer.*
(US Navy)

Second World War fleet sub radar picket conversion Spinax (SSR-489), showing antennas – air search on bridge, height-finder at stern, fighter homing beacon between. (US Naval Institute)

probably the only submarine in history to exceed 30kts on the surface. She is the only Western nuclear submarine with a multiple-reactor powerplant and indeed was barely beaten for the title of the world's first operational multi-reactor ship by the USSR's nuclear icebreaker *Lenin*, which went to sea fifteen days before *Triton* on 29 September 1959, but soon suffered a major radiation leak which left her unusable for years. *Triton*'s powerplant was, like every other detail of this ship, perfectly engineered and wholly successful, and may well remain the most powerful one in any Western submarine. Designed for 34,000hp, she reached 45,000 on her trials, according to Captain Edward Beach, her first commanding officer. Yet she was apparently the first – certainly the first Western – nuclear submarine to be permanently retired from service. Why?

Triton was designed to be, at least in many aspects, a submersible surface ship: able to match the performance of the fastest surface vessels, operate with them, carry out her military functions in support of them, yet avoid attack by disappearing beneath the sea surface as a fully functional submarine. Although Japan, Britain, and France studied close co-operation among surface ships and submarines during the inter-war years and built large high-speed 'fleet' submarines, these were limited in speed by their diesel powerplants. *Triton*'s only genuine predecessors were the British 'K' class steam driven submarines, designed for 24kt surface speeds to permit them to accompany the Grand Fleet in the First World War. Many of the same problems, of operating a submarine on the surface at these unusual speeds, that the 'K's encountered were also experienced by *Triton*. Like them *Triton* was given the lines of a fast surface warship. With her 37ft beam her fineness ratio was better than 12:1, and she was given a knife-like bow with the bulbous forefoot of a large modern surface warship. But unlike the 'K's she was a technical success.

The *Triton* (at least in these aspects of her design) was conceived as a radar picket, operating great distances from the carrier task forces she would support. Positioned between them and the enemy, she would be able, with her powerful and advanced radar, to provide early warning of incoming air attacks, control fighter interception at great distances, and submerge to avoid being attacked herself. Even the relatively slow aircraft of the Second World War had demonstrated the vital importance of radar early warning and radar controlled fighter direction in defeating their attacks. The Japanese *kamikaze* attacks against the US fleet off Okinawa had shown both the deadliness of the 'guided missile' and the great weakness of shipborne radar, its inability to see over the horizon and thus warn of aircraft approaching at extremely low altitudes. The first solution was to position 'radar picket' destroyers at a distance from the fleet; however, they were themselves vulnerable to air attack and suffered grievous losses at Okinawa. This suggested the concept of the radar picket submarine, herself able to avoid attack by submerging. Yet the twelve postwar US radar picket submarines were limited by their diesels to surface speeds little better than 20kts, making them unable to keep pace with the carrier task force they were protecting.

Hence the *Triton* – or at least her radar, air control facilities, and high surface speed. Yet, four months before her launching the Navy announced plans to end its radar picket submarine programme and retire or convert its twelve diesel boats, thus making *Triton*'s chief claim to fame and her sole intrusion on the consciousness of the public – outside of a flap over the Navy's slighting of Admiral Rickover by apparently reneging on a promise that his wife would christen her – her 1960 voyage retracing Magellan's circumnavigation of the world done almost entirely *submerged*, covering 41,500 miles in 83½ days, a substantial improvement on the previous record for distance and endurance submerged. In addition the conventional wisdom, often repeated in published sources, has it that *Triton* was actually built only to test its powerplant, which served as a prototype for multiple-reactor surface ship plants. For example, nuclear engineer Harold Hemond, in his bitterly critical article 'The Flip Side of Rickover' writes:

> The *Triton* (SSN-586) project concentrated attention on how to install twin reactor plants in a twin-shaft ship. The submarine did not need two reactor plants, but Rickover was anticipating the problems he would have installing multiple reactor systems in surface ships. Much effort was also devoted to the development of steam driven powerplant auxiliaries in lieu of electric motor driven auxiliaries with hopes that the on-board electric powerplant could be simplified. But steam-driven auxiliaries were never again used on submarines, and no significant mission could be found for the *Triton*.

Ultimately the authors have come to believe that this view is largely incorrect or at least misleading. But a study of the few published facts on the *Triton* yielded only a succession of question marks. Why was she in fact built? What was the potential value of her designed role as a radar picket? Was she ever tested or exercised in it, and if so, how did she perform? What was her actual performance at high speeds on the surface? Was there any further use made of her high surface speed after the radar picket role was dropped? And what was her performance submerged, in speed, manoeuvrability, and quietness? Was she actually used as an attack submarine, and if so how successful was she in that role? What was the value and significance of her twin-reactor powerplant? What

Triton *in drydock No 4 at Norfolk Naval Shipyard, summer 1989. Note her knife-like upper bow and bulbous forefoot. These surface ship features contributed to her incredible surface speed. However, the lack of bow buoyancy contributed to her wet decks at speed. Note torpedo tubes, retracted bow planes, and 70ft height from keel to bridge.* (Naval Sea Systems Command)

Triton's huge sail contained a conning tower. This was the last time this feature was used in an American submarine. (Naval Historical Center)

were her other unique technical features? What was her contribution to nuclear submarine development? In what roles did she actually function during her decade of active service? How well did she perform them? Did her size, speed, or other unique features make her especially well-fitted for them? Why was she decommissioned when she was? Was it because she was an expensive white elephant with no clear role, or was she a valuable unit?

Seeking answers to these questions yielded much fascinating information but few clear and definitive conclusions. Many historical documents were closed to the authors. Everything regarding nuclear powered ships and active duty submarine operations during this period is classified. Much is secret; the remainder can be declassified by request, a very burdensome process. Thus the authors went directly to individuals connected with her building and career. (The most valuable published sources are similarly based largely on interviews.) These individuals were not free to disclose the answers to some of our main questions – but were happy to enlarge upon her characteristics as a ship, our greatest interest.

Yet here, too, simple answers are few. The questions we have most sought to answer are questions of judgement. What were her proper or potential roles, how successful and valuable was she, why was she built, should she have been built, should she have been retired? These can only be answered by consulting the intentions of those who designed her and the experience of those who operated her. Yet she was surrounded by inconclusive debate during her design and building, and the testimony of her operators is filled with contradictions. One of her

Sailfish *(SSR-572), purpose-built radar picket diesel sub commissioned in 1956, showing height-finder antenna on pedestal aft of sail, and retractable air search on sail.* (US Naval Institute)

commanding officers, for example, began the interview by saying, 'You're going to have to really stretch to portray her a something significant,' but ended with, 'she was a valuable unit . . . she played an important role'.

The Fleet Radar Picket Requirement: The Submersible Surface Ship

Much of the paradox in this evaluation of the *Triton* lies in the fact that her design results from two distinct strands of development: her high speed radar picket characteristics, and her powerplant. The rather fortuitous joining of the two in a single ship resulted in the *Triton*. Both began at a time of tremendous scientific and thus military flux – both the nuclear bomb and the nuclear reactor were novelties a few years old, and the effect both would have on sea power seemed tremendous, if far from clear. The radar picket role, if of less practical significance to the actual building of the *Triton*, is responsible for her most unusual and interesting characteristics.

From the vantage point of 1993, the great lesson of the cold war seems to be the stability of nuclear deterrence, permitting an endemic state of limited conventional warfare. Even though both sides operated under this pattern from nearly the end of the Second World War, it was only genuinely accepted by Western military thinkers in the 1960s, and perhaps never completely accepted by their Soviet counterparts. For almost two decades, the overwhelming power of nuclear weapons, and insolubility of the problem of defence against them dominated military thought. The US Navy early sought to develop its carrier force as a strategic nuclear force. The USSR similarly early perceived nuclear weapons as the 'equaliser' which would enable it to defeat the West's tremendous naval superiority, developing first nuclear warheads for submarine torpedoes, then a series of very large and long-ranged cruise missiles launched from bombers, submarines and surface ships. These began to materialise in quantity by the end of the 1950s and remained the basis of the Soviet anti-carrier force for three decades.

Certainly the USN spent vast amounts of thought and effort of defending itself against this threat and may well have been able to defeat it – if attacked with conventional warheads. But from the moment of the Bikini nuclear tests defending the fleet from air or missile nuclear attacks seemed absolutely vital and perhaps ultimately impossible. This pessimism led to several concepts voiced repeatedly in the defence literature of the day, such as dispersing a fleet over very wide area to avoid its destruction in a nuclear blast, and dispersing its air arm throughout it among small VTOL carriers (the origins of the *Invincible* and the *Principe de Asturias*) or even smaller vehicles, *Spruance* class 'DDHs', SES or SWATH platforms, or even to ordinary frigates via some ingenious aircraft vertical recovery systems. Most of these vehicles or systems were actually built, but a more radical solution, voiced with considerable frequency, was to transfer the functions of the surface fleet to the nuclear submarine. Submarine aircraft carriers, landing craft, freighters, and anti-aircraft vessels were proposed; perhaps the *Triton* and the Soviet 'Echo II' class cruise missile subs represent the closest approach to the realisation of the nuclear submersible warship.

To return to the postwar problem of fleet air defence, the USN converted numerous *Gearing, Buckley,* and *Edsall* class ships to radar pickets. (Britain's four *Salisbury* class aircraft direction frigates were the only such purpose-built ships.) According to Norman Friedman in *Submarine Design and Development* the original concept of the radar picket submarine '. . . Project Migraine, called for the conversion of 24 fleet submarines to operate their SV air search radars at periscope depth. At the war's end four had been completed, and two more were in various stages of construction'. According to military historian James Mandelblatt, currently working on a history of USS *Requin* (SSR-481), preserved as a memorial at Pittsburgh, Pa, the term 'Migraine' was applied to the ten Second World War 'fleet submarine' conversions, which were not intended to operate submerged. All

Note the half-moon shaped well on Triton's sail. Her unusual SPS-26 radar could be retracted, necessary at high submerged speeds. This 3-D radar eliminated the need for a separate height-finder – note raised plastic shield over bridge. (US Navy)

carried powerful air search radars, separate height-finder radars, aircraft homing beacons, and internal fighter direction centres with complete facilities to plot raids and control interceptions. Internal space was required for the control centres and extra personnel; all had their stern torpedo tubes removed. The two 'Migraine I' boats used the stern room for berthing, while the new 'Migraine II' boats, converted before completion, located the air control centre there. These were *Tench* and *Balao* boats, but the 'Migraine IIIs', older 'thin-hulled' *Gato* class boats, were much more extensive conversions, lengthened 24ft to accommodate the amidships air control centre. Possibly they were chosen as less suitable for GUPPY attack sub conversions due to their lesser diving depth. Radars were mounted on a high streamlined sail and raised pedestal aft, possibly to admit operation in an awash condition or reduce problems with electric connections through the hull which gave much trouble in the four earlier boats.

Mr Mandelblatt says *Requin* operated with the Sixth Fleet. He notes that SSRs were operated in pairs so that if one were forced to dive the other could provide continuous radar early warning coverage. This of course was the major flaw with the SSR concept; when she dived to avoid attacks, she ceased to function as a radar picket.

Two new construction diesel pickets were built in the 1950s, *Sailfish* and *Salmon*, similar in most respects to the 'Migraine III' boats. According to Commander Dale Eastep who served on both *Salmon* and *Triton*, they were built at the insistence of the naval air community and funded out of its allocations. He describes *Salmon* as a very valuable unit detecting aircraft hundreds of miles distant, able to transmit the data for display in the carrier's CIC. Her equipment included guidance radar for the Regulus missile. She suffered from a number of serious material deficiencies including her TACAN and new powerful (2400hp) Fairbanks diesels and generators. After the conclusion of the picket programme she was converted to carry and land a 40-man commando force.

Her diesels were still inadequate to match the speed of a surface task force. According to Dr Friedman 'there were several projects for alternative SSR powerplants. One of them became the pressure fired steam engine, which was actually used in two classes of ocean escort (surface) vessels.' This certainly attests most strongly the need felt by the navy's carrier air community for some means of radar early warning at this period; for it seems obvious this steam submarine would suffer from many problems, similar to those of the 'K' class in the First World War. It would require large air intakes and exhausts which would need to be sealed before diving and preclude snorkel operation. Residual heat would need to be dissipated after diving, steam would have to be gotten up after surfacing. And after diving she would be reduced to operating on batteries like any other submarine, losing most of her mobility. And if she could make destroyer speeds on the surface, presumably it would require similar fuel consumption. The steam SSR would presumably be operating at a distance from the carrier group and its fuel supplies, so use of high speed would be severely limited.

Obviously, the nuclear submarine was the answer: steam turbine propulsion with unlimited endurance at top speeds surfaced or submerged. The success of the USS *Nautilus*, 'under way on nuclear power' on 17 January 1955, and the vital need for radar early warning led the USN's carrier aviation community to seek the authorisation for a nuclear radar picket submarine able to match task force speeds – the *Triton*. In 1956 Chief of Naval Operations Admiral Carney's request for appropriations included three nuclear submarines: two *Skate* class and a 'large radar picket using the advanced two-reactor system'.

She was laid down on 29 May 1956. Yet four months before *Triton* was launched on 19 August 1958, the Navy announced plans to end the SSR programme. (In 1959 *Requin* was converted into an attack submarine, still minus her stern tubes.) Why had the support for this programme, so strong two years earlier, evaporated?

The answer is the superiority of airborne early warning. The USN began to take delivery of Douglas Skyraiders carrying the AN/APS-20 radar for AEW in 1948, quickly demonstrating one great advantage over the shipborne radar, its much greater radar horizon covering an area with a radius of hundreds, not tens, of miles. Not blocked by the curve of the earth, a single airborne radar provides more complete low level coverage than a picket line composed of a number of sea-level radars. The AEW aircraft are also much cheaper and more numerous than surface ships and can be more easily defended, operating behind the fleet's fighter defences, than distant picket ships.

Why then were destroyer and submarine radar pickets pursued at considerable expense during the early 1950s? Probably because of their fighter direction capabilities; the AD-5W Skyraider lacked their height-finders, homing beacons, and extensive plotting facilities of their air control centres operated by numerous personnel. The USN's land-based PO-2W Constellation of 1954 could duplicate the air control function of a picket ship, with its height-finder radar and 26-man crew. While the WF-2 (or E-1B) made substantial improvement on the Skyraider in 1958, but what permitted the carrier-borne AEW aircraft to equal the capabilities of the air control centre was the digital computer. Married to a very powerful air search radar, advanced data processing enabled the E-2 Hawkeye to track hundreds of targets and control numerous interceptors simultaneously; this system entered service in 1964.

Before the E-2, only the surface ship, submarine, or land-based aircraft picket could provide an independent control centre operating at considerable distances from the fleet, able to plot interceptions and control strikes, thus greatly extending the effective range of the carrier's aircraft. And submarine pickets could operate even more independently, providing electronic reconnaissance from the heart of enemy-controlled waters. Cmdr Eastep recalls the *Salmon* playing this role in the 1950s, operating off Hainan Island, observing Chinese air and radar activity 'a mile from the beach'.

But another reason for the disappearance of surface ship pickets as a type was the advent of three dimensional radars, able to provide the altitude data necessary for fighter direction, and their provision to all major surface warships. The original 3-D radar, scanning mechanically in azimuth and scanning in altitude by modulating the frequency of the beam, was the SPS-26, tested in 1953 and mounted in the DL-1 *Norfolk* in 1957. *Norfolk* achieved ranges of 110 miles; the standard SPS-52, which followed in 1963, is credited with a range of 60 miles against small high speed targets, much longer ones against aircraft.

The only other ship to carry the SPS-26 was the *Triton*; the large half-moon shaped well permitting it to be retracted into the sail, unlike those of most earlier SSRs, is visible in the accompanying photographs. The authors were very curious regarding the *Triton*'s contribution to the development of this radar but it was apparently not significant. A 3-D radar of course eliminated the need for a separate height-finder antenna, mounted aft of the sail in the older SSRs. The SPS-26 was removed after a few

Triton's *engine room control station – roughly half her 175-man crew served her twin reactor, twin turbine powerplant.* (Naval Historical Center)

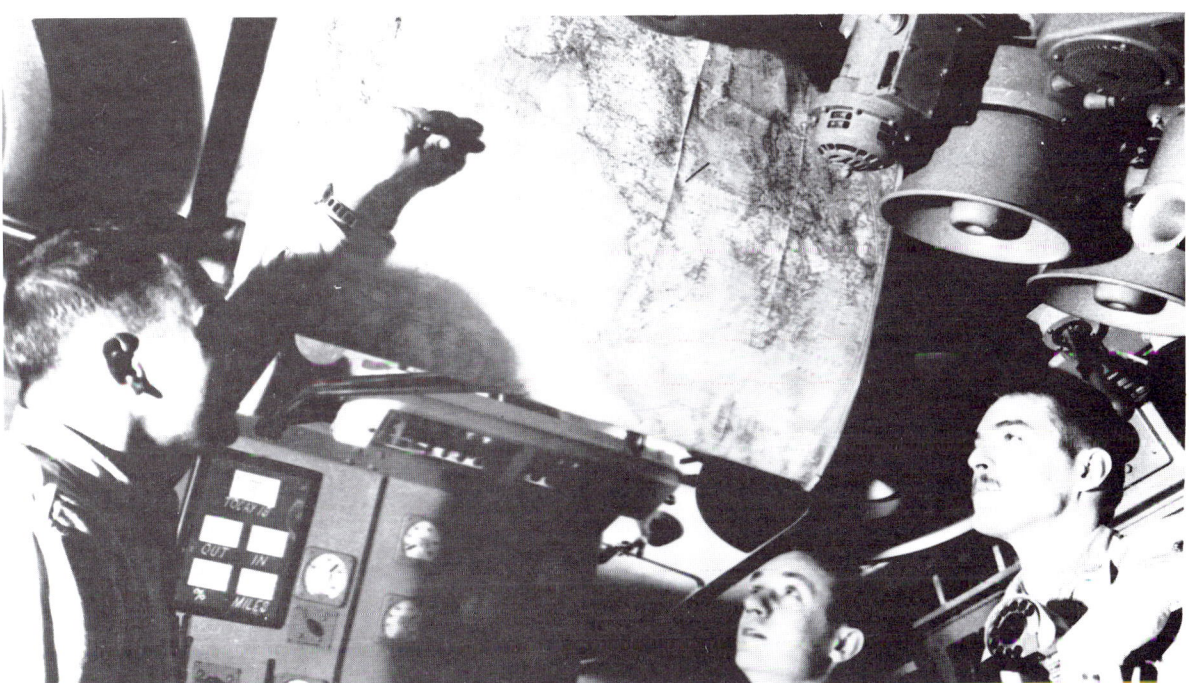

years (before her 1962 overhaul) and replaced with a BPS-2, the same search radar used on the *Salmon*, according to Commander Eastep. She retained this radar and thus substantial air search capability to the end of her career.

Again, this role was only a secondary reason for building *Triton*, and she exercised it very briefly; yet for a time it appeared to be of critical importance. The necessity and weakness of radar picket lines in the absence of AEW was demonstrated powerfully at the Falklands. There radar shaped sea and air tactics in a multitude of ways, including the constant exploitation of its limitations at low levels. The four British nuclear submarines paralleled the SSR tactics of the USN in the 1950s while maintaining their close blockade of the Argentine surface fleet 12 miles offshore. They tracked merchant shipping and radar equipped Argentine aircraft, as well as providing early intelligence of Argentine air attacks. Dr Friedman says: 'presumably they spent most of their time nearly awash, with their radar antenna projecting above the surface'.

The chief source of early warning for the V/STOL carriers however was provided by a picket line of Type 42 destroyers, ideal ships with their area defence radar and missiles. They suffered from the inevitable weakness of vulnerability of the pickets themselves, numbers required for an effective barrier, and its vulnerability to low-level penetration or outflanking. During the first Super Etendard/Exocet attacks, three Type 42s were spread out 15 miles apart, 20 miles west of the carriers. The Super Etendards were detected at sea level 25 miles out by the picket destroyers which failed to vector the Sea Harrier CAP in for an interception or acquire them with their Sea Dart systems. One picket, the *Sheffield*, was sunk by an Exocet, only detected visually a mile from the ship. During the second attack which sank the *Atlantic Conveyor* only one Type 42 picket was available, 25 miles west of the carriers, and was outflanked by the Etendards, attacking from the north. During the third attack, two Type 42s formed the picket line, 20 miles west of the carriers with the *Avenger* passing 10 miles to the south, fortuitously extending the line. *Avenger* attracted the attack, but this time picket *Exeter* detected the Exocet (which *Avenger* claimed to have destroyed with 4.5in gunfire) and shot down one of a group of Skyhawks, led in by the Etendard. Thus the radar picket mission: difficult, dangerous, but vital – in the absence of AEW.

The Submarine Advanced Reactor

Triton owed her fine lines, high surface speed, radar and large air control centre to the fleet radar picket concept. In fact, she owed her existence much more to the conception, design, building, and need to test her seminal powerplant, the so-called Submarine Advanced Reactor (SAR). As we have stated, Admiral Rickover's detractors tend to see the real reason for the *Triton*'s construction as the testing of the twin-reactor powerplant for use in the nuclear-powered surface navy he later sought. Perhaps the SAR, as the first successful multiple-reactor plant, provided proof of the concept; but it was not the prototype for any specific surface ship reactor and it was first conceived in reference to very different concerns.

According to Captain Edward Beach, the first CO of the *Triton* and very early supporter of the concept of the nuclear submarine, the original conception of the SAR was based on advanced physics, the exploitation of high speed neutrons to gain increased power and efficiency. The type of neutron producing fission could be controlled, among other things, by the type and density of the coolant. The pressurised water reactor which became the standard submarine powerplant utilised at so-called 'thermal speeds', and was first known as the STR, Submarine Thermal Reactor; while the sodium cooled reactor was eventually tested in the *Seawolf*, the Submarine Intermediate Reactor (SIR), produced neutrons at intermediate speeds. The Submarine Advanced Reactor would use neutrons at even higher speeds.

Commander David Leighton, project officer for the SAR and long-time collaborator with Admiral Rickover in many areas of the naval and civilian nuclear power programmes, agrees that initially the nature of the SAR was wide open with liquid metal, even gas coolants, being discussed. Both agree, however, that it was very early determined that the goal of the SAR would be higher performance, rather than advanced technology, in the form of a twin pressurised water reactor plant.

Certainly the experience of the *Seawolf* bore this out. Whatever the potential superiority of the sodium-cooled reactor in size and weight to power, this highly reactive

Lord Mountbatten, David Leighton, and Admiral Rickover at the controls of the S3G land prototype of the Triton's *reactor at West Milton, New York, October 1958. (Courtesy of David Leighton)*

substance proved dangerous in use. *Seawolf* apparently suffered serious material defects, sodium leakage and embrittlement of piping, after a year's operation. And of course, the sodium could never be allowed to cool; it would solidify and cease to act as a coolant. Distilled water on the other hand, can be manufactured on shipboard and readily replaced.

The SAR concept embodied the same blend of conceptual foresight and caution in execution which marked Rickover's career. The STR developed by Westinghouse's Bettis Laboratory required much groundbreaking translation of pure science into practical engineering (the production of zirconium for its fuel rods is an example) but it proved an outstanding success: reliable, safe, maintainable, as was the *Nautilus* which used it. Rickover has often been criticised for sticking with the large, noisy, inefficient pressurised water formula – variants of the Bettis design powered all but a handful of US nuclear submarines until the *Los Angeles* of 1974 – but the USN has never had a nuclear accident. On the other hand, Commander Leighton points out that no one could foresee just how reliable it would actually prove to be. From the beginning, Rickover was determined both to see an alternative design under development, in case the Bettis (S1W) proved unsuccessful, and to see that this design embodied twin reactors, in case the nuclear reactor in general proved less dependable than it has. Possibly every nuclear submarine would require two reactors; only experience could tell. Rickover was also concerned to take the initiative and forestall the Navy's requesting a multiplicity of reactor designs, each for a separate application.

Besides, it is often forgotten how revolutionary a warship the *Nautilus* was, in size as well as propulsion. At 4100 tons and 320ft she was a huge submarine, exceeded in displacement only by the *I-400* aircraft carrier submarines, far larger than any attack submarine then built. This flew in the face of the submarine tactical thinking of the day. Traditionally of course, the submarine was an expendable weapon of attrition; the latest conventional designs such as the *Tangs* emphasised reduced length for greater manoeuvrability. Few could appreciate how completely the *Nautilus*' submerged speed and endurance made her an insuperable opponent for the ASW of 1955, in spite of her size and noise. *Nautilus* detected surface ships far beyond the range of their own sonars, overtook them at 18kts, and quickly broke counter-attacking destroyers' sonar contacts with a burst of speed. Nuclear power would change the nature of the submarine, from a barely mobile weapon of opportunity to the primary vehicle for ship-to-ship combat, with capabilities well worth her great expense. At the Falklands, one kill by a British nuclear submarine left the entire Argentine surface fleet with no alternative but to take refuge in port. The US 'maritime strategy' envisioned US nuclear subs preceding carrier strike forces into Soviet Arctic Ocean 'bastions', to find and kill their opposite numbers – playing a role analogous to fighter aircraft in establishing command of the air.

Regaining the advantage of small size in a nuclear submarine is perhaps not a feasible proposition. Shielding requirements remain the same, even when the size of the reactor itself is reduced; power is lost for relatively small

Skipjack (SSN-585), Triton's successful competitor as the pattern for the future. She achieved higher speed with less power with her 'Albacore' hull. The deep troughs behind her bow wave show the amount of energy she wasted to wave making on the surface. (US Naval Institute)

gains in size and cost. Thus while the Navy demanded the small *Skate* class with half the horsepower of the *Nautilus*, only days after *Skate* was laid down, Rear-Admiral Frank Watkins stated she would be too slow to avoid underwater detection.

Commander Leighton says that Rickover saw advantages in high speed and large size from the beginning; *Triton*'s origin lay in his interest in achieving speeds of up to 35kts by means of a three to four times increase in power. Hewlett and Duncan, in their scholarly *Nuclear Navy 1946–1962*, say the design study of this vessel 'an improved SSN embodying very high submerged speed' was authorised by the CNO on 22 October 1951. Yet opposition to the large fast submarine was very strong. Admiral Watkins felt such a vessel fitted neither into present or future tactical concepts. 'It looks like grandstanding and has as its only purpose techniques for gaining the dollar support of the AEC [Atomic Energy Commission].' Several admirals insisted at a Pentagon meeting in 1953 that 'the Navy simply had no use for a submarine with a displacement of more than 4000 or possibly 5000 tons' which would be vulnerable to ASW, hard to manoeuvre, and a large target.

However, both Commander Leighton and *Nuclear Navy* authors agree Rickover had a very important 'political' goal in pursuing the SAR. He was early determined to bring General Electric's Knolls Atomic Power Laboratories at Schenectady, NY into the business of naval reactor design, as a second source of expertise besides Westinghouse's Bettis Lab. But whereas Bettis was committed to submarine reactor design from the start, Rickover encountered considerable resistance at Knolls. Commander Leighton says GE intended Knolls to work on a variety of projects; its people had a bias towards scientific research as opposed to practical engineering; it was Rickover's idea to turn their abilities exclusively to submarine reactor design. Ultimately he was successful but not without much delay and controversy. Knoll's first reactor was the S1G prototype of *Seawolf*'s short-lived, flawed, sodium-cooled S2G. Rickover thus found himself with the complex aim of keeping Knolls in submarine reactor design but converting it to pressurised water technology in the process. SAR was to be the instrument for both. First the Navy, determined that the fleet's nuclear submarines would be small attack boats, resisted. Rickover argued that it was indeed in the Navy's interest to keep the Atomic Energy Commission, which supported SAR, involved. In 1953 he obtained the Navy's agreement to pursue SAR in return for a small sub for production; then-Lieutenant Leighton was named SAR project officer.

The next problem was with Knolls itself. To quote *Nuclear Navy*, 'At Bettis he was largely successful in imposing his principle of full personal responsibility while at Knolls he had only limited success after many years of argument'. The failure with the SAR was not in any defect in the S4G plant carried by *Triton* but in the slowness with which it was achieved. Rickover's intimations of the value of a high speed submarine were vindicated by the single Bettis S5W reactor mounted in the streamlined 'Albacore' hull of the *Skipjacks*. High speed was achieved by the simple, elegant route of decreasing resistance, not increasing power. Still, as Commander Leighton puts it, 'If the S5W had turned out to be a defective design, we would have built plenty of S4Gs – we needed two lines of development'. And eventually a large submarine would be required. But it would not be based on *Triton*; Rickover's problems with Knolls would prevent the SAR/*Triton* from appearing in time to be a serious competitor to the S5W/*Albacore* hull combination.

Thus, it was only determined SAR would be water-cooled in 1954; it was still in the paper stage in 1955. Commander Leighton says Bettis could have produced it with less time and cost but 'Rickover was determined to bring Knolls into water-reactor technology'. Hewlett and Duncan say further resistance from the Navy might have led Rickover to terminate the project except for the need to develop a competitor to Westinghouse and his perception that the large fast submarines would best meet the Navy's future needs.

But what were these needs? Or rather, what immediate justification could be given for the *Triton* as the vehicle which would be available when the needs became apparent? Rickover early proposed she be completed as a guided missile submarine, carrying the new Regulus. Others argued that the Regulus carrier, as a 'ferry boat', would have no need of size or speed – indeed the *Halibut*, the only nuclear Regulus carrier, a converted *Skate*, was much more modest. In 1955 Admiral Carney rejected the concept, stating that nuclear submarines would be confined to the attack, hunter-killer, and radar picket roles. At this point naval air interest in the SSR was at its peak, *Sailfish* and *Salmon* were under construction, and the following year *Triton* was requested as a radar picket. As Commander Leighton put it: 'We were looking for a customer.'

By the autumn of 1955 the SAR project had gained momentum. Rickover approved Knoll's basic design in October. Meanwhile, Electric Boat Co had begun design on the hull and procurement contracts for major components were placed. *Triton* was laid down by 28 May 1956, but 'the pace of development was never fast' and the S3G land prototype twin reactor plant was only operational at West Milton, NY by August 1958, the same month *Triton*'s hull was launched at Groton. As *Nuclear Navy* puts it: Knoll's lack of expertise 'delayed completion of the new reactor design almost beyond its usefulness.'

At this point multiple events began to overtake the *Triton*: the concept of the E-2, AEW plus the digital computer, was proposed, and the radar picket role seemed nearly obsolete. The *Skipjack* was launched, faster *submerged* than *Triton* would ever be, above or below the surface, and on less than half the power. And, as Hewlett and Duncan say: 'the S5W, already in multiple production at Bettis, had preempted any hopes at Knolls that the S4G would be the standard propulsion system of the nuclear fleet'.

Also, a major controversy over strategic missile submarines developed, as the supersonic Regulus II achieved success. Rickover again proposed *Triton* for this role; pressure for the Regulus was intense but brief, ending when it was cancelled in favour of Polaris in December 1958. It had been decided to spare no effort for the Fleet Ballistic Missile, and by March 1957, the basic para-

meters of the system, using the S5W and *Skipjack* hull, were determined. Indeed, *Scorpion* (SSN-589) was cut in two and the 16-missile bay was installed, to complete her as the *George Washington* in 1960.

Triton's completion was slowed by competition for design staff and shipyard personnel, first with *Skipjack*, then *George Washington*. Still, in 1957 Rickover convinced the CNO, Arleigh Burke, that *Triton* should be completed to provide an evaluation of the twin reactor system. He was also concerned that if the ballistic missile sub was required to operate under the Arctic ice cap – which he felt was desirable – it would need the S4G. Two reactors would be necessary to provide reliable power where surfacing would not always be possible in an emergency; twin screws would be necessary for manoeuvring in the tight quarters and high winds of openings in the ice pack. But the S4G was far from proven, pressure was great to develop the Polaris sub, and so when Burke approved her ship characteristics in June 1957, the under-ice capability was deleted.

But how was *Triton* to be used? Polmar and Allen in *Rickover: Controversy and Genius* report an interview with Captain Beach in which he describes four alternative missions proposed by Walter Dedrick (then prospective CO of the *Halibut*): 'advance sonar scout for the surface

Triton *on ways, 19 August 1958 just prior to launching – note propeller and sail cut down to clear ways.* (General Dynamics Corp)

fleet; guided-missile ship; command and staff ship; and high speed minelayer'. The first sounds very much like the role the *Los Angeles* class was originally intended to fulfil: a high speed anti-submarine escort, able to accompany carrier groups submerged. Captain Beach himself urged she be equipped to rescue and tow further nuclear submarines disabled under the ice pack. The command ship role was in fact tested, and *Triton* was prepared for under-ice operation, as will be described.

However, none of these roles developed into anything permanent or important. Was it worth spending $109 million to build the *Triton*? Commander Leighton answers: 'We got what we wanted', which was above all to develop General Electric as a resource, eventually producing the reactors for all nuclear surface escorts except the *Long Beach*, as well as for the *Los Angeles* and *Ohio* classes. She did provide also the first crucial

Triton *after launching*. (Naval Historical Center)

Triton on the surface. Her knife-like bow contributed to her high speed but, lacking sufficient buoyancy, left her decks awash and her bridge wet. (General Dynamics)

experience of actually operating a multi-reactor plant at sea. Captain Beach says concerns over the question of synchronising levels of power from two reactors in parallel were put to rest. The S4G proved uniquely flexible; each reactor could provide steam to a single turbine, each could supply both turbines, or both could supply both turbines at once.

But GE's next service powerplant was the D2G, first used in *Bainbridge* (laid down 15 May 1959). Was this a repeat of the S4G, and the real purpose for which it was built? According to Commander Leighton, no. The *Bainbridge* reactor's prototype, the D1G (Knolls began preparations to build it in 1957) was of a completely different design. And the D2G was at 60,000hp – far more powerful. In fact, the first surface ship plant was C1W of the *Long Beach*, laid down December 1957.

As Commander Leighton points out, however, Bettis and Knolls were not competing in the sense of retaining anything like proprietary information; they continually reviewed and commented on each other's work. While the S4G was not a direct prototype for any other plant (and incorporated such unusual features as a unique system for loading the zirconium-clad fuel elements) the experience gained with it was a valuable contribution to the work of both labs.

Triton: TECHNICAL SPECIFICATION

Displacement:	5940 tons, surfaced
	6670 tons, submerged
	(Actual figure reported be be 7900 tons)
Length:	447½ft oa
Beam:	37ft
Draught:	24ft
Powerplant:	2 General Electric S4G reactors,
	2 General Electric geared turbines
Speed/surface:	27kts at designed 34,000shp
	30kts at +45,000shp, trials
Speed/submerged:	probably 27 to 30kts
Crew:	172 as radar picket, 159 later
Armament:	6–21in TT (4 bow, 2 stern)
	16 Mark 37 torpedoes
Search radar:	originally SPS-26 3-D, later BPS-2
Sonar:	BQS-4
Fire control:	Mark 101

Building, Trials, Performance

Triton was laid down on 29 May 1956, launched 19 August 1958, and commissioned 10 November 1959. Meanwhile *Skipjack* was laid down on the same day, launched 26 May 1958, and commissioned 15 April 1959. *Triton*'s successful competitor as high-speed submarine was quickly overtaking her. And *George Washington*, laid down as *Scorpion* on 1 November 1957, was launched 9 June 1959, and fired her first Polaris missiles in July 1960, supplanting *Triton* in the major role for the very large submarine.

When *Triton* slid down the ways she was a strange combination of the conservative and revolutionary, of perfect engineering and occasionally flawed ship design, and was attended by a strange mixture of excitement at her superlatives and growing doubt as to her utility. For many of her unusual features the best justification that could be made was that future trends were not clear, or at least were not clear when she was planned. She was the last American submarine built with an extensive external superstructure or casing, twin shafts and screws, a conning tower, and stern torpedo room, as well as steam driven engine room auxiliaries.

In his *Around the World Submerged*, Captain Beach tells of his eight weeks training on the original S1W in Arco, Idaho, followed by eight weeks on the S3G at West Milton. 'Already, she had created difficulties because of her great size. Among problems faced by the builders was that her huge bow blocked the space reserved for the railroad which ran just forward of the building ways.' To clear it, the lower part of her bow was cut away and replaced a few days before launching. Likewise, the last 50ft of her stern was constructed on the adjacent ways and lifted into place by crane. Finally, the top 12ft of her huge 20ft × 75ft sail had to be cut off so that she could clear the overhead cranes of the building ways; it was reinstalled after launching.

Captain Beach writes that she stood seven stories above the ground on the ways and describes standing on her bridge at the top of her sail as 'like standing on a three-story building'. He begins his narrative of *Triton*'s world cruise with an inspection of the boat's interior spaces, beginning at the control room beneath the sail, where *Triton* had three decks. The control room was half cylindrical in shape, roofed by the pressure hull, and bisected by the periscope and radar mast wells.

Above the control room was the conning tower, which must have given *Triton* a substantially greater periscope depth. This seems to have played an important part in her operations. Her second CO, Captain George Morin, told us that in her rare use in the radar picket role, she operated submerged, with antennas extended above the surface. She also operated submerged but broached, with the top of her sail at the surface. During her around the world cruises she was twice operated in this mode, once to permit a sick sailor to be taken off by boat, and later a Navy information officer with film of the voyage. Captain Frank Wadsworth, her fourth CO, describes operating her broached with radar and antennas extended. This permitted her to operate at much higher speeds in this mode than she could fully submerged; at these speeds her electronics masts would have been bent over backwards. On the other hand, her very long hull increased the possibility of breaking the surface with her bow or stern at relatively slight up or down angles.

Also, the division of ship control facilities between control room, conning tower, and navigating bridge when surfaced created problems. Captain Beach says the conning tower was particularly inadequate – all fire control equipment was in the control room below for example. When she was redesignated an attack sub on 1 March 1961, his suggestions for making her much more effective in this role included greatly enlarging it, which could easily have been done by removing the huge radar which occupied so much of the volume of her sail. But the SPS-26 was simply replaced with another large (but not 3-D) air search radar. Perhaps there was never much serious intent to convert *Triton* to an attack submarine; we believe electronics dominated her later operational career.

Forward of the control room and beyond a watertight bulkhead was crew's berthing, and before that the forward torpedo room with four tubes, a relatively small complement of torpedoes, sonar equipment and more berthing. Aft of the control room was the officer's berthing compartment containing the captain's cabin. Captain Beach told us this was a reversal of the usual practice. Placing the spaces *aft* of the control room, between it and the radar spaces, meant a constant inconvenient passage of personnel. He says it was done to place the captain's quarters closer to the large CIC/air control compartment, which must have been on the deck below.

Aft of this multi-deck area were the reactor and engine spaces, taking up a very large proportion of the ship's length. First was the number one, then the number two reactor compartments, then the number one engine room 'the largest compartment in the ship, in cubic volume not far from the entire displacement of a World War II submarine' containing the starboard turbine and reduction gear set, then the port engine room.

Captain Beach notes *Triton*'s powerplant utilised such space-consuming features as large de-aerating feed tanks, common to surface ships but normally dispensed with in submarine steam plants. We questioned all those interviewed regarding the unique features of *Triton*'s plant, including her unique steam driven auxiliaries. Many accepted the interpretation that as *Triton*'s plant was intended as a prototype for surface ship plants it was intentionally made consistent with surface ship practice in as many ways as possible. Commander Leighton says it was simply intended as an experimental alternative to the other plants being developed. Beach and Leighton agree that steam driven auxiliaries are in theory more efficient with less power loss than electric motor driven ones, but perhaps the latter were more maintainable and compatible with submarine design.

Finally comes *Triton*'s stern torpedo room, with two tubes. This may have been a consciously conservative feature – Commander Leighton says it was not absolutely clear straight-running torpedoes were obsolete. Captain Beach says the *Skate*s had a stern torpedo room like that of the last twin-screw diesel boats, which contained small tubes intended to fire defensive ASW or anti-escort homing torpedoes. Was *Triton* given a stern torpedo room simply because the hull form with its long narrow stern provided the space, and a full-sized one because her size permitted it?

Certainly she was not designed with much attention to the torpedo attack role. Captain Beach says her maximum load was sixteen torpedoes, very little for her size, when Second World War subs managed to stow as many as twenty-eight. He notes she could have carried double her load of Mark 37 torpedoes in her forward torpedo room if only 18in more space could have been provided by the moving of a single interfering girder. He attributes the presence of a number of such faults in her design to the priority given the *Skipjack*s, to which the best design people working on the *Triton* were reassigned.

In his description of her trials, Captain Beach several times writes vividly of the tremendous power and noise of *Triton*'s engines:

The way she leaped ahead when the power was applied made my heart leap too; we could actually feel the acceleration as we gave her the gun.

I left the engine room and proceeded aft through the remaining engineering spaces, finally reaching the after torpedo room. There all was calm except for the noise of the two huge propellers whirling away just outside . . . only a few feet from me, spinning with violent energy, driving water aft at an unprecedented speed and putting more horsepower into the ocean than any submarine had ever done. I could feel the induced vibration shaking the entire after structure of the ship. The noise of the propellers and the roar of the water as it raced past our hull were almost as loud as the machinery a few compartments forward.

Plainly, performing like a destroyer meant making the same noise, from reduction gears, circulation pumps, turbulence, and propeller cavitation. But what was her actual performance? She is usually described as having a top speed surfaced of 27kts, submerged 20kts. Captain Beach told us it was hoped to get 30kts on the surface but only 27 was achieved at first. He says that Rickover then directed that power be increased by increasing the temperature difference between the hot and cold legs of the reactor's cooling circuit. In the pressurised water reactor, the water serves as coolant, as the medium for heat transfer to the steam generator, and as a moderator of the fission process, reflecting neutrons back into the core to keep the reaction going. The colder the incoming water, the denser it is, and thus the greater the reflection of neutrons and the more intense the nuclear reaction. The temperature difference was increased very carefully in small increments until *Triton*'s powerplant was producing 45,000shp and a speed of 'well over 30kts'.

Unfortunately this never earned her the title of fastest submarine in the world. *Skipjack* had already exceeded this speed *submerged*. However when we asked her top speed submerged, Captain Beach described it as 'about the same' as her surface performance, and indeed one of her COs described her as faster than any other nuclear boats of her day *except* the *Skipjack*s.

Operating at these surface speeds was not without problems, however. It is often forgotten that a characteristic which has contributed much to the submarine's achievements in both world wars is its surface seaworthiness. Their ability to operate in heavy weather, coupled with the endurance provided by their diesels made relatively small boats effective in mid-Atlantic, and capable of patrols of 200 days. Submarines actually had an effect much like that predicted by the *Jeune École* for the steam torpedo boats of the previous century. However dangerous these craft might have been under ideal conditions, they were simply too short-ranged and too fragile to keep the sea. Structurally immensely strong, lacking light superstructures and the need for crews to work and serve weapons on exposed decks, submarines were able to operate on the surface in rough weather and escape the very worst by submerging. Giving submarines much sheer and flare was unnecessary and a source of increased resistance submerged. There was no point in keeping seas and spray off her decks unless there were men and equipment out there.

But suppose there are? *Triton*'s only real predecessors in the role of submersible surface ships were the steam-powered 'K' class. These ships carried, on top of a long, low, submarine hull, a long superstructure with bridge and twin folding smokestacks projecting above it. Originally carrying three guns, above water torpedo tubes, even depth charge throwers at the main deck level, they resembled submersible destroyers, able to move at 24kts and defend themselves on the surface while doing so, and then disappear beneath the surface and act as fully capable submarines when desired. Unfortunately, at their remarkable high speeds, they dug their bows under, inundating their decks, throwing cascades of spray over their upperworks, rendering them almost untenable and their impressive surface weaponry unusable. First the guns and external tubes were removed for use in Q-ships; for a while their surface armament was limited to the 3in AA gun on the superstructure while a variety of modifications were tried. Huge raised bows were added to provide the buoyancy necessary to ride over waves and the flare to direct spray off their decks and bridges. Bridges and stacks were then raised an entire level, and gun armament was finally reduced to a single 4in plus the 3in AA, both mounted on top of the superstructure. The last variant,

British 'K' class steam powered sub on the surface with her original low bow and deck mounted guns – obviously unserviceable due to wetness of deck at this high speed. (Imperial War Museum)

K-26, returned to a three 4in armament, with all guns mounted on the upper deck of the superstructure.

Triton similarly dispensed with all the familiar features of surface ship design intended to keep decks dry at speed, and experienced almost identical problems. Captain Beach told us that as she reached high speed on her trials she immediately drove her bow straight under. 'Her extremely slim bow had most of its buoyant volume well aft, at precisely the point where the maximum hollow of her bow wave occurred at high speed. Thus she lacked buoyancy exactly where needed.' The only exposed men and equipment aboard *Triton* were nearly thirty feet above the sea surface at the top of her sail. But at 30kts, this was not enough. Captain Beach describes conditions on her bridge at full power during her trials:

> Already at full speed – about half power – *Triton* was riding with her bow still a foot or two out of the water. Occasionally, a roll would break over the deck and sweep aft, bursting with a cascade of spray against the bottom of the sail.
>
> The increased drive of the engines began to be noticeable and in a moment the first really big sea hit us. The bow spray spouted above our heads. Water dashed high over the bridge pelting down on top of the lookouts . . .
>
> The forward part of *Triton*'s bridge was fitted with a transparent plastic bubble, and under this van Leonard, Admiral McCorkle, and I huddled for protection . . .
>
> The spray increased, soon there was a steady stream of white water squirting high above our heads. Then with a swoosh, green water swelled up over the sides of the bridge coaming . . . Simultaneously, solid water poured over the top of the bubble like Niagara Falls . . . the lookouts had given up, turning their backs, while Harris gasped for breath, cupping his hands over his eyes in an effort to maintain a lookout ahead . . .

'K' class modified with raised bow and guns on superstructure. Here K-7 shows a temporary modification with a canvas screen placed on top of her bridge. Later, bridge and stacks were permanently raised to a higher level. (Imperial War Museum)

The immediate solution was to increase the volume of the bow buoyancy tank; that is, the space within the casing enclosed to form the tank was extended without altering the external form of the bow. Beach recommended more extensive changes – at the very least the size of the tank should have been increased even more, and ultimately he feels she should have been given a flared bow – but he says it was judged not to be worth the cost.

Another solution Captain Beach suggested was to mount a small hydrofoil or bow plane under her keel to lift her bow higher at speed. He made the experiment of rigging out her bow diving planes at ¼° up angle while maintaining high speed on the surface. This brought her

Modified 'K' class at speed showing behaviour with her raised and flared bow – note dryness of guns and bridge watch in their new higher positions. (Imperial War Museum)

Captain Edward L Beach on Triton's *bridge at the end of her circumnavigation. Note the retracted Plexiglas bubble, height above the sea, extreme sharpness of bow.* (Courtesy of Captain Beach)

bows right up; unfortunately, as speed increased the bow would rise up enough to lift the planes right out of the water. Then, deprived of their lift, Triton would fall back with a tremendous crash, threatening to damage the planes. Captain Beach says he wished he had worked this out in time to have tried it on her trials; he could not get this adopted either. However, it seems to have been a simple and straightforward, if unusual, solution. It seems to have been followed on the AAVP-7 (or LVTP-7) amphibious vehicles. Designed to transit 10ft seas and heavy surf, they were given an extendible bow plane, to counter their own tendency to push their noses under.

Around the World Submerged

On her trials Triton experienced a single mechanical deficiency: a spring bearing on her long starboard shaft was not provided with sufficient cooling and overheated at maximum rpm. She also experienced a single handling problem, when the submerged emergency power reversal was tried. Submarines tend to be unstable in this manoeuvre, and to sink deeper at the stern as the ship backs down submerged. This instability proved unusually great, perhaps because of Triton's long narrow hull; she began to oscillate badly and sink stern-first. A quick return to two-thirds ahead regained control.

Captain Beach says Triton's technical perfection gave him 'a new lease on life'; his last experience in submarines had been in the new *Tang* class *Trigger* which suffered disastrous breakdowns of her novel lightweight radial diesels. But whereas *Trigger* was broken down and drifting on her shakedown cruise to Rio de Janeiro, *Triton* went around the world on hers. Captain Beach surmises that the immediate purpose of the trip was to impress Premier Khrushchev at the summit meeting planned for May 1960, which was of course called off due to the shooting down of a U-2 reconnaissance plane over Soviet territory. Captain Beach says presidential press secretary, James Hagerty, told him on his return that 'of all the demonstrations we planned, yours was the only one that came through'.

Our chief interest here is what role *Triton*'s unique characteristics played in the voyage, and its possible contribution to future submarine development, for example the extended patrols of the ballistic missile subs. Certainly it was a substantial advance. In 1950, the USS *Pickerel* covered 5194 miles in 21 days of continuous snorkelling, an average of a little better than 10kts. *Nautilus* covered 1381 miles in 90 hours at an average of better than 15kts on her shakedown cruise. This was

Captain Beach being taken off by helicopter at the end of the circumnavigation. The large hump apparently contains the radio communications buoy used by the SSBNs, tested first by the Triton. *This seems to be concealed by touching up in some earlier photographs.* (Captain Beach)

reported as the first time a submarine had made 16kts submerged for more than an hour and the distance was more than ten times that ever made by a submarine submerged without snorkelling. Later she did 1397 miles submerged at an average speed of 20kts. In 1958, *Seawolf* spent 60 days submerged, covering 13,761 miles. *Triton's* 16 February to 10 May 1960 voyage covered 41,500 miles in 83 days 10 hours submerged, largely at a steady 21kts. (Her officers suggested that at ½kt more they could do it in 80 days, tipping their hats to Jules Verne as well as Magellan.)

Soon, the first ballistic missile submarines went into service, performing patrols of similar length on a regular basis. A recent television programme filmed aboard an *Ohio* class states they ship provisions for 105 days. *Triton* was designed to carry provisions for 75 days, and actually carried enough for 120, for a complement of 183. There was some concern for the effect on the crew, which proved groundless, so perhaps a psychological barrier was broken. *Triton* was first to carry exercise machines, to become standard equipment in the missile boats. *Triton's* real contribution here was her demonstration of the ease with which it was done, with minimal planning and preparation and without the slightest mechanical or psychological problems; and at steady 21kts it was a far stiffer test of the ship and powerplant than even longer missile boat patrols carried out at a very silent few knots.

Triton demonstrated not only the absolute technological reliability Rickover demanded from his engineers and builders but the ferocious standards of competence and education he laid down for officers and men. These are things the Soviets seem to have consciously sacrificed, taking major risks in their efforts to equal or surpass American submarine capabilities and numbers. Their numerous accidents and problems in keeping their fleet operational show the cost of this policy.

Operational Career

Triton joined the second Fleet in August 1960, and was deployed in European waters to participate in NATO exercises. She continued operational patrols and training with the Atlantic fleet in 1961. She lost her radar picket designation on 1 March 1961, with her complement reduced from 16 officers and 156 men to 13 and 146. Captain Beach was relieved by Captain George Morin, who commanded *Triton* to September 1964. She received a major overhaul, including the refuelling of her reactor, at Portsmouth, New Hampshire, lasting from June 1962 to March 1964, when her homeport was changed from New London, Connecticut, to Norfolk, Virginia. In April 1964, she was designated flagship of the Atlantic Submarine Force; she was replaced in this essentially ceremonial role by the *Ray* (SSN-653) in June 1967. Meanwhile Captain Robert Rawlins took over as her third CO in September 1964, serving till November 1966, when he was replaced by Captain Frank Wadsworth, CO until her decommissioning. In June 1967, she was again homeported in New London. Her second major overhaul and refuelling, scheduled for 1967, was cancelled and from October 1968 to May 1969 she underwent preservation and inactivation and was decommissioned on 3 May 1969. On 6 May she left New London under tow for Norfolk Naval Shipyard where she remains today.

Much of this information can be found in the ship's history contained in the Naval Historical Center's *Dictionary Of American Naval Fighting Ships*. However this not only contains some serious inaccuracies but failed to answer several major questions. Was *Triton* actually used, tested, or exercised in the radar picket role? Was any further use made of her high surface speed? Was she actually operated as an attack sub? If not, what was she used for? How effective or valuable was she? Why was she decommissioned?

Captain Beach says he exercised her radar picket equipment in 'one little problem' but she was never actually employed in this role. Under Captain Morin, her air control facilities were used, not for defence, but for strike control. *Triton* was operating between the carrier force and its target – submerged not on the surface. The striking aircraft homed upon the *Triton* which then vectored them in to the target. The SPS-26 and air control equipment were removed well before her 1962 overhaul.

Captain Morin says her high surface speed was not used in any tactical situations. He says it was useful for keeping up when operating with the fleet and recalls that whenever she went over 25kts her bows dug in, her bridge was wet, and her decks for her entire length were under 2ft or 3ft of water. Vice-Admiral Henry Chiles, Commander Submarines Atlantic today, who joined her during her

View forward from helicopter over Triton's *after deck.* (Captain Beach)

Triton *broached with sail above water. She used this manoeuvre throughout her career in a variety of interesting ways.* (US Navy)

1962 overhaul as a young officer, describes her surface speed as useful for reducing time spent entering and leaving harbour. Captain Wadsworth says it was 'fun running around on the surface'; as we have said, he used the tactic of operating her semi-submerged, with her sail above the surface, to put her antennas into the air at high speed.

Captain Rawlins described an incident during his command when she was ordered to surface and proceed to the area where a small civilian aircraft was down off Puerto Rico. She 'cranked up to full speed – greater than 20kts', reached the area, began a search pattern, found the survivors in poor condition, and again used full speed to get them to medical attention at Roosevelt Roads. He says she was a 'fairly stable platform' on the surface, but notes that all submarines roll in a heavy seaway.

How was she in the role of attack submarine? 'Quite honestly, not very good,' according to Captain Beach. He says she had fantastic manoeuvrability in spite of her great length – her control surfaces were immense – and banked like a plane, leaning over at 30°. 'When you gave it full rudder that gyro-compass would spin like a top.' But he only exercised her in the attack role once. In 1961, he made an attack on a destroyer at 10,000yds, holding his bow on her to minimise the size of the target he was presenting. But when he fired the torpedo the destroyer steered straight for it and signalled a hit, showing, in other words, that he was aware of every detail of the attack. He had her on sonar the whole time, heard her torpedo tube outer door open, heard the torpedo fired.

Captain Morin says he operated her as a normal attack submarine. He says she was awkward in this role, 'a poor excuse for an attack sub compared with others we were building at the time'. At periscope depth a 2° down angle would put her stern on the surface. But Admiral Chiles remembers Captain Morin, 'taking on the best US ASW facilities, and showing just how aggressively you could operate a big, big sub'. He says also that she was extremely manoeuvrable with her huge rudder and her bow planes far from the surface were very effective for depth keeping.

Captain Rawlins remembers her as 'an excellent submarine as a submarine' with 'tremendous capabilities' though lacking quieting up to the standards of the *Thresher*. Captain Wadsworth comments on her manoeuvrability, 'You could whip it around like a fighter plane'. He spent hours on an exercise eluding a destroyer by 'whipping around on his stern, where he couldn't track us'. He notes also that much of *Triton*'s equipment was not sound-mounted. 'Operating near a passive array or a new sub the risk of detection was high.' He says her perceived lack of quietness – he never found it a problem – may have played a role in the decision to decommission her.

But was she an expensive, inefficient white elephant, or a valuable unit? The answer was usually unequivocal. Captain Rawlings: 'She did a tremendous amount of operational work which was of great value . . . every day was different; we did all kinds of things.' Captain Wadsworth: 'She proved to be one of the most valuable units we had', although her unusual equipment made her expensive to maintain. Admiral Chiles says that, 'she was a white elephant designed for a role not of her times, but she was a very valuable submarine . . . she took part in many interesting operations'. (He was 'surprised and chagrined' when he found himself assigned to her; he was hoping for one of the new fast attack boats. But he found service aboard her one of the most fascinating parts of his career.)

But what were these operations? All the officers with firsthand knowledge flatly refused to say. 'She served on many special operations that are still classified.' 'She possessed huge extra spaces, a huge CIC like a surface ship's. It was never really used as a CIC but provided a lot of flexibility for putting other electronics in, even double sets, as other cold war missions were found for her.' 'I don't think you're ever going to find out about the really fascinating part.'

Plainly the *Triton* was used for some truly important service in the latter half of her career. Unfortunately we are reduced to speculation as to what it was. Before taking this up, we did learn of a number of roles for which *Triton* was tried or considered during the middle part of her career.

Some of the accompanying photographs show a large hump near the aft end of her casing deck. This is the housing for the radio communications buoy which became standard on the SSBNs. It was tested on *Triton*, by Captain Beach, first; a door on the hump opened, with television cameras mounted to observe how it floated off. It occasionally tangled in a screw; the solution was to release it while turning sharply.

Triton also tested an early 'SINS', ship's inertial navigation system, required by the SSBNs for accurate positions for missile launch when cut off from the surface. But *Triton*'s was a very large and unsuccessful Sperry version. A room 15ft × 15ft was built, in the pump room below the control room with the device 'a huge globe – sitting in the middle of it'.

Both devices were removed during Captain Morin's command. During her overhaul she was also readied for under-ice operations, but never actually used for them. Commander Eastep discussed ideas for surfacing through the ice: they planned to make careful contact with the under surface of the ice with the sail (built of HY-80 steel), blow the after tanks slightly to get a 5°–7° down angle, then blow the forward tanks to gently break through. The value of twin screws for reliability and manoeuvrability in tight spots was often mentioned, but Captain Morin considers the latter not important. He notes her screws were too close together to generate much turning moment; he points out he always used tugs to bring her alongside the pier.

One role for which she was actually tested was command ship. This has generated much confusion; she is frequently described as being considered as a command post for the President during a nuclear war, in print and by her own officers. But according to Captain Morin, the role for which she was actually tested was as a possible wartime command post for the commander of the Atlantic submarine force. The huge former air control space was equipped with communications equipment for COMSUBLANT and his staff. One exercise was carried out with part of the staff aboard; they tried to control a group of subs but it proved quite awkward and difficult to communicate with them. (The Soviets appear to believe strongly in the tactical value of co-operation between submarines and to be determined to make it work in spite of the difficulties.) *Triton*'s ceremonial flying the flag of COMSUBLANT also confuses the issue; he never commanded from aboard.

But what was the role in which *Triton* proved so valuable between 1964 and 1968? It was plainly based on electronic equipment housed in *Triton*'s former air control centre. This huge compartment was divided in two, with

Triton's Air Control Centre. After the radar picket role was dropped this large space was readily adapted to many other uses including, the authors surmise, electronic intelligence gathering. (Naval Historical Center)

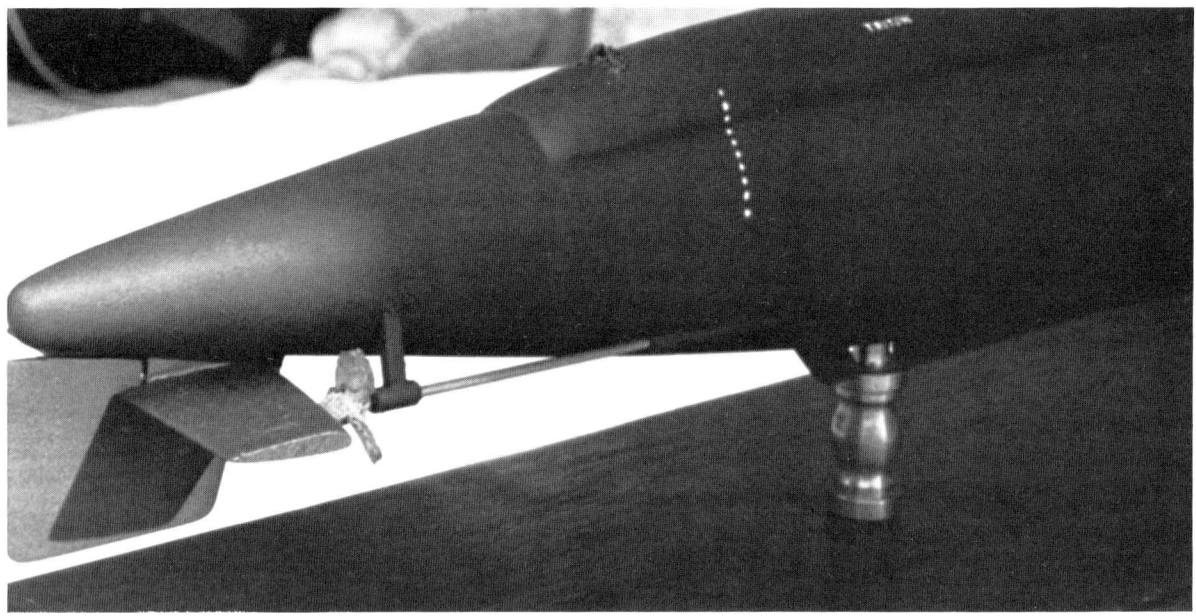

Triton's long narrow stern, twin 11ft screws, stern diving planes and huge rudder. (Authors)

the smaller part serving as the ship's CIC-radar room. The larger was devoted to the classified function, with electronic modules in racks, which could easily be removed and replaced, so the ship could be specially configured for a particular patrol.

This is purely speculation but we guess that at least some of these missions involved signals intelligence, monitoring the range of electronic emissions of a foreign power for many purposes: radio communications for cryptanalysis, radar frequencies for electronic warfare purposes, etc. Many ships and aircraft have been used for this, but in the late 1960s the US did employ specially modified craft such as the mercantile *Pueblo* and *Liberty* with large interior spaces and numerous antennas devoted to this function. Perhaps *Triton* was similarly equipped at the time. The *Pueblo* and *Liberty* were of course easily attacked when North Korea and Israel took offence at their activities, even in international waters; but *Triton* could easily have avoided detection even in close proximity to Soviet bases. We can imagine her operating much like *Salmon* off Hainan a decade before, or *Conqueror* off Patagonia fifteen years later. Except instead of advertising her presence with her powerful low-frequency radar, she would have operated submerged, antennas extended, listening passively and close in, able to pick up even short range and highly directional high frequency links.

The *Permit*s of the early 1960s were apparently handicapped in this role by their small sails, minimised for lessened underwater resistance, with reduced space for masts and antennas. The *Sturgeon*s which appeared later in the decade remedied this problem and perhaps made *Triton*'s special facilities unnecessary. Certainly the specialised intelligence gathering surface ships also disappeared shortly afterwards.

It seems to the authors that 'Desert Storm' played a crucial role in disposing the Soviet military to acquiesce in *glasnost* in spite of the tremendous decline in defensive strength this eventually entailed. The United States had finally demonstrated a decisive, qualitative military superiority based on advanced technology – not in nuclear weapons, but in electronics. Underlying the political conflict of the Cold War lay a technological competition of unprecedented magnitude, a phenomenon unique in human history. In Desert Storm, American electronics capabilities (sensors, missile guidance, electronic warfare, etc) countered every Iraqi threat and rendered their military blind, headless, helpless. *Triton*, we guess, played a serviceable part in the long effort to achieve this.

This electronic sophistication made *Los Angeles* submarines able to strike pinpoint land targets hundreds of miles away in Iraq. In them, the submarine has become an all-round warship, bearing out Rickover's early intimations of the value of size, and speed, in future submarines. And they use a GE reactor.

But where the multi-purpose submarine warship was the role of the future, the submersible surface warship was the blind alley. Only the maintenance of continuous radar operations and active radio communications require the submarine to stay above the surface. The 'Echo II' class had to remain there to prepare and launch their long range Shaddock missiles, operate the electronic gear necessary to communicate with Bear-D targeting aircraft, and give mid-course guidance to the missiles. Probably *Salmon* would have used her BPQ-2 Regulus missile guidance radar in a similar way. *Triton* showed such craft could be given the performance of surface ships, at great expense. Yet such missions mean discarding the invisibility of the submarine, at least temporarily. *Triton* only operated in the air control mode submerged, which she could do because all air control antennas were mounted in her sail, not on her deck like earlier radar pickets. Perhaps this shows what was really most important.

Details of control surfaces. From a model in Captain Beach's possession made by Ralph Lawton. Bow planes extended, antennas, bridge, bulbous forefoot. (Authors)

Decommissioning, Reserve, Disposal

In spite of her valuable service and many good qualities, there were obvious reasons for her early decommissioning. As a unique ship with many unusual systems, the cost of spare parts and maintenance was high. Captain Wadsworth notes *Triton*'s engineering repair and training manuals were far from the standards of the late 1960s and upgrading them for a single ship and powerplant would have been expensive.

Also *Triton*'s reactors were designed with a unique system of loading its zirconium-clad fuel elements. Although designed to be refuelled from a tender, like other nuclear submarines, this would have been a complex process, and in fact she was only refuelled in dockyard by cutting away the hull above the reactors. Meanwhile *Forrestal* had just suffered a disastrous fire, causing $35 million damage, and Captain Beach believes that the decision was made to cover this cost with the funds originally intended for *Triton*'s overhaul and refuelling, even though her new fuel elements were ready.

This decision cost her six or seven years of operating life. But was there ever any thought of returning her to service? Again, her unusual systems and lack of spare parts would have made this 'a horrendous job' as Captain Beach put it. Admiral Chiles says, 'The hope that she would continue in service was, in all honesty, wishful thinking on the part of people who served aboard her.' She did generate tremendous loyalty on the part of officers and crew, partly perhaps because of her technical perfection, partly the variety and interest of her missions. As Captain Wadsworth said, 'Every day was different . . . we spent 60% of our time at sea . . . we never had a commitment we didn't meet.'

Admiral Chiles says that, 'At first there was some feeling we might want her back.' But by the 1970s any serious thought of it was gone. He inspected her some time ago and, 'it was not a pleasant experience'. Apparently much equipment has been removed and her interior spaces are in chaos. What condition is she in now? 'Good enough to be towed to the breaking yard. But not in any condition to serve as a fitting monument to her era.' Indeed her scrapping has been delayed by the debate over the disposal of the radioactive components of submarine reactors. Now her scrapping is a year or two away – a sad, if familiar, fate for this unique, anomalous, remarkable ship.

Sources

The authors would like to thank Captain Edward Beach, Lieutenant-Commander David Leighton, Vice-Admiral Henry Chiles, Mr James Mandelblatt, Captain George Morin, Captain Robert Rawlins, Captain Frank Wadsworth, Lieutenant-Commander Dale Eastep, Dr Gary Weir, Ralph Kennedy of the *Triton* veterans' association, and Lieutenant-Commander Dave Morris and the other members of Admiral Chiles' staff who helped make this article possible.

Printed sources cited:

Captain Edward L Beach, *Around The World Submerged* (1962).
Norman Friedman, *Submarine Design and Development* (London and Annapolis 1986).
Harold Hemond 'The Flip Side of Rickover', *US Naval Institute Proceedings* (July 1989).
Hewlett and Duncan, *Nuclear Navy, 1946–1962* (Chicago 1974).
J L Mooney (ed), *Dictionary of American Naval Fighting Ships* (Washington 1991).
Norman Polmar and Allen, *Rickover: Controversy and Genius* (New York 1982).
Norman Polmar, *Ships and Aircraft of the US Fleet* (14th ed, Annapolis 1987).

WARSHIP NOTES

This section comprises a number of short articles and notes, generally highlighting little-known aspects of warships history.

THE 2750IHP CORVETTE ENGINE

John Harland, the author of a recent book on whalecatchers, describes the simple but effective machinery that drove 'Flower' and 'Castle' class corvettes, as well as the LST (3)s.

In 1936, Smith's Dock of Middlesbrough engined *Southern Pride*, a large whaler of special design, with a forty-cylinder triple expansion machine, which developed 2300ihp (indictated horse power). Engines of similar design, but different dimensions, were used the following year by Bremer Vulkan to power two more big catchers, *Southern Gem* and *Unitas I*. These 'scout catchers' were bigger and faster than the average whaleboat of the day and, bunkered with 400 tons of oil, had twice their fuel capacity and endurance. Intended to roam far and wide from the factory ship to determine in which sector the most whales were to be found, they allowed the hunt-leader to maximise the effort of the other boats in the expedition's hunting group. The first catcher built for this purpose, the *H J Bull* of 1935, was also fitted with a four-cylinder reciprocating engine, the 'Fredriksstad Steam Motor'. This extremely efficient double-compound engine was fitted in a few whaler-type patrol vessels built to German order during the war, and subsequently powered a great many postwar whalers. (For more on the engining of whalers, see the author's, *Catchers & Corvettes: The Steam Whalecatcher in Peace and War, 1860–1960*, Rotherfield 1992).

The 'scouts' built in British and German yards needed a bigger engine than the three-cylinder triple expansion engine used in the ordinary whaler. Getting the extra size and power from a scaled-up 'three-legged' engine would have demanded a low pressure (LP) cylinder of unmanageably large diameter, and the difficulty was obviated by replacing the single LP with two identical cylinders of more practical size. In 1938, rather than design a completely new engine for the 'Patrol Vessel of Whaler Type', later to be called the 'Flower' class corvette. It was decided to save time by powering these utility coastal escorts with a version of *Pride*'s machine, for which the casting patterns were available. The designed revolutions and cylinder diameters were upped slightly, increasing the indicated horse power of the corvette engine to 2750, but otherwise it was identical with that fitted in *Southern Pride*. The cylinder dimensions were 18.5in, 31in, (2)38.5in with a 30in stroke. Its appearance in plan and elevation is shown in Figures 1 and 2. Later on, the same engine was used in the 'Castle' class corvette, and subsequently a matched pair drove the twin-screw reciprocating engined frigates and the Landing Ship Tank(3).

About a thousand examples of this engine were built between 1941 and 1944, a remarkable record for something that started life as the main propulsion unit of a humble steam whaler! The total given in *Selected Papers on British Warship Design in World War II*, Conway Maritime Press (London 1983), page 116, is actually 1150, but we have had difficulty in reconciling this figure with calculations made in other ways.

Using figures in Peter Elliott's *Allied Escort Vessels of World War II*, Macdonald & Jane's (London 1977), we can calculate that 314 single-engined corvettes (including 'Castles') were built in the United

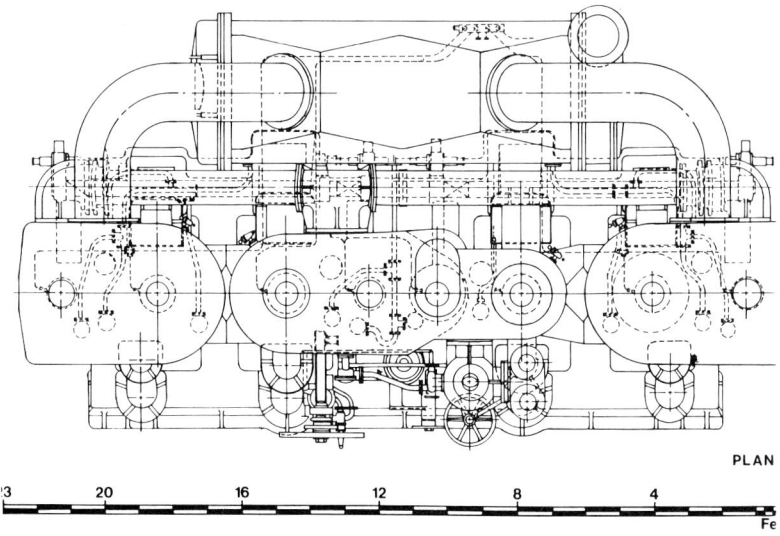

Figure 1: Arrangement drawings of the corvette engine. (John McKay).

Figure 2: Arrangement drawings of the corvette engine. (John McKay).

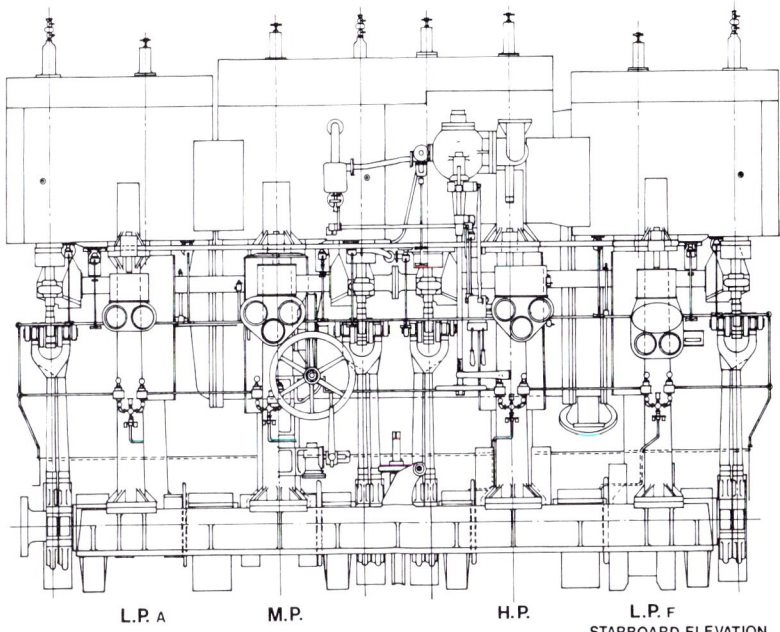

L.P. A   M.P.   H.P.   L.P. F
STARBOARD ELEVATION

Kingdom, Canada and France. A further 290 twin-screw frigates of 'River', 'Loch', 'Bay', 'Colony' classes, including *Tacoma* patrol frigates, were built in the United Kingdom, America, Canada, and Australia.

With the assistance of Jim Colledge, we broke this down further as follows:

Twin-screw frigates
'River' class built in UK 57 (of which five had turbines):	52
'Loch' class built in UK 36 (of which two had turbines):	34
'Bay' class including four survey vessels & two flag-tenders	26
'River' class built in Canada 70:	70
'River' class built in Australia 12:	12
'River' class (*Tacoma* class PF) built in the USA 75:	75
'Colony' class built in USA and transferred to RN 21:	21
Sub total	290

Single-screw corvettes
'Castle' class; all built in UK:	44
'Flower' class built in UK:	135
'Modified Flower' class built in UK:	10
'Flower' class built in Canada:	79
'Modified Flower' class built in Canada:	42
'Flower' class completed in France:	4
Sub total	314

314 single-engined corvettes	314
290 twin-engined frigates	580
Total engines in frigates and corvettes	894

The same engine was fitted in the British designed twin-screw Landing Ship Tank(3), a design decision reflecting Britain's inferior wartime diesel engine production capacity *vis à vis* Germany and America. If we accept that engines were constructed for every one of the 45 LST(3)s built in the UK plus 36 (or possibly 37) ordered in Canada, we have a total of 81 ships and 162 engines. This raises the total number of 1056. On page 29 of *Selected Papers on British Warship Design in World War II* it is claimed that a further 36 LSTs were ordered in Canada, and later cancelled. If we made the very dubious assumption that all 72 engines were pre-built for these cancelled vessels, we get a total figure of 1128.

Richard Robinson of Kengray, South Africa, who has collected an immense amount of information about requisitioned trawlers, says that the engine builders who turned out machinery for trawlers were told to work 'at capacity', and perhaps something similar happened here, more engines being built than were actually needed. Some 2750ihp machines were installed in a few catchers built for United Whalers built by A & J Inglis, a subsibiary of Harland and Wolff after the war, but I am reasonably sure these were salvaged from corvettes.

The layout of the 'Flower' class engine room is shown in Figure 3. A. Air Pump, B. Evaporator. C. Generators. D. Auxiliary feed-pump. E. Condenser. F. Bilge pump G. General service pump. H. Hotwells. As with the classical three-cylinder triple expansion engines, the HP (high pressure) cylinder was closest to the boiler, that is to say immediately forward of the IP, with the LP cylinders at either end.

The appearance of the starboard side of the engine as installed in the 'Flower' class, can be seen in Figure 4. Although various types of single eccentric radial gear were found in whalers built by other yards, those built in Middlesbrough favoured the classic double eccentric Stephenson system, and the same arrangement was used in the corvette. The gear was of the 'All-around' type, that is to say whichever direction the reversing wheel was turned the machine shifted continuously from Ahead to Stop to Astern, and back again. In small engines, this gear was operated by the large hand wheel, but in the corvette, as in most whalers, the wheel was used in emergency, and the reversing

Figure 3: Layout of corvette engine room. (John Harland).

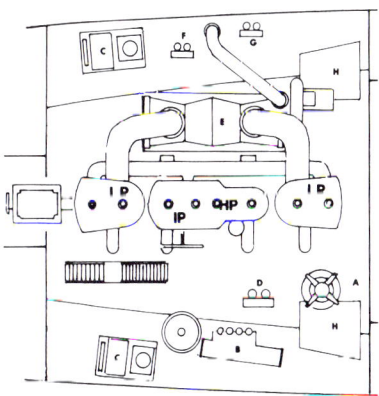

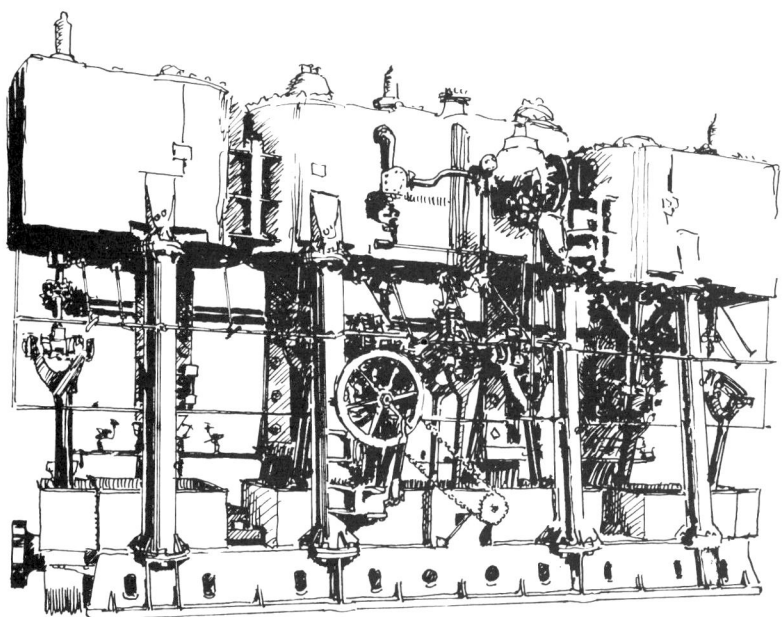

Figure 4: Sketch of starboard side of corvette engine. (John Harland).

gear was actually worked by a lever below the handwheel controlling a single-cylinder reversing engine. Pushing the lever forward got the engine turning ahead. A heavy chain connected the reversing engine to the 'Turning-' or 'Jacking-gear', which could be used to turn the machinery over slowly in the absence of main steam.

All the controls were on the starboard, or 'round pillar side' of the engine, within reach of the engineer on the starting platform, while the heavy square engine pillars carrying the crosshead guides, and the condenser, were on the engine's port side. The stop-valve, which controlled the admission of steam from the boilers, was of the 'double-seat' type, a pattern which facilitated operation of the valve against steam pressure. This was operated by a wheel and vertical shaft toward the forward end of the engine. Beside this were a pair of levers, one operating the starting valve, and the other the throttle, which was of the 'butterfly' type. The starting valve allowed the engineer to bypass steam directly to the Intermediate stage, and kick the HP piston off dead-centre, if the machine had stopped in that position.

Steam entry to the HP stage was controlled by a piston valve, while IP and LP used double-ported slide valves of the Andrew & Cameron 'Matchbox' type. A piston valve was better suited to the high pressure/low volume stage, while slide valves were preferred for the second and third stages, where the volume of steam was greater and the pressure lower.

The double-ported pattern was preferred to a simple 'D'-slide valve of traditional pattern, because in this type the length of travel of the valve rod was halved and face friction markedly reduced.

The engine was placed on the centreline of the engine space, and as in virtually all small single-screw ships with a right hand propeller, the condenser lay to port, and the starting platform to starboard. With this configuration the crosshead guides were in plain sight of the engineer on the starting point.

For reasons which need not concern us here, the propellers of a twin-screw ship work best if they turn 'outward', that is to say the starboard screw is right-handed, the port one left-handed. In the ordinary way, the engines are mirror images of each other, but in the twin-screw frigates and LST(3)s, production was greatly simplified by using identical engines. These were built with a connecting flange at both ends of the crankshaft, and the decision to designate the engine port or starboard, could be left to a fairly late stage.

This interchangeability required a

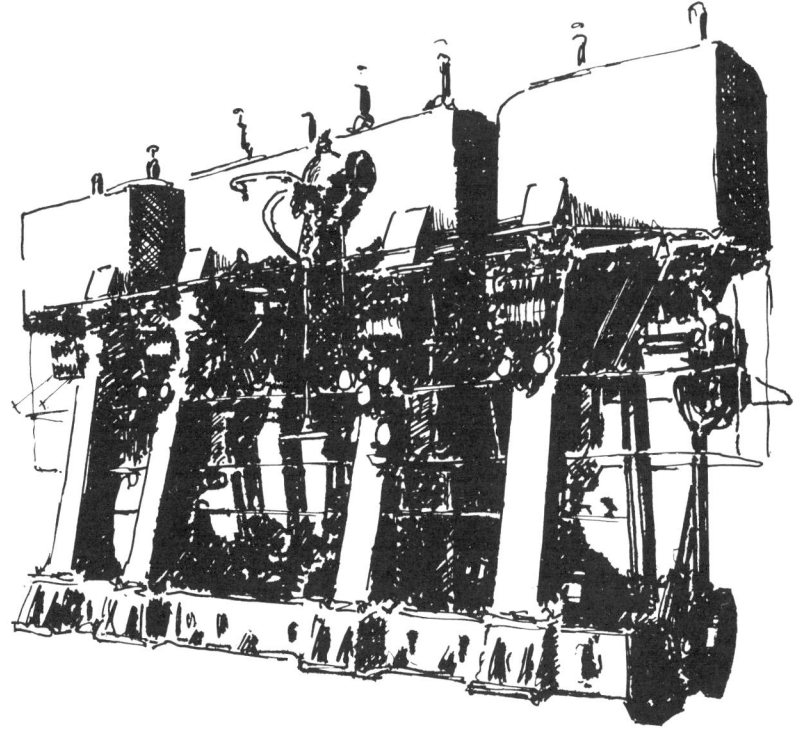

Figure 5: Sketch of inboard side of a frigate engine. (John Harland).

change in the relationship of the condenser to the engine. In the corvette, the eduction pipes carrying steam from the LP cylinders to the condenser were on the port side. In the twin-screw frigates, the condensers and eduction pipes had to be shifted to the outboard sides of the engine, and this required appropriate change in the entablature of the cylinders. Basically the four cylinders, considered as a unit, were picked up and swung around on the engine columns, so their order was reversed end for end, bringing the LP eduction pipes onto the outboard side, and throttle and starting valve onto the (square pillar) side of the engine.

Considering the starboard engine: since this drove a right-hand propeller, the ahead guides and box-columns which carried them were placed on the engine's port side, as in the single-screw version. The HP had been swung around so it lay abaft the IP, and the steam controls were on the inboard or port side. The reversing gear was modified by shifting the operating lever onto the port side just below the wheel for the stop-valve. On the port, or left-hand screw side, the same engine arrangement was followed, but everything was turned end for end, once again placing the controls onto the square-pillar or inboard side, with eduction pipes and condenser outboard. Figure 5 is a sketch of the inboard side of the starboard engine of a frigate.

The single right-hand screw of the 'Castle' class corvette was driven by what was in effect the starboard engine of a frigate, and so offers an exception to the general rule that in single-screw steamers the condenser is to port of the engine. In 'Castles' and the frigates, the condensers were sufficiently distant from the engine, to allow space for the oilers keeping an eye on the crosshead guides, and a check on bearing temperatures. The engine room layouts are shown in Figures 6 and 7.

A splendid example of the corvette engine, modified to allow its being turned over by an auxiliary engine, is to be seen aboard the HMCS *Sackville*, which is maintained by the Canadian Naval Historical Trust in Halifax, Nova Scotia. A 1/10th scale model of the 'double-ended' engine, built by

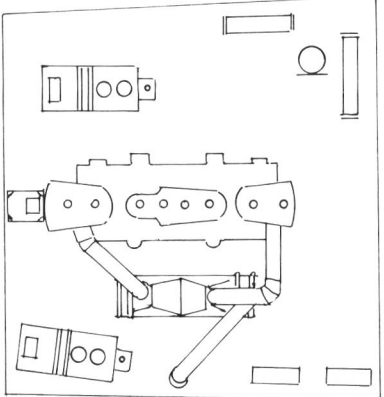

Figure 6: Layout of 'Castle' class corvette engine room. (John Harland).

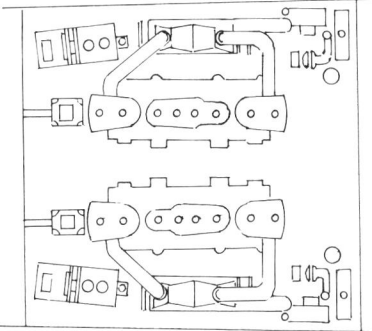

Figure 7: Layout of frigate engine room. (John Harland).

the apprentices at Smith's Dock in 1945, is on view at the Museum of Science & Industry, in Newcastle. Unfortunately this is disfigured by the addition of a ratchet mechanism, intended to turn it over by hand.

The intention of the corvette's designer, William Reed, was to fit them with water-tube boilers, but with hostilities imminent, this relatively sophisticated equipment was earmarked for 'real' warships, and the first 'Patrol Vessels of Whaler Type' were fitted with a pair of oil-fired Scotch boilers. These sat back to back in separate stokeholds, and used Howden 'closed ashpit' forced draught. In later corvettes the firetube boilers were replaced with three-drum water-tube boilers, and 'closed stokehold' forced draught. *Southern Pride*, by the way, had a unique arrangement for a whaler, namely a Scotch boiler in the after stokehold, and a water-tube pattern in the forward fire room.

From 1925 or so, for logistical reasons all catchers were designed for oil-firing, and the same plan was followed in the Patrol Vessel of Whaler Type. Whalers working in the Antarctic needed to refuel at sea every few days, a task that was much easier and quicker with liquid fuel. In 1938, trawlers were universally coal-fired, and since the corvettes were intended as 'coastal' escorts, coal fuel would have been a perfectly reasonable choice. In the event, these little vessels were used as mid-ocean escorts, a task for which they were not designed. Had the planning authorities opted for coal-firing, their job would have been made even more difficult.

THE LAST VOYAGE OF U-234
David Miller looks at the curious voyage of a U-boat dispatched to Japan with an odd cargo of 'strategic materials' on the very brink of Germany's defeat

The boat. *U-234*, a Type XB minelayer, was constructed at Germaniawerft, Keil. Laid down in 1941, her launch was delayed due to damage incurred during a USAF bombing raid on 14 May 1943 and eventually took place on 23 December 1943. She was commissioned on 2 March 1944 under *Kapitänleutnant* Fehler.

After working-up, *U-234* operated in the Baltic and at one point Fehler was briefed for a resupply voyage to the German garrison at Lorient, but this was cancelled. Then, after only six months in service, *U-234* returned to Germaniawerft (5 September 1944) for conversion to a transport. The major work involved removing the twenty-four mine tubes located in the saddle tanks to create two large, rectangular stowage spaces. The six forward mine tubes were, however, retained and special containers were fabricated to fit inside them, which were held in the place by the normal mine-release mechanism. In addition, two saddles were welded to the deck either side of the conning tower, on which were mounted two large steel cylinders some 6m (19.7ft) long and 1.2m (3.94ft) in diameter.

Some routine work was also done, which included the installation of a schnorchel, a Balkon sonar and a revised oil system, while the starboard propeller, which had proved very noisy in service, was also replaced. This work completed, *U-234* left the yard in December to conduct trials and schnorchel training. Loading, which included removing the outer keel plates so that the keel duct could be filled with cargo, started in January and lasted until March. All this was done in strictest secrecy, but *U-234*'s destination seems to have been widely known, at least among the U-boat community.

Cargo. So important was the mission that a special body – *Marine Sonder Dienst Auslands* – was set up to decide what should be taken. This group co-operated with *Kapitän zur See* Souchon, head of the *Marine Attaché Abteilung* in *Berlin Oberkommando Referat Japan*, as well as with the officers of the Japanese Naval Mission in Berlin.

The eventual load in the various tubes consisted of military equipment, metals, documents and ammunition, all of them carefully selected as being of special interest to Japan. Equipment items included: injection pumps, aircraft warning devices, a direction-finding set, a multi-band receiver, a fire control computer, coils, various ammunition fuses, vacuum tubes (valves) and planospheric lenses. There were twenty-three cases of ammunition, included recoilless anti-tank rounds (*Panzerfaust*), while the metals carried were thallium (106kg), steel (6110kg), lead (11,151kg), zinc and mercury (1926kg). There was also a huge amount of documents and plans (including aircraft drawings from Junkers and Messerschmitt) and some miscellaneous items: benzyl cellulose (3000kg), silk ribbons, and the drug atabrin (465kg).

In the six forward tubes were ten cases of uranium oxide (U_{235}) to which we shall return later, while in the keel duct were machine-gun belts (38kg), 1474 bars of lead (55,758kg) and 564 bottles of mercury (22,168kg). The grand total of the cargo was estimated by the U-boat's officers to be some 240 tonnes, although the manifest subsequently put together by the US authorities accounted for only 162 tonnes.

Crew. In command was *Kapitänleutnant* Johann Heinrich Fehler (34), who had served aboard the raider *Atlantis* during 1940–41. He then trained for U-boats and on qualifying was assigned to his first boat, *U-234*, as commanding officer! There were also five other officers (three deck officers, an engineer and a doctor) and 39 ratings.

Passengers. The passengers aboard *U-234* were a mixed lot. The most senior was *General der Flieger* Ulrich Kessler, who was on his way to become air attaché in Tokyo, for which he had been under training from March 1944 to April 1945. Other Luftwaffe officers were the former head of AA defences at Bremen, *Oberst* (Colonel) von Sandrart, who was to instruct the Japanese in tactical air defence, and a communications specialist, *Oberstleutnant* (Lieutenant-Colonel) Menzel.

There were four naval officers. *Fregattenkapitän* Falck, a specialist in ship design and construction, was destined to be technical aide to the German Naval Attaché in Toyko (Admiral Wennecker) and had attended several secret courses before departure. Other line officers were *Kapitänleutnant* Bulla, a specialist in air-sea co-operation, and Hellendorn, a specialist in naval air defence. Rather more sinister was *Oberstleutnant-Geschwaderrichter* (Lieutenant-Colonel – squadron judge) Nieschling, a naval judge, who was going to Japan to purge German elements there (in particular to follow-up the Sorge spy scandal) and also to keep an eye on other passengers during the voyage.

Dr Ing Schlicke was a top-ranking electronics specialist, who was given a reserve commission in the rank of captain for this voyage. His main task was to help the Japanese to set up a radio/navigation station, to be codenamed 'Goldweber'.

Finally, there were two civilians, both from the Messerschmitt factory, whose reluctance to go to Japan had only been overcome by a direct order. The senior of these was August Bringenwald, whose task was to put the Me 262 turbo-jet fighter into production in Japan and to prepare for the production of the Me 163 rocket fighter. The second was Franz Ruf, who was to supervise the building and equipping of the factories which would produce the two fighters. There were also two Japanese passengers, returning home after service with the military mission in Germany: Lieutenant-Commanders Hideo Tomonaga and Genzo Shoji, both of the Imperial Japanese Navy.

The Voyage. On the afternoon of 25 March 1945 eight German and the two Japanese passengers embarked and *U-234* left Kiel in company with *U-516* and one Type VIIC. Two days later the three boats arrived safely at Horten and from then until 3 April Fehler carried out schnorchel trials, during which he collided while submerged with *U-1301* (Type VIIC/41). Both boats sustained minor damage and Fehler proceeded to Kristiansund, Norway, arriving on 5 April. In the event, the crew had to do the repair work themselves and on completion they did the final loading of fuel and food, and the remaining passengers also joined.

U-234 sailed on 15 April, even though the crew were convinced that they would never reach Japan, proceeding at schnorchel depth until 1 May, when Fehler decided to proceed on the surface for about two hours each night and submerged at about 40–100m for the rest of the day. On May 4 they heard fragments of an English broadcast announcing that Dönitz had been appointed Hitler's successor and on 10 May they heard the broadcast ordering all U-boats at sea to surrender and proceed to the nearest Allied port.

The crew's first reaction was to make for neutral Eire but Fehler continued slowly southwards while he decided what to do. General Kessler tried to persuade him to head for Argentina, an idea which had already occurred to Fehler, but having heard other U-boats signal their surrender he felt obliged to follow suit and he broadcast their position on the 600m band.

At this point the two Japanese officers told Fehler that they would commit suicide rather than surrender and after failing to persuade them to change their minds he agreed to leave them undisturbed in their cabin. The two Japanese then distributed numer-

Luftwaffe General Kessler had travelled for a month aboard U-234 *'hot-bunking' with two engineers and was twice transferred between vessels in an open boat, but when he came ashore at Portsmouth Navy Yard he still looked every inch the general, complete with jackboots and monocle!*

U-873 *a Type IX D2 enters Portsmouth (New Hampshire) harbour, wearing the United States flag, 16 May, 1945.*

ous gifts to German crew members before taking a massive overdose of luminol (sleeping tablets); 36 hours later the doctor discovered that they were still alive, which caused some alarm among the Germans, but they eventually died and were buried at sea on 11 May.

U-234 received her first Allied message at 0800 hours on 12 May and later that day was ordered to proceed to Halifax, broadcasting details of her position and course hourly as she did so. Also on 12 May, USS *Sutton* (DE-771), which, with USS *Scott*, was escorting the recently-surrendered *U-1228*, was ordered to capture *U-234* and she proceeded on an intercept course. At 0500 hours on 14 May *Sutton* was joined by two RCN warships, but they parted again in the afternoon and at 2141 hours *Sutton* found *U-234* (47° 07'N, 42° 25'W). The two vessels remained in company overnight and just after 0800 hours *Sutton* sent a boarding party to take possession of the prize. The captain and most of the passengers and crew were transferred to *Sutton* and a small party under the orders of the prize crew took *U-234* into Portsmouth, arriving 16 May.

In the United States *U-234*'s arrival caused great excitement, especially in the Press. She was the fourth such prize to arrive in the space of eight days, her predecessors being: *U-805* (Type IX C/40), 14 May; *U-873* (Type IX D2), 16 May; and *U-1228* (Type IX C/40), 9 May. The Press watched the prisoners come ashore with General Kessler creating the greatest stir, as his leather greatcoat, tall peaked cap and monocle fitted in so exactly with the popular, contemporary image of the 'typical Nazi general'. Rumours abounded and at one time the Press suspected that one of the two civilians was Hitler in disguise and the other was Messerschmitt, while Himmler and Ribbentrop were also reported to be aboard. There were even reports that the bodies of the two Japanese were in the torpedo tubes!

The passengers and some of the officers were flown to Washington, DC for interrogation, while the crew were taken to a Portsmouth naval prison. Unfortunately, some US personnel appear to have misbehaved and many 'souvenirs' were taken from the boat and the crew, including pistols, canned goods, parts of uniforms, dextrose pills and virtually all the liquor.

The US authorities were interested in all the stores aboard the *U-234*, but most of all the radioactive uranium

The four surrendered U-boats alongside at Portsmouth Navy Yard. The workmen engaged in removing stores from the forward mine chutes of U-234 (lying outboard of U-873) were almost certainly ignorant of the high radioactivity of the uranium oxide in one of those tubes.

oxide, which was contained in boxes marked 'U-235: For Japanese Army'. Published interrogation reports indicate that nobody aboard admitted to any knowledge of the uranium oxide, other than that they knew it to be on board. Nieschling, for example, told his interrogators in July that all he knew was that it was valuable and suggested that either Falk or Fehler would know more about it, which both of them subsequently denied.

The U_{235} was eventually unloaded and taken away. Where it went, what happened to it and to what use it was put has never been revealed by the United States' authorities. Was it used in the American A-bombs dropped on Japan, or did the simple fact of its existence suggest that the Japanese might have A-bombs and thus lead the Americans to pre-empt them by using their own bombs against Hiroshima and Nagasaki? Those who know the answer to those questions have kept very quiet for some fifty years.

An enduring mystery. Precisely who put the U_{235} aboard and why remains a mystery. Since it was such a valuable commodity and as the loading list had been so carefully prepared, it is clear that both the Germans and the Japanese thought it to be worthwhile. Further, someone aboard must have known why it was being sent and for whom it was intended. One possibility is that the two Japanese officers were 'in the know' and that they may have committed suicide to ensure that they kept their secrets from the Americans.

U-234 sailed from Kiel on 23 March, at a time when Germany was on the very brink of defeat. The Allies had crossed the Rhine (22/23 March) and the Russians had reached Danzig (23 March), and it was absolutely clear that Germany would be defeated. When *U-234* sailed from Norway on 15 April the situation was even worse, but despite that someone still considered the voyage to be essential. All the items aboard were important and valuable, but none was crucial and justified such a long and hazardous voyage – except, perhaps, the U_{235}.

Postscript. The former *U-234* was eventually towed to sea, where on 20 November 1947 she was torpedoed and sunk by USS *Greenfish* (SS-351) off Provincetown, Massachusetts (Lat 43° 37'N, Long 69° 33'W). She sank immediately, but the ripples from her voyage and her cargo linger on.

Sources:
1. US Navy Op-16-Z dated 27 July 1945.
2. USS *Greenfish* memo dated 25 September 1948.
3. USS *Sutton* Action Report serial 29 dated 29 May, 1945.
4. *The Portsmouth Periscope* (7 June 1945).
5. US National Archive RG-38 Box 13.

SEAKEEPING AND ADDED WEIGHT

A technical note from David K Brown, RCNC and Dr Adrian J R M Lloyd about a matter of importance to a warship's seakeeping.

It is often said that adding weights near the end of a ship will increase pitching motions but, while this is true in principle, the actual increase in pitch is very small. However, extra weight will reduce freeboard and, if near the bow, will change trim, further reducing freeboard, and this may increase the chance of green seas sweeping the ship.

Until modern seakeeping theory was developed and powerful computers became available to put numbers to actual ships, views were inevitably subjective. The belief that weights near the end increased pitch was the reason that *Warrior*'s armour did not extend to the bow, and has also been said to explain the poor seakeeping of the German *Z-35* class destroyers when a twin 5.9in was mounted forward, and also that of the US *Sumner* class.

Problems first became apparent with the *Z-23* and *Z-24* as originally completed with a single 5in forward when they met bad weather in March 1941 and were said to lack buoyancy forward. During Operation 'Zarin' in the Barents Sea in September 1942, after the twin 5.9in had been fitted, these fears were confirmed; the extra weight of the turret resulted in the bows diving into the seas and water being thrown over the bridge, leaving the turret surrounded by surf and sticking out like a half-tide rock. Vice-Admiral Bey recommended that the turret should not be fitted in later ships.[1]

Recently the second author studied this problem, calculating the effect of adding 50 tons 15 per cent (16.5m) abaft the bow of a 3000-tonne ship, 110m in length. The changes in motion were calculated for the ship steaming at 20kts into Sea State 5 (significant wave height 3.25m, modal period 10.9 seconds*).

The changes in ship geometry are fairly small:

	Initial	Added Weight
Displacement, tonnes	3000	3050
Longitudinal centre of gravity abaft amidships, metres	2.00	1.34
Radius of gyration/ length	0.225	0.228
Draught forward, metres	4.19	4.41

The effect of these changes on the motion of the ship is not large. Computed values of motion are given below and it should be noted that the accuracy of the methods used has been confirmed by full scale trials.

Average (significant) motions

	Initial	Added Weight
Pitch, degree	2.60	2.62
Heave, metres	1.246	1.286

The earlier Z-23 showing that the 1936 A type was always wet forward, even though they carried only a single 15cm gun forward. (CMP)

The German destroyer Z-38 of the 1936 A(Mob) type which was said by many sources to suffer from the addition of the large twin 15cm (5.9in) turret forward. (CMP)

Heave acceleration m/sec²	1.408	1.468
Vertical acceleration at gun mount m/sec²	3.46	3.48

It is unlikely that any of these changes in motion would be noticed by the crew. What will change and will be noticed is the frequency with which green seas come on board. If the initial freeboard was 6m, the new freeboard will be 5.78m and the probability of a single wave encounter causing water to come on board will increase from 0.76 to 1.2 per cent. The increased wetness is due almost entirely to the reduction in freeboard though there is a small contribution from the increase of pitch relative to the wave.

Almost certainly, destroyer sailors would notice the increased wetness and blame it on pitching. The captain, observing more green seas on the forecastle would reduce speed. Reports from sea are of great value to the naval architect but, as in this case, they can be quite difficult to interpret.

*Definitions
Significant – mean of the ⅓ highest motions which closely represents perceived values.
Modal – The value which occurs most often.

[1] M J Whitley, *Destroyer! German Destroyers in World War II*, Arms and Armour Press (London 1983).

A CAREER IN THE EDWARDIAN NAVY: HENRY KEPPEL GARNIER, RN

Using the photographs from a recently discovered family album Andrew Lambert outlines the career of a British naval officer in the Edwardian age.

Garnier's career was largely unremarkable, and ended at the rank of retired Commander in 1929. However, it should be stressed that this, and not active flag rank, constituted a 'normal' career. Indeed, the expansion of the Royal Navy after 1900 led to a rapid increase in the size of the officer corps, which, after 1919, was simply too large for the needs of the nation.

After passing through HMS *Britannia* at Dartmouth, at that time still a floating establishment made up of two old wooden battleships, the aptly named Henry Keppel Garnier went to sea in 1905. His first appointment as Midshipman in July 1905 was a posting to HMS *King Edward VII*,

The view over the quarterdeck of King Edward VII; Commonwealth *and* Africa *lead the rest of this homogeneous squadron. The amount of water being shipped over the bows in a moderate sea explains why this class was considered wet forward.*

WARSHIP NOTES

HMS King Edward VII *as flagship of the Atlantic Fleet 1905. She is seen leaving Portsmouth harbour.*

Vice-Admiral Sir Charles Beresford hoisted his flag as Commander-in-Chief Channel Fleet aboard King Edward VII *on 16 April 1907. Here he poses with his staff and his* alter ego *under the after 12in turret in June. From the left the staff are: seated Captain Doveton Sturdee, Chief of Staff, and Captain Henry Pelly, Flag Captain. Among those standing are the Secretary, John Keys, Flag Commander Fawcett Wray and Flag Lieutenant Herbert Gibbs.*

The torpedo boat destroyer Lightning, *one of the first four such craft ever built. Although completed in 1895 she was still in service in 1908, as a tender to the depot ship* HMS Hecla *at Portsmouth.*

then flagship of the Atlantic Fleet. In April 1907 *King Edward VII* became flagship of the Channel Fleet, under the larger than life Vice-Admiral Sir Charles Beresford. In August Garnier qualified as a Sub-Lieutenant, and in the following year moved into the armoured cruiser *Minotaur*, serving in the First Cruiser Squadron. Rear-Admiral Stanley Colville flew his flag in the old armoured cruiser *Drake* with three battlecruisers. During this time Garnier was temporarily assigned to the destroyer HMS *Moy* and the torpedo boat *Lightning*, a tender to the torpedo depot ship

The armoured cruiser HMS Minotaur *in 1908–9, Captain W Boothby. She formed part of the First Cruiser Squadron with the first three battlecruisers and the older armoured cruiser* Drake. *The funnels were raised by 15 in in 1909 to improve draught.*

Spithead 1909; the Fleet Review with the new Russian armoured cruiser Rurik, *built by Vickers, passing through the British cruiser fleet. Her newly installed foremast should be noted.*

Hecla. After *Minotaur* Garnier joined the ill-fated armoured cruiser *Monmouth*, on the China Station in March 1910, being promoted to Lieutenant the following month. In June 1912 he joined the 'Improved County' class armoured cruiser *Roxburgh*, part of the 2nd Fleet in Home Waters. After the outbreak of war he remained with *Roxburgh*, in the Grand Fleet, until September 1915, when he joined the 'Town' class light cruiser *Dartmouth*, eventually taking over as Gunnery Lieutenant. He was removed into the older light cruiser *Boadicea* in November 1917, and promoted Lieutenant-Commander early the following year, remaining there until the end of the war. Be-

tween 1919 and 1924 Garnier served on the Headquarters staff at Malta, for maintenance duties, and ended his career with two periods at Pem-

broke Barracks, Chatham, supervising the disciplinary training of ratings for disposal, and later as assistant to the Drafting Committee. He was retired with rank of Commander on 5 October 1929, after a career lasting twenty-five years.

The Channel Fleet at sea in 1907, Britannia *astern of the flagship.*

The battlecruisers and armoured cruisers at the Spithead Review of 1909. Indomitable, Cochrane *and* Black Prince *are prominent. This view was taken from the* Minotaur.

HMS Roxburgh, *Captain Cole W Fowler, entering Plymouth Harbour in June 1912. An 'Improved County' class cruiser,* Roxburgh *mounted a mixed battery of 7.5in and 6in guns, the larger pieces being in four single turrets, fore, aft and on the forward quarters.*

The First Cruiser Squadron steaming out of Berehaven for manoeuvres; Drake *leads* Invincible, Inflexible, Indomitable *and* Minotaur, *from which the picture is taken. Garnier's caption reads 'burning North Country Coal', a reference to the copious quantities of smoke being produced by* Indomitable.

One of the most powerful surface escorts of the present-day Royal Navy, the Batch III Type 22 Cornwall. *Beyond her is the RFA* Olmeda, *a fleet tanker, and in the distance the Brazilian frigate* Defensora *and the Danish minelayer* Sjaelland.

THE BATTLE OF THE ATLANTIC REVIEW

As part of the Liverpool-based celebrations to mark the fiftieth anniversary of the turning point in the Battle of the Atlantic, in May 1993 there was a large assembly of warships from many countries off the coast of North Wales. Perhaps appropriately, the Force 8–9 conditions were reminiscent of the weather often faced fifty years ago by the convoys, their escorts and their opponents. Although it was not ideal for photography, we are grateful to Leo van Ginderen for supplying these atmospheric shots of some of the ships involved.

Very ill at ease in the rough North Atlantic, the modern 'corvette' Danaide *represented the Italian navy at the review.*

WARSHIP NOTES

An unusual visitor to the review was the South African multi-purpose auxiliary Drakensburg. *The largest warship so far constructed in that country, she was also locally designed and can function in the replenishment, support, or disaster relief roles.*

Roughly the size of a 'Flower' class corvette, the British coastal survey ship Bulldog *gives a fair impression of what life must have been like serving in small ships in the North Atlantic in bad weather.*

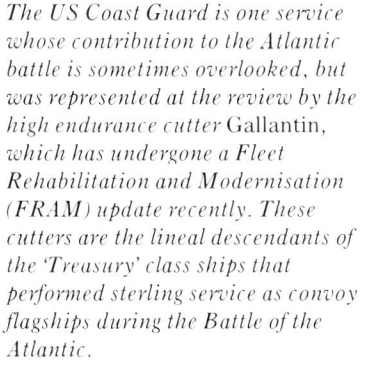

The US Coast Guard is one service whose contribution to the Atlantic battle is sometimes overlooked, but was represented at the review by the high endurance cutter Gallantin, *which has undergone a Fleet Rehabilitation and Modernisation (FRAM) update recently. These cutters are the lineal descendants of the 'Treasury' class ships that performed sterling service as convoy flagships during the Battle of the Atlantic.*

Specialist minelayers are now a rarity in most navies, but the Royal Danish Navy still operates a handful, although Sjaelland *(N83) now functions as a support ship for submarines. The design is a Nato type dating from the 1960s.*

Life aboard the smaller ships, like the 'Hunt' class MCMV Middleton *must have been very uncomfortable, and some of the minor craft were eventually allowed to run for cover.*

The fleet was inspected by HRH The Duke of Edinburgh from the royal yacht Britannia, *with the Type 21 frigate* Active *serving as guardship.* Britannia *is flying his personal standard from the main.*

Britannia, *trailed by* Bulldog *and* Active *disappear into the murk at the conclusion of the Duke's inspection.*

The sight of Russian military hardware at Western events is no longer a novelty, but it is still interesting to see one of the relatively new Sovremennyy *class destroyers at close quarters.* Gremyashichiy *here flies the pre-revolutionary Russian ensign and jack.*

NAVAL BOOKS OF THE YEAR

The last twelve-month period has not been a very fruitful one for publishing in this journal's field of interest, so a few titles of more general naval historical relevance have been included. As usual this section is divided into reviews proper, short descriptive notices, and a list of books announced or not actually received. In all sections order is alphabetical by author and place of publication is London unless otherwise specified.

Jean Boudriot, La Frégate: Étude historique, 1650–1850, *published by ANCRE (Paris 1992).*
310 × 230mm, 348 pages, 300 illustrations. No price quoted.

Best known for a series of beautifully illustrated monographs on historical ship types, Jean Boudriot is France's leading technical historian of the sailing ship, and for many years virtually the *only* serious student of such matters. For the historian rather than the modelmaker, these books are somewhat frustrating in that the all-too-short introductions give a brief taste of the author's undoubted knowledge of the design background before passing on to the specifics of the ship concerned. Put together, these introductions will one day form the basis for a technical history of the French navy in the age of sail, but for the cruising ship this has already been achieved by the volume under review.

Needless to say, it is thoroughly illustrated, but mostly with contemporary draughts, models and prints rather than the author's own characteristic line work. However, the real value of the work lies in the text, an attempt to describe the French frigate as a coherent development from the middle of the seventeenth century to the end of sail. Given the fragmentary nature of the French sources, this is no easy task and the book sometimes gives more emphasis to a particular surviving document than it might warrant. Various proposals, for example, are mentioned, but it is never clear how seriously some of these often radical suggestions were treated. Although policy may have been more frequently discussed than in Britain (or more frequently committed to paper, which is what matters to the historian), in France what has survived is often the product of historical accident. Nevertheless, the main features of frigate development are quite clear, the author choosing to organise his chapters around the main deck gun calibre, which usually defined the frigate of a particular generation (although there was always some overlap). Each chapter includes a list of relevant vessels, which are the best so far available, and differ in some respects from those published by the author in earlier monographs.

The Admiralty 'as captured' draught of the French Topaz. *According to the new study of the French frigate by Jean Boudriot, the ship was designed by J M B Coulomb and built at Toulon in 1789. She was armed with twenty-six 12pdrs and six 6pdrs.* **(NMM)**

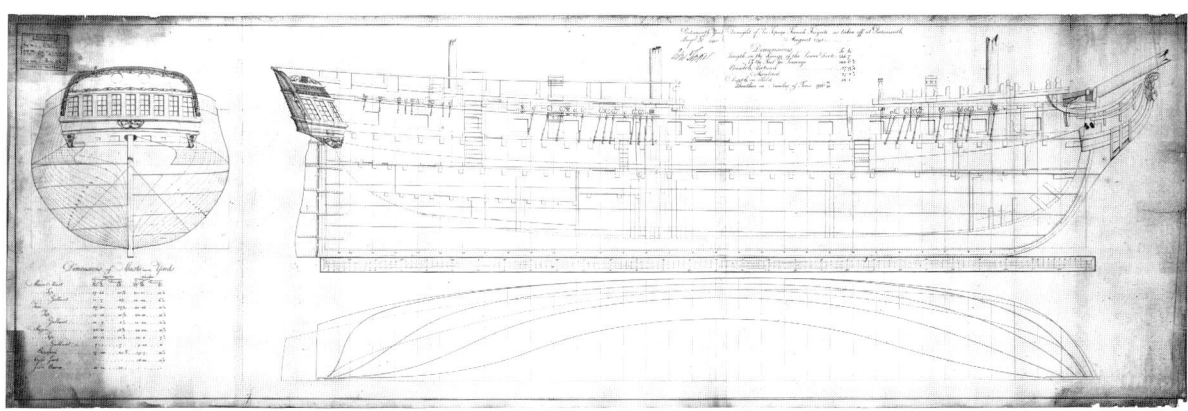

The chapters of general history are followed by a couple on more general topics, including the evolution of the internal arrangements, armament, decoration, and masting, concluding with a pictorial section featuring contemporary frigate models. Some information on performance under sail is quoted, and it is interesting that they were not universally as fast as many of their Royal Navy opponents made out – in fact, some captured ships seem to have been faster after modification in British dockyards.

If there is one major criticism that can be levelled at the book, it is the lack of any sense of the 'enemy'. Reading British sources it is never long before there is some reference to what is supposed to be going on across the Channel; possibly it was different in France, but a French shipbuilding report on the British and Dutch navies dated 1737 – published recently under Boudriot's English imprint as *Eighteenth Century Shipbuilding* – suggests the opposite. Admittedly, many of the best illustrations of French frigates come from Admiralty 'as captured' draughts, but any comparison with British practice from French sources would have been valuable.

This said, the book is an excellent survey that is far ahead of anything so far published. In an ideal world, it would be quickly followed by a similar study of the French line of battleship.

Robert Gardiner

Ian Buxton, Metal Industries: Shipbreaking at Rosyth and Charlestown, *published by the World Ship Society (Tynemouth 1992). 104 pages, 89 photographs. £8.00 Paperback.*

The end of a ship's life is inevitably sad, both to those who served in her and to those who designed her, which may partially account for the lack of readily available accounts of shipbreaking. This book, by Dr Buxton, Reader in Marine Transport at Newcastle University and Vice President of the World Ship Society, tells the story of shipbreaking at Rosyth and nearby Charlestown from 1923 until it finished in 1963.

It is largely an economic history and a fascinating one at that since it was not easy to make a profit and quite ingenious approaches were needed. In the early years the cost of oxygen was such that it was cheaper to dismantle a ship by cutting off the rivets with hammer and chisel. This led the company to explore better and cheaper processes for generating and transporting oxygen – and they also made a profit out of selling this expertise.

A large proportion of the profits came from armour between the wars and after the Second World War from the non-ferrous content. Figures are given for HMS *St Austell Bay*:

	Tons	Sale value (£)
Ferrous	1107	12,384
Non-ferrous	109	15,135

The company made £17,350 out of this ship but the profit was not easily earned. Non-ferrous scrap took 50 man-hours/ton to win compared with 14–18 for ferrous.

The photographs are outstanding: well selected, mainly of warships (including many of the German ships scuttled at Scapa Flow), with unusual views showing the arrangement of structure. The captions are comprehensive and accurate – even the ships appearing small in the background are identified. The reproduction is amongst the best I have seen in halftone and reflects very great credit on the printers. It is an unusual topic, the text is interesting and the photos fascinating; strongly recommended.

D K Brown, RCNC

Lt-Cdr John R P Lansdown, With the Carriers in Korea 1950–1953, *published by Square One Publications (Worcester 1992). 489 pages, 95 photographs. £18.50.*

The Korean war of June 1950 to July 1953 is remembered by few people today and even fewer will be aware that the RN and RAN operated a light fleet carrier on the west side of Korea for virtually the whole of the war. The ships concerned were:

Triumph	June–September 1950
Theseus	September 1950–April 1951
Glory	June–September 1951; February–April 1952; November 1952–May 1953
Sydney	September 1951–January 1952
Ocean	May–November 1952; May–October 1953

All were supported by the maintenance carrier *Unicorn* throughout the war. Though she spent much of the war in Japan she was also the only carrier to bombard an enemy coast, on 22 September 1951.

The carriers operated very intensively throughout with 9–10 days in the operating area during which they would average some 60 sorties a day, weather permitting, from a complement of 32 aircraft, a figure which compares well with that achieved by the bigger fleet carriers in 1945. *Glory* and *Ocean* shared the record with 123 sorties in a day. The task was 'interdiction', stopping the supply lines. It was not easy as the North Koreans could use large numbers of people to rebuild bridges and relied on ox carts for transport.

The final total was 23,000 sorties, 15,000 bombs dropped, 58,000 rockets and 3,300,000 rounds of 20mm fired; an achievement all the more remarkable as many of the aircrew had only just completed their training. Casualties to aircrew were quite high, 26 Fleet Air Arm and 1 RAF killed in action and 7 more in flying accidents. There is also a harrowing story of life as a North Korean prisoner.

This war saw the last operational flight of the Seafire but the Firefly and Sea Fury bore the brunt of the action, the latter being the only piston-engined aircraft to shoot down a MiG-15 jet. The first rescues were by the amphibian Sea Otter but this was soon replaced by American manned helicopters whose bravery brought back many pilots.

This is a story which needs telling and the author is well qualified, having been AEO *Glory* from September

The light fleet carrier Theseus *on her way home from Korean War service in May 1951, with Firefly and Sea Fury aircraft on deck. Details of the ship's operations are available from a new book by Lt-Cdr J R P Lansdown reviewed here.* (CMP)

1952 to May 1953. The book is not easy to read due to an excess of detail and to the unfamiliarity of the place names, though two good maps are provided. The photographs seem from the squadron line books and are relevant, exciting and largely new, though the reproduction could be better. The problems met seem relevant to future limited wars and this book is worth studying.

David Brown, RCNC

M J Melvin, Minesweeper: The Role of the Motor Minesweeper in World War II, *published by Square One Publications (Worcester 1992).*
234 pages, 68 illustrations.
£14.95.

The author was a wartime stoker in an MMS, moving on to the RAF after the war before becoming a clergyman and leading the restoration of *MMS 191*. There are some interesting accounts of wartime life and the problems of keeping the engines turning. Very detailed tables list the disposition of every MMS and BYMS both geographically and by flotilla. Their postwar careers and fates are also listed – including one which became a Belgian barquentine.

The author is frequently muddled over the general background, attributing the design to HMS *Vernon* – they were designed by DNC department under W J Holt, RCNC. He claims that insufficient attention was given to MCM before the war, which may be true but he is wrong in blaming the War Office for the shortage of ships. His wider claim that the German mine offensive was never defeated seems invalid as the Normandy landing penetrated a major barrier with losses which, though painful, were far too few to affect the operation. Only the pressure mine was unsweepable.

The illustrations are interesting but not well reproduced.

D K Brown, RCNC

J Rohwer and G Hummelchen, Chronology of the War at Sea, *revised edition published by Greenhill Books, 1992.*
268 × 210mm, 446 pages.
£35.00.

This book provides brief notes on all operations at sea, in all theatres, from 19 August 1939 to 30 November 1945; a sad note on which to finish, the wreck of HMCS *Merittonia*. The format is not quite day by day: a complete operation, such as a convoy is fully covered before moving on to the next time and place. The entries, though short, are surprisingly readable and very comprehensive, listing all the vessels and other units involved.

It is a much updated version of a two-volume chronology published 1972–74 and the authors claim, apparently with justice, that the new version is both corrected and amplified particularly as regards Sigint.

The operations of Soviet naval forces are very well covered, which will be of great interest since there is no readily available and comprehensive English language history. There is a slight trace of the book's German origin as every sortie of their ships seems to be listed while the occasional use of the word enemy refers to the British. Overall, though, the balance is very good.

Both the authors are well known historians with reputations for accuracy, giving confidence in the book and they acknowledge the help of a long list of advisers from all countries involved. A considerable number of checks revealed only one trivial error – and many of these checks were 'trick' questions.

A book of this sort needs a comprehensive and reliable index and in fact it has six: warships, merchant ships, army units, air force units, personnel, miscellaneous (convoys, mine barrages, operations, listed by code name). All seem accurate. There are some 10,000 warship names, 1500 merchant ships, almost 1900 units, 1200 convoys, 400 other operations and 3250 lucky individuals listed. Being greedy, one might ask for a 'weapons' index so that the use of, say, HF/DF, FIDO (Mk 24 mine/torpedo) may be examined.

It is a most interesting and valuable book and an essential reference for any study of the war.

D K Brown, RCNC

Paul Schmalenbach, Die Geschichte der deutschen Schiffsartillerie, *third edition, published by Koehlers (Herford 1992).*
240 × 165mm, 193 pages, 42 text illustrations, 56 photographs, DM 49.80.

For those with a working knowledge of German, this posthumous edition is a sound account of the development of guns and fire control as installed in ships of the German navy. The text illustrations are clear and the photographs are generally excellent. Relatively little material which would be new to advanced students of the subject is provided. More tables giving details of the ballistics of German naval guns, and further information on propellants, in which the German navy were usually in advance of their rivals, might have been given at the price of a few more pages. Some notes on guns for projected ships such as the 42cm SKL/45 for the 1918 capital ship programme, and the huge 53.3cm Gerat 36 would have been of interest.

N J M Cambell

SHORT NOTICES

'Anatomy of the Ship' series:
John McKay and John Harland, The 'Flower' Class Corvette Agassiz
240 × 254mm, 160 pages, 30 photographs, 350 line drawings.
£24.00
Al Ross, The Escort Carrier Gambier Bay
240 × 254mm, 112 pages, 30 photographs and 200 line drawings.
£20.00
Both published by Conway Maritime Press, 1993.

The latest contributions to this well-known series cover ship types so far ignored by 'Anatomy' authors. *Gambier Bay* will always be famous for her heroic action against Japanese heavy units off Samar in 1944, but is in fact typical of the fifty *Casablanca* class ships, all of which were retained by the US Navy for employment in the Pacific. The text to the volume includes transcripts of firsthand reports on the loss of the ship.

The 'Flower' class corvette also needs no introduction, although the prototype chosen may not be familiar, being one of the early short-forecastle Canadian-built examples (appropriately, for a collaboration between two authors from British Columbia). John McKay, the draughtsman responsible, is one of the most accomplished artists contributing to the series, and this is both his most extensive and most finely detailed to date. While

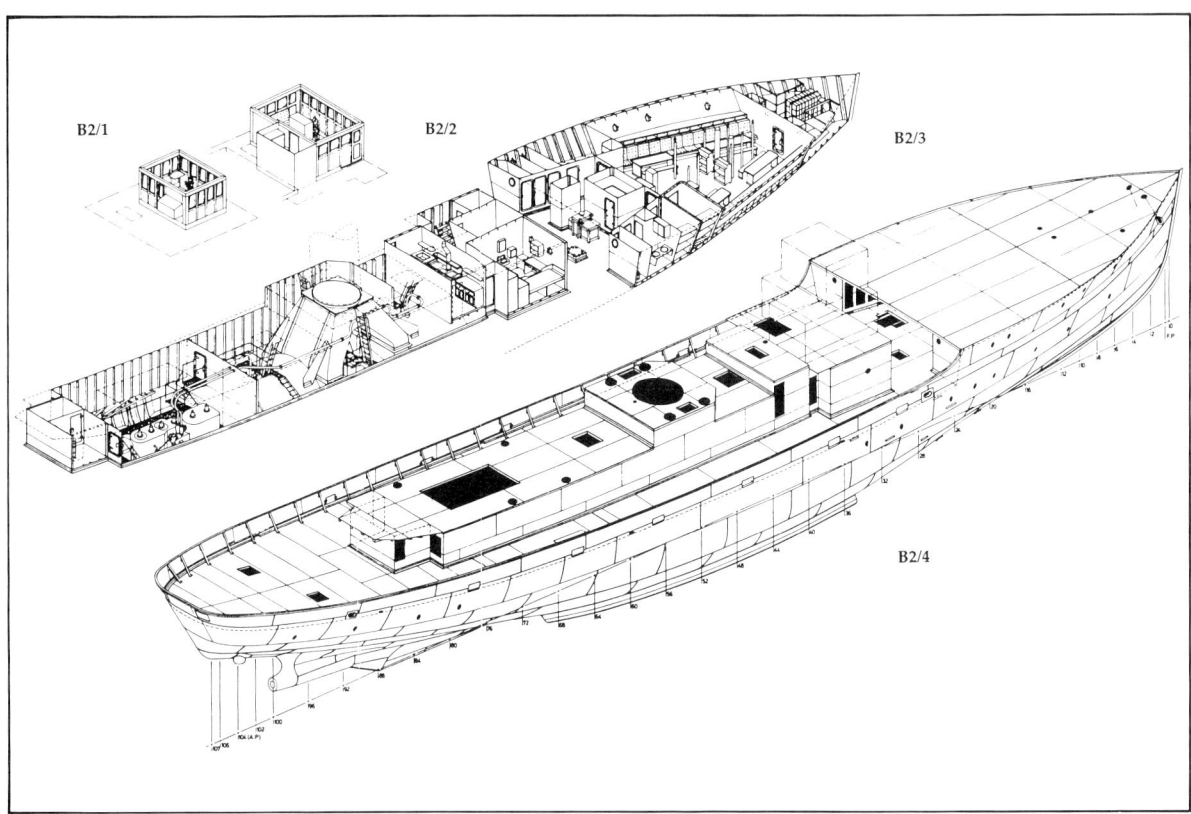

One of John McKay's characteristic isometric drawings, from his new 'Anatomy of the Ship' volume on the 'Flower' class corvette Agassiz.

the volume concentrates on its name ship, there are also drawings of the main modifications and developments of the class, concluding with a plan of the succeeding 'Castle' class.

M P Cocker, Mine Warfare Vessels of the Royal Navy – 1908 to Date, *published by Airlife Publishing (Shrewsbury 1993). 212 × 140mm, 224 pages, 200 photographs. £22.95.*

It is difficult to see what this book is meant to achieve. It lists all the Royal Navy's vessels, purpose-built and converted, that were devoted to mine warfare, but in no greater detail than can be found in any readily available standard reference. There is little comment on each type or class, and what there is betrays a lack of familiarity with recent publications. One would not expect original research from a book pitched at this level but it is no longer excusable to claim that the *Abdiel*s 'could make 40kts with no pressing of engineroom power', while the assertion that the stern of the *Adventure* was rebuilt 'for better positioning of the minelaying doors' is at best a half-truth (for David Brown's analysis of the baleful hydrodynamic effects of the transom, see *Warship 5*). Where it does have interesting information, such as the Navy's 1982 minelaying experiments with the ferry *Ailsa Princess*, the book fails to explain anything of the circumstances. In similar vein, the potted history of mine warfare is superficial and unrelated to specific design developments.

The best feature of the book is the large number of well printed photos, but readers will have to decide whether that is a worthwhile return for such a high cover price.

M Duffy, S Fisher, B Greenhill, D J Starkey and Joyce Youings (eds), A New Maritime History of Devon, Vol I From Early Times to the Late Eighteenth Century, *published by Conway Maritime Press in association with the University of Exeter (London and Exeter 1993). 295 × 248mm, 256 pages, 100 photographs and 60 line drawings. £35.00.*

A compilation of local history essays would not usually find a mention in *Warship*, but this one has a number on naval topics: four are dedicated to various aspects of Plymouth (Devonport) Dockyard and there are more general chapters on naval strategy and the south-west, and Devon men and the navy.

John English, Amazon to Ivanhoe: British Standard Destroyers of the 1930s, *published by the World Ship Society (Kendal 1993).*
244 × 186mm, 144 pages, 120 photographs. Paperback.

This is another of the World Ship Society's excellent little monographs, and since Edgar March's monumental *British Destroyers* is such an unsatisfactory work it is well worth taking another look at the interwar vessels. The author has clearly read all the design papers in the Covers, but no very profound analysis of British destroyer policy emerges. It may be that the principles were so well understood that there was barely any real discussion, but the result is that development from class to class seems a matter of 'tinkering' rather than conscious desire for improvement. If the design background is thin, the book scores more heavily on service history, with short biographies of the lives of all the vessels, including outlines of major refits. It is also well illustrated and at a time when specialist books seem to be very expensive indeed, it is a pleasure to note the relatively low price of the work.

William M. Fowler, Jr, Under Two Flags: The American Navy in the Civil War, *published by W W Norton (New York and London 1990).*
234 × 152mm, 352 pages, 22 photographs, 2 maps.
£15.95/$30.00.

Although written by an academic, this is intended to be a readable general account of the naval side of the war between the states. In this it succeeds very well, but it is likely to appeal more to those who know little or nothing of the conflict rather than those with any depth of knowledge. The author is not much concerned with technical matters, so the significance of so many warship developments is barely addressed.

Robert Gardiner, The First Frigates: Nine-Pounder and Twelve-Pounder Frigates, 1748–1815, *published by Conway Maritime Press, 1992.*
267 × 267mm, 128 pages, 100 photographs.
£25.00.

The first of a new series called 'Conway's Ship Types' designed to use the huge and under-exploited collection of ship plans in the National Marine Museum to illustrate monographs covering some categories of vessel previously neglected in print. This title may be outside the interest of most *Warship* readers, but the next will be more appropriate: in *Paddle Warships*, due for publication in October 1993, D K Brown will tackle the earliest steam-powered warships. Each volume includes detailed data and construction tables, a general design history, and short chapters on specific features, such as armament, rigging, performance, and so forth. Besides the original draughts, there are illustrations of contemporary models, paintings and prints.

Jan Glete, Navies and Nations: Warships, Navies and State Building in Europe and America, 1500–1860, *published by Almqvist & Wiksell (Stockholm 1993).*
242 × 172mm, 2 vols totalling 752 pages, numerous charts, diagrams and tables.
Paperback.
No price quoted.

The cumbersome title disguises one of the most important books on the era of sailing navies published in recent years. Long-standing readers of *Warship* will remember Jan Glete's extremely useful navy-by-navy listing of warship sources in issue 45, and a careful compilation of statistics on ship size and type forms the basis of his present *magnum opus*. But building on this data he has constructed a detailed history of the development of all the major navies in the period covered, showing not only the degree by which they grew or declined but also the reasons for the change. In passing he has much to say on matters of ship design – including some valuable information on the more obscure types like surviving oared craft, and the lesser known naval powers. However, the book is an excellent example of the new 'vertically integrated' style of naval history, encompassing virtually every aspect of the subject, from the political factors ruling the funding of navies, through the strategy of their deployment and its influence on tactics, right down to how that affected choice of ship type and shaped design.

It is a long book, and not easy reading, but has so much of interest to say on so many questions that it really must be read by anyone with a more than passing interest in sailing navies and their ships.

Basil Greenhill and Ann Giffard, Steam, Politics and Patronage: The Transformation of the Royal Navy, 1815–1845, *published by Conway Maritime Press, 1993.*
234 × 156mm, 256 pages, 40 illustrations.
£25.00.

Further evidence to support the growing consensus that the Royal Navy was far from backward in its attitude to new technology in the nineteenth century, this book leans heavily on the private papers of a naval officer ancestor of one of its co-authors. As such it gives a clearer view of the attitudes and motivations of the navy's officer corps than a purely technical or political history could achieve.

From Beardmore Built, *one of the author's excellent illustrations comparing a plan of the Dalmuir yard with a very instructive bird's-eye perspective.*

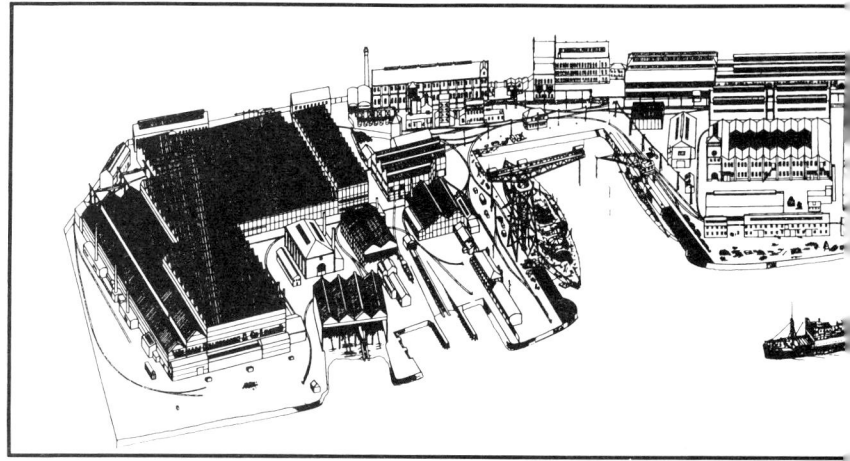

'History of the Ship' series: *Norman Friedman (ed), Navies in the Nuclear Age: The Warship since 1945, published by Conway Maritime Press, 1993. 295 × 248mm, 208 pages, 250 photographs and line drawings. £28.00.*

This series was mentioned in last year's 'Short Notices' so there is no need to repeat the main features. The above title is the only one of this year's quartet devoted to a naval theme, but since it was one of the first general histories to benefit from the flood of information now coming out of the old Soviet Union, it is particularly significant. Soviet ship design is beginning to make sense and the policy and strategy shifts that drove it are now confirmed by Russian sources. The individual chapters are principally arranged around ship types, but there are more general essays on such important matters as the influence of nuclear weapons on navies, and the arcane practices of electronic warfare.

Ian Johnston, Beardmore Built: The Rise and Fall of a Clydeside Shipyard, *published by Clydebank District Libraries & Museums (Glasgow 1993). 270 × 210mm, 192 pages, 175 photographs, line drawings and maps. £14.99 hardback; £9.99 paperback.*

Beardmores had a short but very significant history as warship builders (1900–1930), although their military–industrial activities extended to ordnance, tanks, aircraft and airships (they built railway locomo-

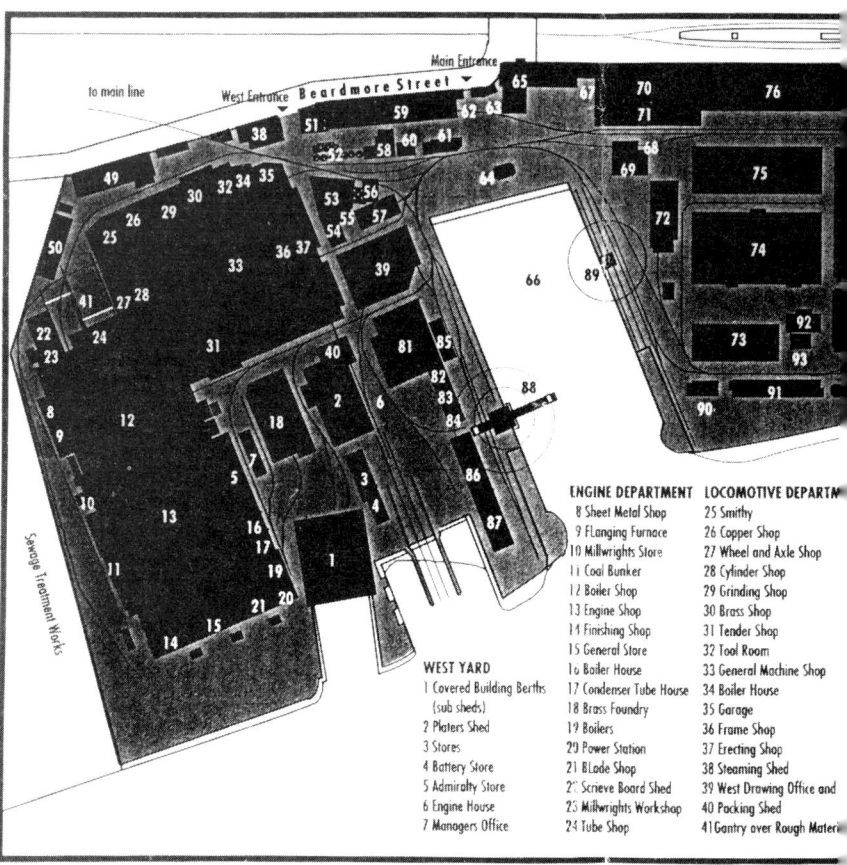

tives and other heavy engineering items as well), all of which are covered by this interesting book. The company became shipbuilders by taking over the distinguished Govan yard of Robert Napier but then proceeded to lay out the new and purpose-built Naval Construction Works at Dalmuir. This served them well in the expansionary phase of the pre-1914 'naval race', but was too large and specialised a facility to survive in the postwar atmosphere of financial austerity and treaty restrictions. However, Beardmores produced some important vessels. They were early advocates of naval aviation and it is fitting that they were responsible for *Argus*, the world's first true flush-decked carrier; they were also awarded one of the abortive 'G3' battlecruisers.

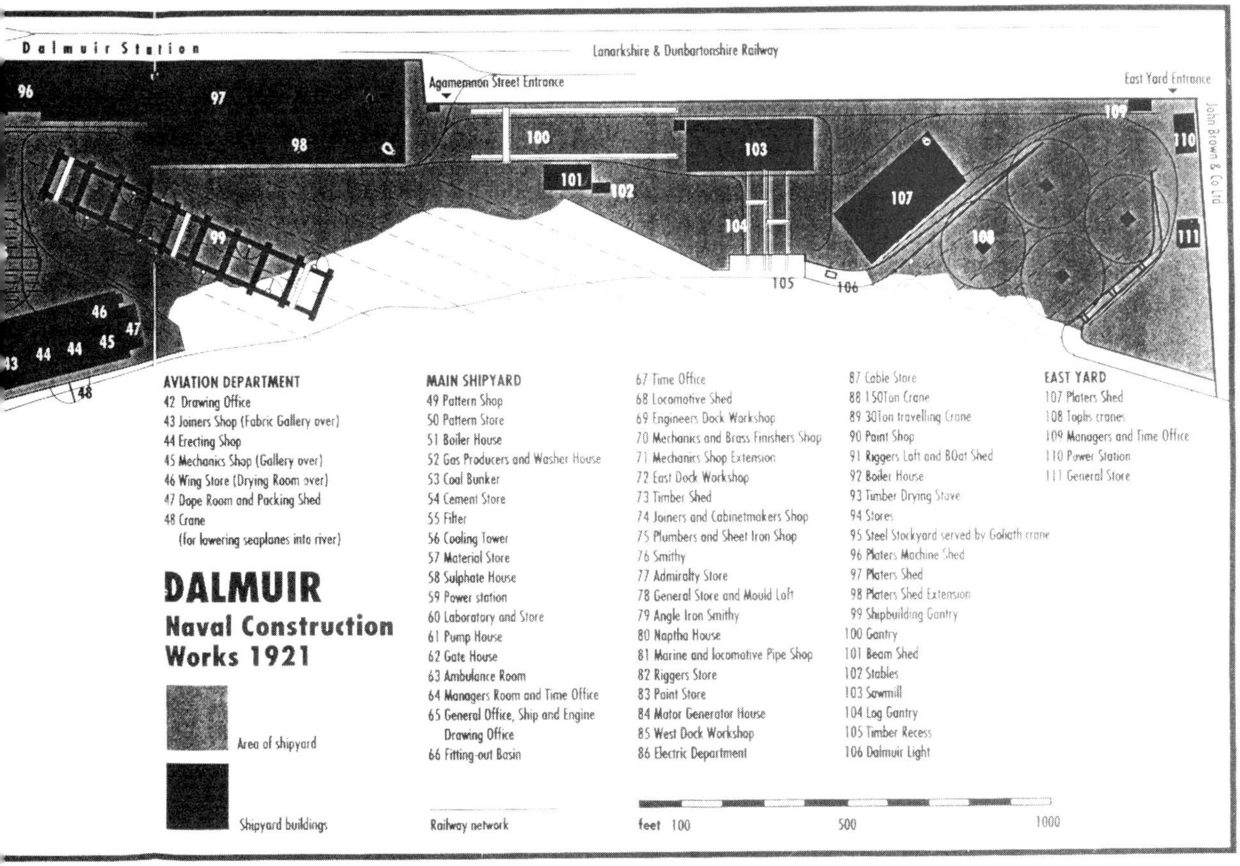

Dalmuir Station
Lanarkshire & Dunbartonshire Railway
Agamemnon Street Entrance
East Yard Entrance
John Brown & Co Ltd
96
97
98
99
100
101
102
103
104
105
106
107
108
109
110
111
46
47
45
44
44
43
48
AVIATION DEPARTMENT
42 Drawing Office
43 Joiners Shop (Fabric Gallery over)
44 Erecting Shop
45 Mechanics Shop (Gallery over)
46 Wing Store (Drying Room over)
47 Dope Room and Packing Shed
48 Crane
(for lowering seaplanes into river)
DALMUIR
Naval Construction Works 1921
Area of shipyard
Shipyard buildings
MAIN SHIPYARD
49 Pattern Shop
50 Pattern Store
51 Boiler House
52 Gas Producers and Washer House
53 Coal Bunker
54 Cement Store
55 Filter
56 Cooling Tower
57 Material Store
58 Sulphate House
59 Power station
60 Laboratory and Store
61 Pump House
62 Gate House
63 Ambulance Room
64 Managers Room and Time Office
65 General Office, Ship and Engine Drawing Office
66 Fitting-out Basin
Railway network
67 Time Office
68 Locomotive Shed
69 Engineers Dock Workshop
70 Mechanics and Brass Finishers Shop
71 Mechanics Shop Extension
72 East Dock Workshop
73 Timber Shed
74 Joiners and Cabinetmakers Shop
75 Plumbers and Sheet Iron Shop
76 Smithy
77 Admiralty Store
78 General Store and Mould Loft
79 Angle Iron Smithy
80 Naptha House
81 Marine and locomotive Pipe Shop
82 Riggers Store
83 Paint Store
84 Motor Generator House
85 West Dock Workshop
86 Electric Department
87 Cable Store
88 150Ton Crane
89 30Ton travelling Crane
90 Paint Shop
91 Riggers Loft and BOat Shed
92 Boiler House
93 Timber Drying Stove
94 Stores
95 Steel Stockyard served by Goliath crane
96 Platers Machine Shed
97 Platers Shed
98 Platers Shed Extension
99 Shipbuilding Gantry
100 Gantry
101 Beam Shed
102 Stables
103 Sawmill
104 Log Gantry
105 Timber Recess
106 Dalmuir Light
EAST YARD
107 Platers Shed
108 Toplis cranes
109 Managers and Time Office
110 Power Station
111 General Store
feet 100
500
1000

Although only a portion of this book is actually devoted to warships, it is worth consideration for its integrated description of the activities of a major industrial concern. The author is remarkably at home with all the company's products – a list of acknowledgements to many of the best authorities suggests careful checking of fact and interpretation – and it is excellently illustrated, employing both professionally-taken photographs from the company's archives and the author's own maps, plans and line drawings.

Paul Kemp, Convoy Protection: The Defence of Seaborne Trade, *published by Arms & Armour Press, 1993. 234 × 156mm, 124 pages, 95 photographs. £17.99.*

This relatively short book is a comparison between the German cam-

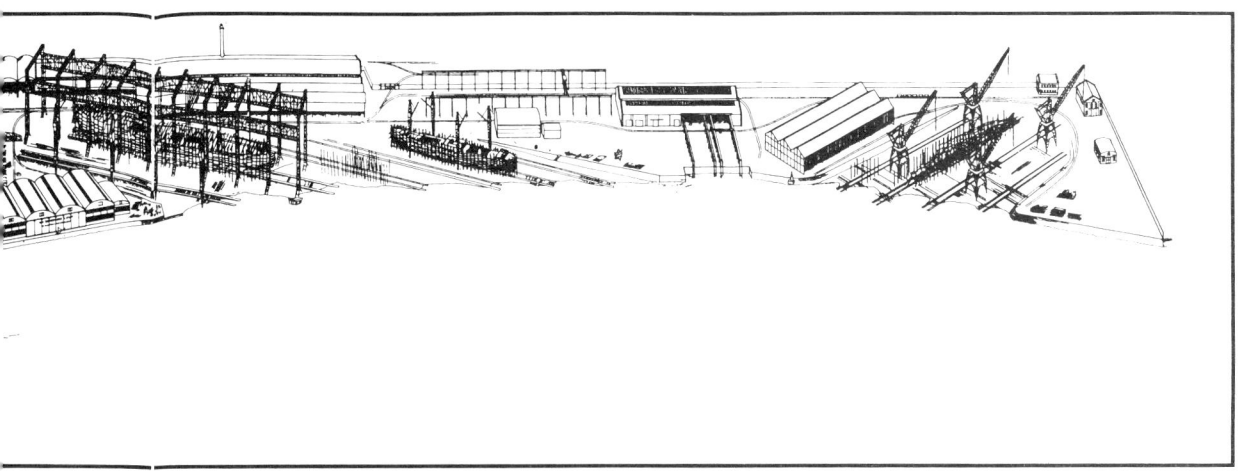

paign against Allied shipping in the First World War, which failed (just), and the highly successful American assault on the Japanese merchant fleet in the Second. The author obviously knows a lot about submarines and that understanding is always in the background, but the book seems entirely dependent on secondary sources and comes to no startling or original conclusions. Perhaps it was never intended for the real specialist, and at the level of an introduction to the subject it provides interesting reading on the mechanics of trade defence. However, most of the quotations that bring the narrative alive are drafted in from other works, and the analysis is unconvincing on such crucial issues as why the Royal Navy became so opposed to convoy during the nineteenth century.

John Lambert and Al Ross, Allied Coastal Forces of World War Two, Vol II Vosper MTBs and US Elcos, *published by Conway Maritime Press, 1993.*
296 × 248mm, 256 pages, 200 photographs, 700 line drawings.
£35.00.

Those familiar with the first volume need only be told that there is even more material in this second tranche (a third is in prospect). Like a dozen or more 'Anatomy of the Ship' volumes grouped together, this is essentially a collection of drawings – of complete boats, variations, modifications, fittings, machinery and weaponry – but with a good leavening of photos and a text which covers the main points of design, construction and operation. The theme of this volume is the short hard-chine boat, the major designs being the British Vosper and the US Elco. Lambert was responsible for the Vospers and Ross the Elcos.

David Loades, The Tudor Navy: An Administrative, Political and Military History, *published by Scolar Press (Aldershot 1992).*
216 × 138mm, 328 pages.
£35.00.

An academic history of the Tudor navy is overdue, given the amount of new work that has been prompted by

The bridge controls of a Vosper MTB, an example of the detail available from Volume II of Allied Coastal Forces. *(Vosper Thornycroft)*

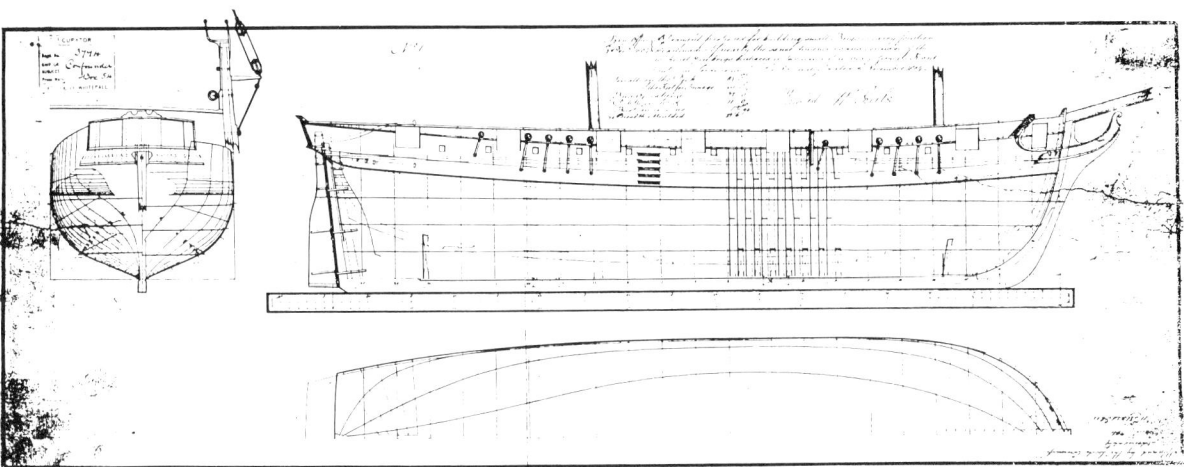

The illustrations to David Lyon's Sailing Navy List *concentrate on smaller, less well known craft, like this gun-brig the* Confounder *of 1805.* (NMM)

events as diverse as the Armada celebrations or the raising of the *Mary Rose*. It is not surprising considering the target readership that this book does not make a special feature of technical discoveries about ships and guns, but the author is at least familiar with their significance. In the modern style, it is an 'integrated' history, bringing together the political and economic background to the navy's development with a description of its operations.

D J Lyon, The Sailing Navy List: All the Ships of The Royal Navy – Built, Purchased and Captured – 1688–1860, *published by Conway Maritime Press, 1993.*
295 × 248mm, 352 pages, 300 illustrations.
£60.00.

As the most successful force of the sailing era the Royal Navy has been the subject of intense study, yet before this book there was no authoritative listing of the ships themselves. Colledge's dictionary-like *Ships of the Royal Navy* was very useful for discovering the main data on a named ship but could give no sense of the development of a ship type, let alone the navy as a whole. What has been needed for years is a fleet list, divided by category and class as for modern warships, with principal dimensions, armament and building data. For the navy's purpose-built vessels, all this is provided by Lyon's monumental work – in accurate and authoritative detail – and yet this is only a small proportion of the book. It also includes the vast number of captured vessels, those purchased and hired, and others that served in paramilitary forces like the Revenue and Excise. Small craft, so often ignored because of difficulties with the sources, find their place, right down to yard craft; many vessels enjoyed long careers after final decommissioning as harbour service vessels and hulks, and these are also listed, with relevant dates.

The quantity of the information is matched by its quality, being largely taken from official sources. The author spent much of his working life as curator of the incomparable collection of Admiralty plans at Greenwich, and this book derives from the needs of cataloguing that archive and answering a plethora of public queries about its subjects. Probably its greatest contribution is to sort out the design aspects: each Rate is listed chronologically, with ships built to the same draught grouped as classes. The work of each Master Shipwright and Surveyor is identified, and the whole forms a skeletal design history of the Royal Navy in the age of sail. Against the significance of the information itself, the illustrations are gilt on the gingerbread, but the contemporary draughts, models and prints have been chosen to reflect all the main types, but there is a refreshing emphasis on some of the lesser known and smaller vessels.

Few books are genuinely indispensible, but if you have any interest whatever in the sailing navy, this is one.

L A Ritchie (ed), The Shipbuilding Industry: A Guide to Historical Records, *published by Manchester University Press (Manchester 1992).*
213 × 138mm, 206 pages.
£35.00.

There is a company-by-company listing of what records are held and where. For the serious scholar this is a very useful starting point, although more information about the plans collections (type of draught, and of which ships, for example) would have made it even more valuable to the technical historian. Nevertheless it does point out which collections are not at the National Maritime Museum, and gives full addresses for the holding institutions. On a more general level, each of the shipbuilders included is treated to a short history, with dates for changes of name and status, which is interesting information in its own right, although possibly not enough to justify purchase at such a high price.

Robert C Stern, The Lexington Class Carriers, *published by Arms & Armour Press, 1993.*
245 × 190mm, 160 pages, 175 photographs.
£19.99.

An extremely full and comprehensive monograph on the largest of the 'first generation' carriers, this book uses the ships as a paradigm for the development of carrier aviation down to the end of the Second World War. It includes all the usual subjects now *de rigeur* in such technical works – design, machinery, weapons, protection, etc – but also the crew and habitability considerations, and operational factors. On matters of fittings and equipment, it is detailed enough to satisfy the most ardent 'rivet-counter', without losing sight of the more sophisticated interplay of technology and policy that makes the study of warships so fascinating to the more sophisticated. An additional plus is the selection of photographs, which in the majority of cases truly illustrate points made in the text – unlike so many contemporary books where their relevance is marginal. It is, however, a very crowded book with text and pictures crammed into what seems too small a format, both in terms of page size and extent; but if this was done with the laudable intention of keeping the price down, then few readers will complain.

A typical crowded spread from Robert C Stern's highly detailed monograph on Lexington *class carriers.*

M J Whitley, German Coastal Forces of World War Two, *published by Arms & Armour Press, 1993.*
245 × 245mm, 192 pages, 120 photographs and line drawings.
£30.00.

This is the latest in a series of books on *Kriegsmarine* warships from this author and publisher, and the familiar combination of design and operational history is continued. Apart from the obvious *Schnellboote* (E-boats to the Allies), this study also includes their depot ships, minesweepers, and other small craft like R- and MZ-boats, *U-jägers* and even the curious F-boats; one perhaps unexpected inclusion is the *Kleinkampfverbände*, the midget submarines introduced towards the end of the war, so the 'coastal forces' of the title should be understood in its widest sense.

This allows the inshore war to be covered in full, and although the emphasis is on the Channel and North Sea, chapters are also devoted to the Baltic, Mediterranean and Black Sea campaigns, some of which are not at all familiar to English-speaking readers. The technical background is less fulsome, but given the vast range of craft encompassed, this is probably inevitable – and extensive appendices list the essential data. Compared with previous volumes the photographic collection is a little disappointing (in content and quality), but this may simply reflect the paucity of wartime images of small craft.

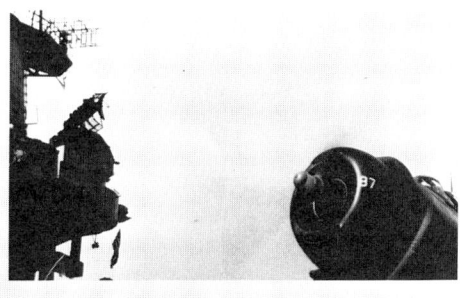

NAVAL BOOKS OF THE YEAR

BOOKS ANNOUNCED

C K S Aylwin, Maritime Forces Supporting Malta during the Siege of 1940–43, *published by the author, 1992, £10.00.*
Very detailed research into all ships and aircraft involved.

K Jack Bauer and Stephen S Roberts, A Register of Ships of the US Navy, 1775–1990, *published in the UK by Greenwood Press, 1993, £59.95.*

C Chant, Destroyers and Frigates, *published by Brassey's, 1993.*

R E and T N Dupuy, Collins Encyclopaedia of Military History, *published by HarperCollins, 1993, £40.00.*
From 3500 BC to the present; 1700 pages.

Norman Gelb, Desperate Venture: The Story of Operation Torch – the Allied Invasion of North Africa, *published by Hodder & Stoughton, 1992, £18.99.*

Jim Mitchell's atmospheric painting of Schnellboote *at sea forms an attractive cover illustration to Mike Whitley's new book on German Coastal Forces.*

Edwin Gray, Hitler's Battleships, *published by Leo Cooper, 1993, £17.95.*
Small in number but big in appeal, German capital ships of the Second World War are treated to another general account of their activities.

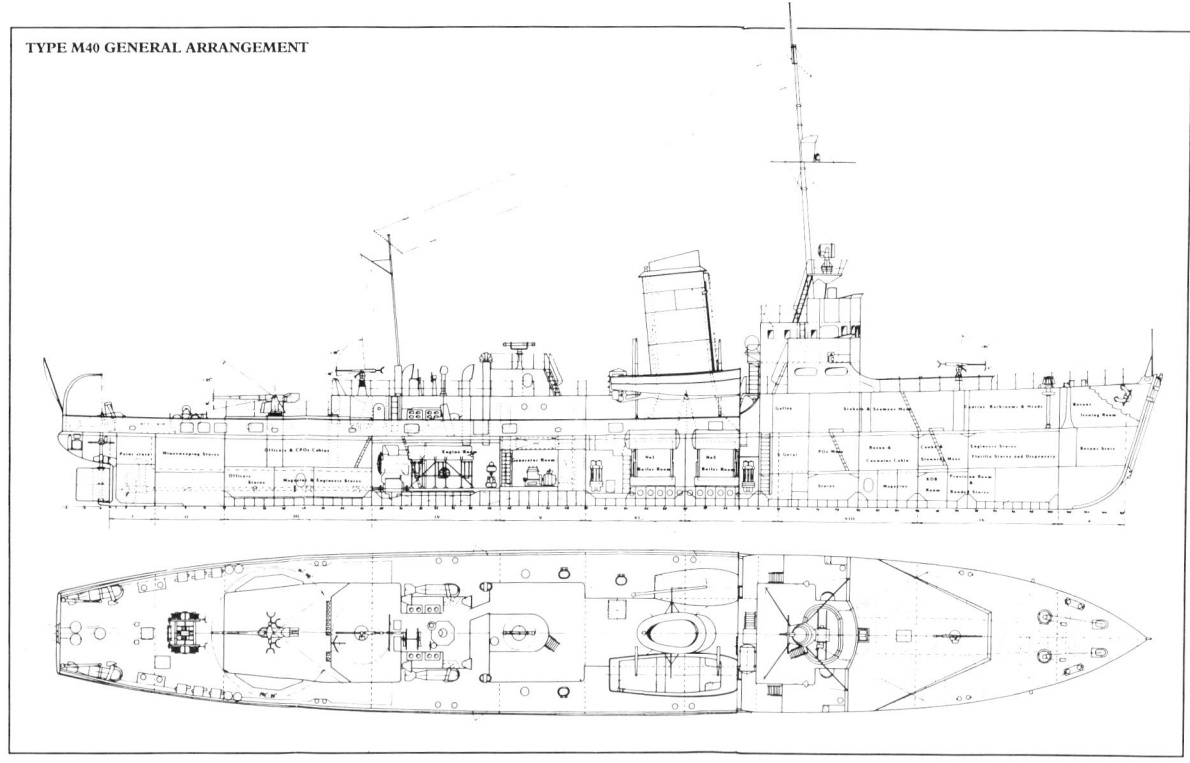

A general arrangement of the M40 Type minesweeper, from German Coastal Forces.

Michael J Hadley and Roger Sarty, Tin-Pots and Pirate Ships: Canadian Naval Forces and German Sea Raiders, 1880–1918, *published by McGill-Queen's University Press (Montreal, Quebec 1991).*
An apparently obscure subject but actually a very useful contribution to the early history of the RCN.

F C Lane, Venetian Ships and Shipbuilders of the Renaissance, *published by Johns Hopkins, £13.00.*
Paperback reprint of 1934 classic.

Edward C Myers, Thunder in the Morning Calm: The Royal Canadian Navy in Korea, 1950–1955, *published by Vanwell (St Catherines, Ontario 1992).*

Heinz J Novarra, Grey Wolves of the Sea: German U-Boat Type VII, *published by Schiffer (West Chester, Pa 1992).*
Pictorial collection, translated from German.

Edward Smithies with Colin John Bruce, War at Sea, *published by Constable, 1992, £14.95.*
Series of interview with participants.

Ray Sturtivant and Gordon Page, Royal Navy Aircraft Serial and Units 1911–1919, *published by Air-Britain Publications, 1992, £28.50.*
List of over 15,000 aircraft by serial, with units, bases and fates.

G Till, Coastal Forces, *published by Brassey's, 1993.*

J Gordon Vaeth, Blimps and U-Boats: US Navy Airships in the Battle of the Atlantic, *published by Naval Institute Press (Annapolis, Md 1992).*

Ben Warlow, Shore Establishments of the Royal Navy: being a List of the Static Ships and Establishments of the Royal Navy, *published by Maritime Books (Liskeard 1992), £40.00.*
Useful, if ferociously expensive, little book detailing some 2700 'stone frigates', with their functions and dates of operation.

Denis and Peggy Warner with Sado Seno, Disaster in the Pacific: New Light on the Battle of Savo Island, *published by Naval Institute Press (Annapolis, Md 1992).*

John Winton, For Those in Peril: Fifty Years of Royal Navy Search and Rescue, *published by Robert Hale, 1992, £25.00.*
A well written account of a somewhat neglected area of naval activity.

THE NAVAL YEAR IN REVIEW

The events covered by this review stretch from approximately
May 1992 to May 1993, with some reference before and after.
Compiled by Ian Sturton.

A. INTRODUCTION

The year was noteworthy for the survival of Russia intact and under constitutional rule, for setbacks to European unity and for the spreading Bosnian tragedy. Reductions in the armed forces of NATO members and countries in the former Warsaw Pact continued; NATO naval forces were reduced by 10 per cent between 1990 and 1993, and are not at present due to fall much further.

Increased defence spending in the Far East is linked to economic advance and territorial rivalries. China's offer to shelve the issue of sovereignty over the uninhabited but potentially oil-rich Spratly Islands was rejected by ASEAN foreign ministers. Although world defence spending in 1991 was down by 3 per cent, the third successive annual fall, China was a big exception, while, according to figures published by SIPRI in Stockholm, between the early 1980s and 1991 the countries of South-East Asia increased defence spending from 15 per cent of the world's total to 25 per cent.

The strengths of the major naval powers are listed in Table 1.

B(i). THE STRATEGIC BALANCE

The future strategic environment continues uncertain; in addition to political and economic instability throughout Europe, there is concern about the proliferation of nations with ballistic missiles and the ambitions of threshold nuclear powers. In June, the NATO Council agreed on a NATO peacekeeping (and potentially peacemaking) role, outside its traditional area if members agree, but as there is no political consensus on aims, extending peacekeeping to peacemaking is problematic. Events in the former Yugoslavia have highlighted Europe's, and NATO's, disunity, and have shown that there is

The Netherlands frigate Willem van der Zaan, *second unit of the* Karel Doorman *class*. (RNethN)

Table 1. MAJOR WARSHIP TYPES OF PRINCIPAL NAVIES, 1 APRIL 1993

Type	USA	Russia	UK	France	China	India	Japan	Italy
CV (large)	14	1	–	–	–	–	–	–
CV (medium)	–	3	–	2	–	1	–	–
CV (small)	–	–	3	–	–	1	–	1
Battleship	–	–	–	–	–	–	–	–
Cruiser (helicopter)	–	1	–	1	–	–	–	1
Cruiser (missile)	50	23	–	–	–	–	–	–
Destroyer	38	36	12	15	16	5	42	3
Frigate (fleet)	59	41	26	3	39	10	20	14
(escort)	–	108	–	20	–	9	–	13
SSBN	21	54	3	5	1	–	–	–
SSGN	–	30	–	–	–	–	–	–
SSN	85	65	13	5	5	–	–	–
SS (all types)	–	76	5	8	c30	17	15	8
MCMV (ocean and coastal)	17	172	32	16	c110	12	31	12

Note: *The fourteen US carriers include one in overhaul. Russian totals include many ineffective or unserviceable ships.*

no prospect of a common European defence policy. Precipitate disarmament has made even peacekeeping difficult, while the UN flight ban over Bosnia, enforced by over 50 NATO planes from April 1993, has had no effect on the ground war.

The Italian frigate Orsa, *last of four* Lupo *class built between 1974 and 1980. Four similar ships built for Iraq were taken over in 1992 and should enter Italian service as offshore patrol vessels in 1993–94.* (Fincantieri)

B(ii). DISARMAMENT

The post-START agreement of June 1992, providing for further cuts in strategic nuclear forces in a two-stage process to 2003, was signed as the START 2 Treaty (January 1993), but cannot take effect until the original START 1 is ratified. Ukraine, claiming that its sovereignty would be infringed if strategic forces nominally under Commonwealth of Independent States (CIS) control were in reality entirely in Russian hands, has not yet agreed to return strategic weapons to Russian soil, and has delayed deciding START 1 ratification until late 1993.

If START 2 were fully implemented, Britain's maximum Trident capacity of 512 warheads would be about 15 per cent of the allowed US total and about 30 per cent of its submarine warheads; similar figures would apply to France. Both countries will probably reduce missile and warhead acquisitions.

B(iii). THE ENVIRONMENT

Up to 200 nuclear submarines, many with two reactors, are due to be scrapped in the next ten years, causing major radioactive waste disposal problems. In September, sixteen countries bordering the Atlantic signed the Paris Convention, imposing a 15-year total ban on dumping radioactive waste in the ocean; Britain and France had held out for dumping larger items such as submarine reactors. The Soviet Government declared in 1989 that it 'did not dump, does not dump or plan to dump radioactive waste at sea', but official Russian reports published in 1993 show the declaration to have been entirely untruthful. Twenty-one nuclear reactors from ten submarines and the icebreaker *Lenin*, seven still with fuel rods, were dumped in the Barents and Kara seas (eight in only 20m of water), and two more in the Sea of Japan between the early 1960s and 1991; 11,090 containers of low-level radioactive waste were also dumped in the Kara Sea. It is feared that within the next three years there will be an 'uncontrolled' and 'impulsive' leak of plutonium from the nuclear warheads of the submarine

The French light frigate La Fayette *after launch, 13 June 1992. Development of the new submarine deterrent force is greatly reducing the money available for new destroyers, frigates and hunter-killer submarines.* (DCN)

The Spanish Antarctic research ship Hesperides, *completed in 1991 as support ship for the Livingstone Island base.* (Bazan)

Komsomolets, which sank in the Barents Sea in April 1989.

The explosion at the nuclear reprocessing plant near Tomsk on 6 April 1993 further highlighted storage problems. The Tomsk-7 plant is believed to be the main fuel reprocessing centre for the Russian navy, and the explosion is thought to have occurred during the first stage of the process. The 235 Russian nuclear-powered ships have 407 reactors; a backlog of 140 cores with spent fuel has built up, but the navy has storage facilities for only three.

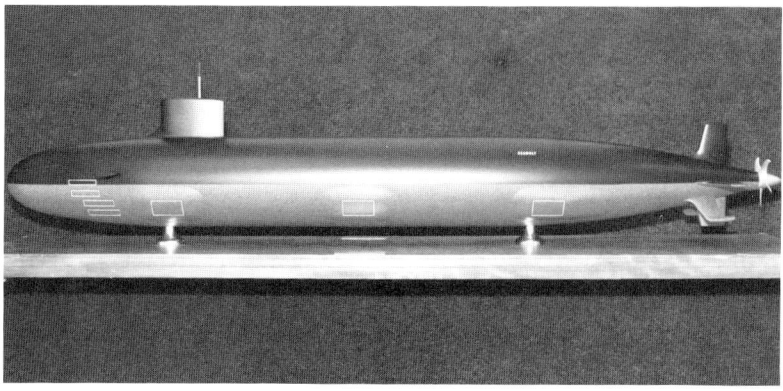

A model of USS Seawolf (SSN-21); construction of the second unit, Connecticut, was eventually approved, July 1992, and fabrication started on 9 September. (USN)

C. BUDGET PROPOSALS AND NEW PROGRAMMES

C(i). USA, NATO and Allies

Major NATO Navies

(a) United States. The defence spending (authorisation and appropriation) bill for FY93, beginning 1 October 1992, was finalised at $274.5b; the Clinton administration proposes spending $1360b on defence in 1993 through 1997, about $60b less than planned by the outgoing administration; the USN was ordered to make $3.0b cuts out of an armed forces total of $10.8b in Bush's proposed $266b for FY94. The Pentagon request for FY94 is $263.4b, $85b for the navy, avoiding major programme cuts. Details of new construction shipbuilding proposals are given in Table 2; the drastic fall in new orders is noteworthy.

The base force plan provided for the force total to be cut by 30 ships to 413, including 12 carriers and 11 active naval air wings. Various options for the future navy include withdrawing by 1999 all non-AEGIS cruisers and destroyers, all frigates and all A-6 aircraft – these last would be replaced by modified F-14s, and by deploying Marine Corps F/A-18s to carriers. Air wings might be reduced from 60 to 50 strike/air defence aircraft pending arrival of the A/F-X in 2015. Carriers will be withdrawn as new ones enter service, and there may be no dedicated training carrier. Construction of large ocean surveillance ships with the SURTASS towed array may be delayed, as modifications against more local submarine threats are desirable.

The future role of the navy is being very thoroughly reviewed. Carriers might be used as platforms for joint operations, with Marines and their helicopters embarked, where previously the carrier would have provided close air support for the Marine mission. Another policy document has suggested that the navy focus on 'littoral warfare', downgrading the carriers' strike role by leaving long-range and deep-strike missions to the USAF.

The USN finally left Subic Bay Naval Station, its largest overseas base, on 24 November, most facilities being moved to Guam. Other overseas stations are to be closed – although the Bush Administration studied financing a docking upgrade at Haifa – and in the continental United States the navy may be concentrated in a few 'megaports'.

Table 2. USN SHIPBUILDING PROGRAMMES, 1991–1994

New Construction	Approved (authorised and funded)			Proposed (subject to amendment)
	FY91	FY92	FY93	FY94
SSBN *Ohio*	1	–	–	–
SSN-21 *Seawolf*	1	1	–	–
CVN	–	–	–	–
DDG-51 *Arleigh Burke*	4	5	4	3
MHC *Osprey*	2	3	2	–
LHD *Wasp*	1	–	1**	–
LSD (cargo variant)	1	1*	–	–

Notes: *Proposals for FY94 provisional; no programme for FY95 and beyond has been announced. $829m for long-lead items for CVN-76 were approved in FY93. * LSD-52, authorised in FY92, was funded in FY93. ** LHD-6 was partly funded in FY93; funding should be completed in FY94. Conversion of* Inchon *as Mine Warfare Command Ship requested in FY94.*

HMS Iron Duke, *the fifth Type 23 frigate, commissioned on 20 May 1992.* (Yarrow)

The Army will continue to preposition stores and equipment afloat, and contracts were awarded for conversion of five merchant ships to large, medium-speed ro-ro ships for strategic sealift

(b) United Kingdom. Defence spending for FY92–93 was £24.2b ($46.0b), with projected spending for 1993–94 and 1994–95 cut to £23.5b and £23.8b respectively by the 1992 Autumn Statement; real defence spending is to be reduced by 10.5 per cent in the next three years, falling from 4 to 3.75 per cent of GNP. A wide-ranging defence cost-cutting exercise

USS Arleigh Burke (*DDG-51*). *Later units of the class will be modified with a hangar and two helicopters (Flight IIA).* (USN)

to meet reductions under this Statement and under 'Options for Change' is in progress; with cuts determined by financial considerations and expediency, a publicly-debated defence review is long overdue. Portland Naval Base will close by 1996, Sea Training being transferred to Plymouth. The RNR is to be reduced by 1200, to 3500, and the RNXS disbanded; eleven of twenty-four reserve training centres will close, and the 'River' class minesweepers will be withdrawn. A nucleus of 500 reservists will be trained aboard sea-going ships. Both Devonport and Rosyth Dockyards will remain open; on 24 June 1993, Devonport was awarded the £5bn contract for Trident SSBN refits, while Rosyth was given refit responsibility for some eighteen major surface ships for twelve years. New orders were limited to the fourth Trident SSBN and the long-awaited helicopter carrier (LPH); invitations to tender for the next batch of frigates and for the new amphibious assault ships are expected in 1994. The first

Batch 2 *Trafalgar* may be ordered in 1996 and a further batch of *Sandown* class SRMH is expected. The Royal Yacht will remain in service but is no longer considered as a possible hospital ship. The 1993 Defence White Paper, 'Defending our Future', published on 5 July 1993, confirmed the long-predicted decision to lay up the four new *Upholder* class diesel submarines – which may be offered to Canada – and to reduce the destroyer and frigate total to 35; personnel will be reduced by 2500.

(c) Canada. The 1993–94 defence budget was set at $9.6b, an increase of 3 per cent before inflation; for the next five years, defence spending is planned to show zero real growth (against the previously proposed 1.5 per cent) for the next five years, saving $4.6b; the equipment budget will be 23.8 per cent of the total in 1993–94, against 21.1 per cent in 1991–92, the increase being financed by personnel cuts. By 1997, the new maritime coast defence vessels will be

An artist's impression of the new Canadian Maritime Coast Defence Vessel (MCDV); first of class is due to be built by Halifax-Dartmouth Industries between 1993 and 1995. At 962 tons full load displacement and 55m overall length, the class is designed for general patrol duties with some MCM capability. (Canadian Maritime Command)

entering service and the patrol corvette and replacement submarine programmes should be well under way, but there will be only two replenishment ships.

(d) Germany. According to *Bundeswehrplan 94*, the equipment budget will be reduced by a further $15.3b between 1994 and 2006 – this is additional to the reductions announced in January 1992. In this time, the navy will receive four Type 124 AAW frigates, the first of four Type 212 submarines and (from 1998) two Type 702 replenishment ships; 40 per cent of its strength will be contributed to NATO maritime forces.

(e) Italy. Uncertainty caused by economic recession and major political corruption scandals was reflected in procurement indecision. The S90 submarine order has been delayed while a new light carrier, perhaps with an amphibious capability, remains a high priority project; Italy joined the Anglo-French Future Frigate Programme.

(f) Netherlands. A cut of 3.9 per cent left a final figure of $8.4b, causing programme delays; the second new replenishment ship was put off from 1996 to 1998. The main reductions proposed in the January 1993 Defence White Paper *'Prioriteiten Nota'* were in the army, but three more *Kortenaer* class frigates and the two *Zwaardvis* class

The launch of Brandenburg, *the first German Type 123 frigate, on 28 August 1992.* (FGN)

submarines are to be sold in 1995-96, reducing the fleet to the planned sixteen frigates and four submarines ten years earlier than previously intended. The tripartite Belgian-Dutch-Portuguese monohull coastal minesweeper programme was cancelled; three *Alkmaar* class will get remotely controlled sweeps instead.

The South-Eastern Flank. Greece spends 6.2 per cent of GNP on defence and in recent years has received about $1.1b of foreign military aid; surplus US and Dutch tonnage is being acquired, the ex-US ships as 'cascaded weapons' under the military aid package. **Turkey** will also get ex-US ships, and existing frigate and missile corvette programmes are being expanded.

The Netherlands frigate Karel Doorman, *with air/surface search radar installed.* (RNethN)

The Greek frigate Makedonia, *ex-USS* Vreeland *(FF-1068), commissioned 25 July 1992.* (Hellenic Navy)

Lesser NATO Navies

Norway. The 1993 defence budget of $3.8b was a real 2.4 per cent less than the 1992 figure; the catamaran MCMV programme is unaffected and new coastguard ships are planned, but the two training corvettes have been deleted, and one frigate and one minelayer laid up.

Danish defence spending ($2.6b in 1993) is to be reduced by 5 per cent over the next two years; the navy will get fourteen, not sixteen, Stanflex 300 patrol craft. **Belgium's** defence budget will be frozen at $3.2b annually until 1997; there will be two active frigates, eleven MCMV and two support ships.

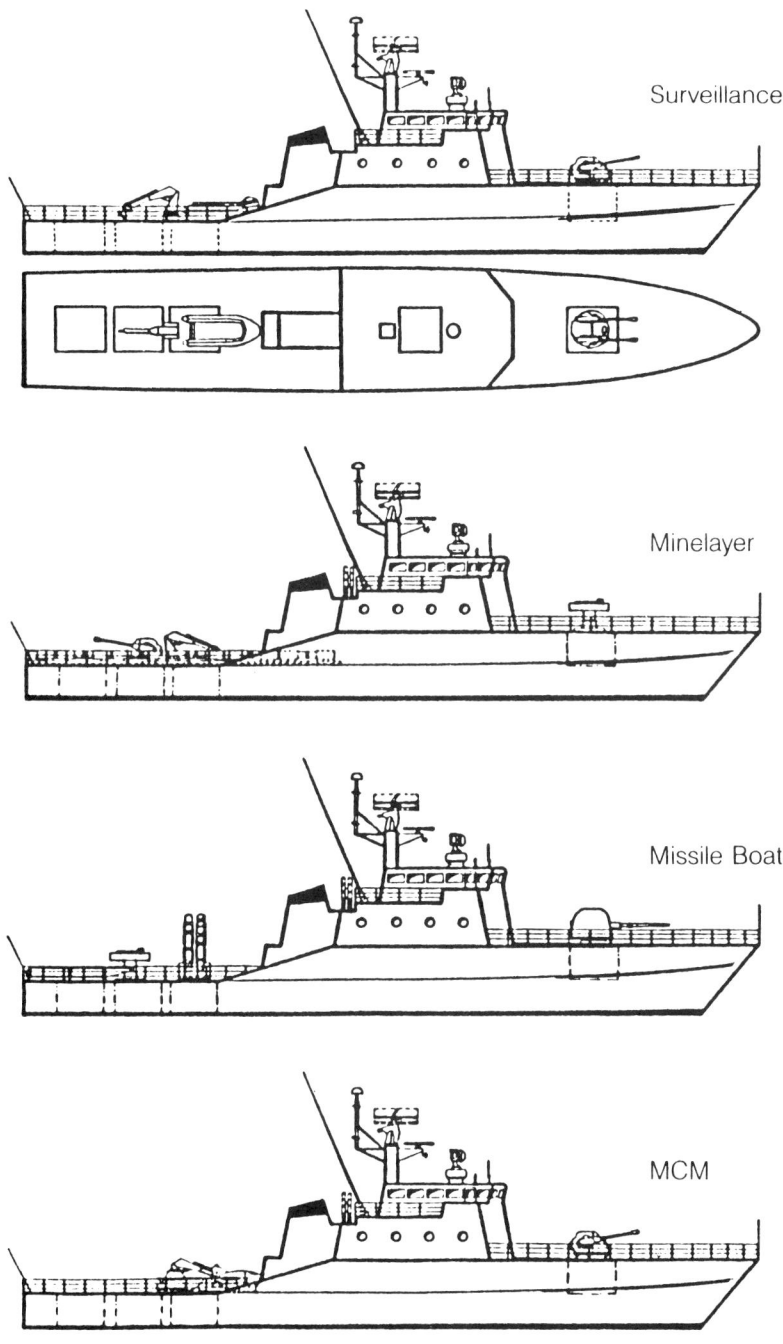

The four prime configurations of the Danish Flyvefisken *(Stanflex 300) large patrol boat class. Configurations can be changed in 48 hours; it is proposed to fit four of the class with modular SAM systems.* (Vickers PLC)

warships are getting older and few replacements are in sight.

Spain. Austerity measures reduced the 1993 defence budget to $7.5b, a real cut of ten per cent, despite the 1991 decision to increase defence spending to 2 per cent of GNP. The cuts affected four new GRP minehunters and the Harrier II Plus radar.

Major US Allies

Japan. Defence expenditure, fixed at around 1 per cent of Japan's burgeoning GNP, continues to grow by about 2 per cent annually after inflation, and the expected full-scale defence review is being brought forward. Equipment spending is expected to fall by about 25 per cent over the last three years of the 1991-95 plan; contracts will be slowed but not cancelled. JMSDF requests for FY93 include the fourth AEGIS destroyer, the first of a new class of diesel submarine, possibly including an AIP system, and a new 8000-ton LST which, designed with a through deck and starboard island, is arousing interest as a precursor of Japan's eventual carrier.

Australia. Defence spending has been cut further. The 1993–94 defence budget will show zero real growth, and spending will fall by a real 0.5 per cent annually over the following three years. The submarine and frigate programmes continue; three consortia were shortlisted for the four plus two coastal minehunter competition.

C(ii). Neutral European Nations

(a) Finland. The fast attack craft and minelayer new construction programmes are complete; ex-USSR 'Osa' class fast attack craft are being converted to light minelayers.

France and Spain

Both countries cooperate with NATO but are not full military members

France. The 1993 draft defence budget was for $39.5b, an increase of 1.35 per cent but an actual fall after 3 per cent inflation; defence spending is to be 3.1 per cent of GNP by 1994. The overall equipment outlay remained unchanged from 1992; the navy's share is scheduled to increase by 5 per cent per year between 1992 and 1997, while spending on the nuclear deterrent will fall from 30 per cent in 1990 to 25 per cent in 1993. Development of the submarine-based M5 continues, but there will be four new SSBN instead of six. The order for the second nuclear-powered carrier was confirmed, but in general

THE NAVAL YEAR IN REVIEW

The Japanese submarine Harushio, *first of a class of six; Japan aims to keep 14–16 diesel submarines in service.* (JMSDF)

HMAS Swan *in 1991, with Ikara and Seacat removed (all RAN Ikara and Seacat launchers have been landed). The remaining 'River' class will be replaced by the ANZAC class from 1995.* (RAN)

The Japanese AEGIS destroyer Kongo, *commissioned in March 1993; the four of this class should be in service by 1998.* (JMSDF)

The Swedish submarine Nacken, *as modified in 1988 to operate with a prototype Stirling closed-circuit diesel motor.* (Kockums)

(b) Sweden. The main recommendations of the defence review were approved by Parliament. The total defence appropriation was $6.5b, of which $966m was for the navy, and equipment spending should rise by 1.5 per cent annually over the next five years. In addition to an increased naval ASW capability and the three new submarines, existing submarines are to be updated with AIP systems.

C(iii). Eastern European Nations

Russia, CIS and Succession States. Russian defence forces were established on 4 April 1992, and in July the Northern, Baltic and Pacific Fleets were declared Russian, the St Andrew's Cross being hoisted on 24 July. Russia and Ukraine agreed at Yalta in August on joint control of the Black Sea Fleet for an interim period of three years; Georgia will be a minor partner.

There was no overall increase in defence spending between 1992 and 1993, but equipment spending increased by 10 per cent, although still 60 per cent less than in 1991. Government arms orders fell by 12 per cent in the nine months after Soviet break-up. Russian experts predict industrial production will fall by 8–10 per cent in 1993, against 18–20 per cent in 1992, and GNP by 5–8 per cent, against 25 per cent in 1992.

Press reports during the year indicated that the material state of the

THE NAVAL YEAR IN REVIEW

fleet was bad and morale very low; major fleet operations seem no longer possible. There have been no operational ballistic missile submarines in the Baltic since early 1991, and 'Kilo' and 'Tango' classes are being withdrawn from Liepaja to St Petersburg, while major surface units in the Northern and Pacific Fleets have been laid up or sold for scrap. Malnutrition-linked infections are reported to have caused four deaths in the Pacific Fleet training base on Russky Island. A sailor in the carrier *Novorossiysk* set fire to a lampshade for light in a power cut; the fire spread, setting fire to the compartment, whereupon the sailor left, closing the door. The carrier was disabled. Explosions in naval munitions stores at Vladivostok and Kaliningrad apparently occurred while personnel were emptying shell casings to sell on the black market.

Russia wants to regain use of the naval base at Cam Ranh Bay – the Soviets withdrew, leaving only communications and SIGINT facility – but talks with Vietnam were suspended because of the Russian arms trade with China.

Succession States. Before agreeing joint control of the Black Sea Fleet, **Ukraine** commissioned the intelligence collector *Slavutych* (flagship), intended for the Russian Northern Fleet, and the 'Petya II' class light frigate *SKR-112*, which defected from the Black Sea base of Sevastopol on 21 July. Russia supplied secondhand tonnage for the new

The Russian frigate Neustrashimy *fitted with funnel extensions or heat deflectors, November 1991.* (BMVg/MoD, Bonn)

The Indian Type 1500 submarine Shankush, *second of two built by HDW at Kiel. The first Indian-built Type 1500 was completed in 1992.*

*The Egyptian FAC (missile)
Ramadan; missile capacity may be
doubled by installing lightweight
OTOMAT or Harpoon*
(Vosper Thornycroft)

navies or coastguards of **Lithuania**, **Estonia** and **Azerbaijan**; **Latvia** preferred to obtain ships elsewhere.

C(iv). Middle East

Iran. Threatening acts in Gulf islands and the arrival of the navy's first submarine have kept the country and its multi-billion dollar arms procurement programme in the headlines.

The South Korean frigate Che Ju, *last of seven Ulsan class, has Marconi S1810 search radar and ST 1802 fire control; Breda/Bofors 40mm/70 guns replace the Emerlec 30mm fitted in early units of the class.*
(GEC-Marconi)

THE NAVAL YEAR IN REVIEW

HMNZS Canterbury *as modernised in 1988 with LW-08 radar, RCA gunfire control and extended flight deck.* Canterbury *and* Endeavour *represented New Zealand at the Battle of the Atlantic commemorative Fleet Review, May 1993.* (RNZN)

Israel's 1993 defence budget was for $10.4b, a reduction of $80m; cuts will mainly hit equipment and R&D funding, and a third new submarine is ruled out.

Oman. The two missile corvettes in hand may be followed by four OPV and possibly a multi-role frigate. **UAE** has expressed interest in an ASW 'package' for an eventual programme of up to eight warships of frigate/corvette size. **Qatar** ordered

HMNZS Endeavour, *a replenishment oiler completed in 1988.* (RNZN)

231

four 56m FAC (missile) from Vosper Thornycroft, for delivery in 1996–1998. **Saudi Arabian** defence spending in 1993 was put at $16.4b, 13.7 per cent up on the previous year. Canada gave formal approval for the acquisition of three *Halifax* class frigates – the Saudi outline agreement with France for three improved *Lafayette* class AAW frigates is still unconfirmed – and the first of the second batch of *Sandown* class SRMH may be ordered soon. **Egypt** hopes to acquire Type 209 submarines, assembled by Ingalls, but US approval is awaited for their export.

C(v). 'PACRIM' and Indian Ocean

(a) China. The economic growth rate

The Brazilian frigate Liberal, *the second of two GP versions built in Britain. Contracts for modernising the pair are being negotiated.* (Vosper Thornycroft)

An artist's impression of the Frigate 2000 design ordered by Malaysia from Yarrow; the first unit was laid down in December 1992. (Yarrow)

reached 14 per cent in mid 1993. According to SIPRI, Chinese military spending is between two and four times as high as Peking admits; naval plans continue to move towards a 'blue-water navy' to control China's territorial claims to much of the South China Sea, including the Spratly Islands. Pilot training facilities are being expanded on the basis that 'there will be deployment of an aircraft carrier in 1997'; a small purpose-built ship or mercantile conversion is more likely than a large secondhand ex-Russian purchase.

(b) **India.** Indian defence spending for 1993–94 was $6.4b, up by a real 2.5 per cent and about 13 per cent of government spending. The proposed nuclear-powered submarine continues to take priority over the future carrier, and surface ships in hand have been delayed by problems with the supply of Russian equipment.

Lesser Navies

South Korea's Defence White Paper proposed enhanced ASuW and ASW capability for the navy, with greater emphasis on protecting sea lines of communication; the submarine total will increase from six to nine and an eventual twenty new destroyers over the next fifteen years will include some of a larger 8000-ton AAW type. **Taiwan** was rebuffed by Germany in an attempt to purchase submarines and frigates; following arms sales by America and France, the Dutch Government may reconsider its decision not to sell submarines. The total French frigates order may be for as many as sixteen hulls. **The Philippines**, with much lower per capita defence spending than its neighbours and a drop in US support, is trying to modernise its navy; the orders for fast attack craft from Spain and Australia have yet to be confirmed. **Malaysia**'s decisions on the eighteen-unit OPV programme and on naval helicopters is expected in 1993, while an interest in an amphibious landing ship and a logistics ship is reported. **Singapore** is investigating a possible submarine force. **Indonesia** has invested heavily in secondhand tonnage, and may purchase more Type 209 submarines. **Pakistan** may acquire ex-French or new construction *Agosta* class submarines to replace existing boats, and has expressed interest in British Type 21 frigates to replace the ex-US frigates on which lease agreements are expiring. **New Zealand** may reconsider the ban on nuclear-powered ships in its ports that effectively terminated its membership of the ANZUS Pact.

C(vi). Latin America

Financial constraints continue to affect Latin American navies. **Argentine's** local construction and refit programmes progress very slowly, **Brazil** continues to give plans to build a nuclear-powered submarine high priority, while **Chile** may acquire British Type 21 frigates.

C(vii). Africa

South Africa. The defence budget for 1992–93 of $2.9b showed a real fall of 14.1 per cent. There are plans to replace the 'Minister' class FAC with a helicopter-equipped OPV; weapons and sensors would be transferred from the older type.

D. WARSHIP BUILDING

D(i). New Designs and Principal Orders

Multinational. Italy entered the Anglo-French Future Frigate Programme; it was renamed the Common New Generation Frigate Programme. Contracts to the BAe/SEMA consortium for the new combat design system, and to EMPAR for a multifunction radar as part of the AAW

Table 3. NEW DESTROYER TYPES

Country	China	India	Russia	USA
Class	Luhu	Delhi	Udaloy II	DDG-51 Flight IIA
No in class	1+?3	0+?4	0+?1	0+?
Builder(s)	Jiangnan	Mazagon DY, Bombay	Yantar, Kaliningrad	Bath IW Ingalls
Building dates	1988–?	1987–2000 I	1989–?1994	1994–?
Displacement (max)	c5000t	6200t	9000t	9217t
Lxbxd(max), metres	150 × 16 × 5	160 × 17 × 6.5	163 × 19 × 5	155.3 × 20.4 × 6.3
Missiles	8 YJ-1 SSM Crotale	4 SS-N-22 SA-N-7	8 SS-N-22 8 SA-N-9 2 CADS-N-2	90 VLS for Tomahawk, SM-2ER, ESSM and ASROC
Guns	2–3.9in/56 8–37mm/63	1–3in/60 4–30mm/65	2–5.1in/70	1–5in/54
ASW	6–12.75in TT 2 FQF-2500	6–12.75in TT 2 DC rails	10–21in TT 2 RBU 6000	6–12.75in TT
Aircraft	2 Dauphin	2 Sea Kings	2 Ka-27	2 SH-60B/F
Machinery	CODOG	CODAG	COGAG	COGAG
Max shp/bhp	55,000	64,000	74,000	105,000
Speed (kts)	30	28	30+	32

Table 4. NEW FRIGATE TYPES

Country	China	Russia	UK/France/Italy
Category	Frigate	Frigate	Frigate
Class	Jiangwei	Gepard	Common New Generation
No in class	2 + ?4	1 + 2	?12 UK + 4 France + 6 Italy
Builder(s)	Hudong	Zelenodolsk	National Shipyards
Building dates	1988–?1995	c1990–?	?1996–
Displacement (max)	c2600t	1900	c6000t
Lxbxd (max), metres	115 × 14 × 4.9	102 × 13.6 × 4.4	144 × 19 × 5
Missiles	6 YJ-1 SSM	8 SS-N-25	8 SM or VLS
	HQ-61 SSM	1 SA-N-4	VLS SAM
			PDMS
Guns	2–3.9in/56	1–3in/60	1 medium
	8–37mm/63	2–30mm/65	2–30mm
ASW	2 RBU 1200	1 RBU 6000	4 TT
		2 DC rails	
Aircraft	1 Dauphin	–	1 medium
Machinery	Diesel	CODOG	CODLAG
Max shp/bhp	14,400	–	–
Speed (kts)	25	26	30

Note: *All details of Common New Generation Frigate speculative.*

system were reported. Invitations to tender for a new SSM are expected in 1993.

(a) United States. Construction of the second and third SSN-21 submarines was approved in July, and SSN-22 *Connecticut* was begun in September. Cost constraints on Centurion, the SSN-21 replacement, will make it 25 per cent less capable than *Seawolf*. Later units of the DDG-51 class will have the flight IIA configuration, improving AAW capability at the expense of ASW performance, but including a permanent helicopter and hangar. Proposals for an enlarged MCMV class and a new heavy lift ship to transport them overseas were dropped.

(b) United Kingdom. The fourth *Vanguard* class SSBN was ordered on 7 July 1992 and laid down on 1 February 1993; it will cost an estimated £550m, the cheapest of the four in real terms. Part of *Vanguard*'s £250,000 towed array sonar broke off during trials and could not be retrieved. The House of Commons Select Committee on Defence was given an official figure of about £5.5b for the cost of the four Trident submarines over their expected lifetimes, including decommissioning.

An artist's impression of the Royal Navy's new 20,000-tonne LPH Ocean. *The crew will number 600, and 700 commandos and 12 helicopters will be embarked.* Ocean *will restore the British helicopter assault capability lost when* Hermes *was sold in 1986. (VSEL, courtesy Mike Smith)*

THE NAVAL YEAR IN REVIEW

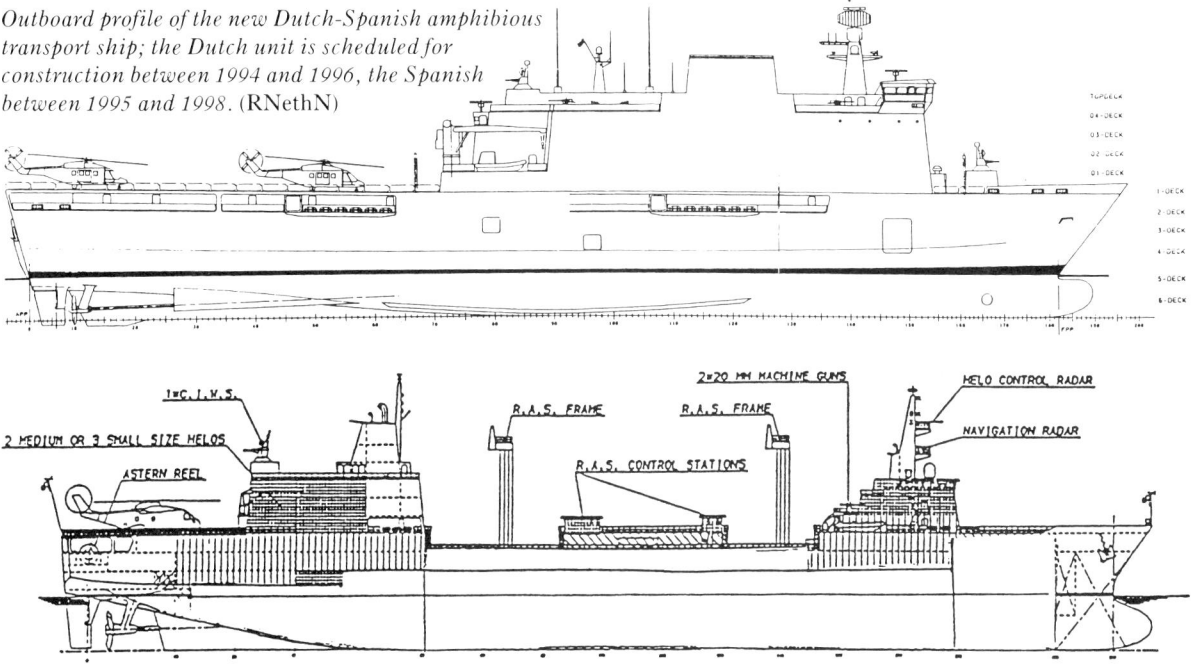

Outboard profile of the new Dutch-Spanish amphibious transport ship; the Dutch unit is scheduled for construction between 1994 and 1996, the Spanish between 1995 and 1998. (RNethN)

Schematic drawing of the new Dutch-Spanish fast combat support ship (Project NL-SP AOR '90). The Dutch Amsterdam *is due to be launched in June 1993, the Spanish* Mar del Sur *one year later.* (RNethN)

Westminster, the eighth Type 23 frigate, will be the first fitted with the Dowty-SEMA SSCS command system. The long-awaited LPH, to be named *Ocean*, was ordered from VSEL and Kvaerner Govan (hull) on 11 May 1993 for £170m; Swan Hunter, the losing bidder, called in the receivers two days later, and the contract is to be investigated by the Select Committee.

(c) Netherlands. The two projected air defence and command frigates, eventual replacements for the *Tromp* class, will be of an enlarged M type, having a commonality of systems with the future German Type 124 class frigates.

(d) Turkey. Two more Meko 200 frigates were ordered at a cost of $510m, one to be built by Blohm & Voss at Hamburg and one at Golcuk between 1994 and 1998. These Track 2B ships will have a modified control system and a vertical-launch Sea Sparrow system. Three more *Yildiz* class missile corvettes were ordered, for a total of five.

(e) France and Spain. DCN and Bazan are to develop a new 2000-ton diesel submarine designated *Scorpene*, to be built at Cherbourg and Cartagena to replace Spain's *Daphne* class from 2003.

(f) Russia. Three nuclear- and two diesel-powered submarines are expected to be launched in 1993, the same totals as in 1992; one diesel submarine in each year was for Iran. No further submarines will be built in the Far East. Chinese interest in purchasing the 80 per cent complete carrier *Varyag*, fitting out in the Ukrainian shipyard of Nikolayev and

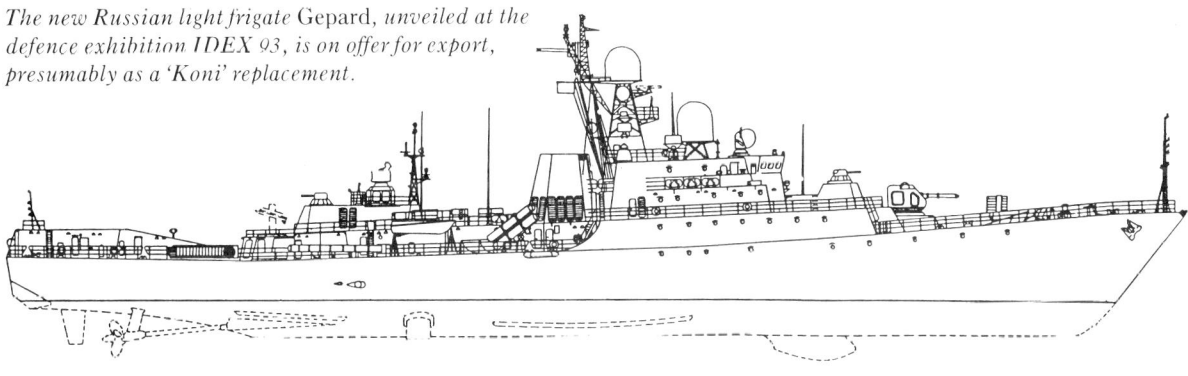

The new Russian light frigate Gepard, *unveiled at the defence exhibition IDEX 93, is on offer for export, presumably as a 'Koni' replacement.*

The Japanese destroyer Yugiri, *third of eight* Asagiri *class. The first of an improved type is to be laid down in 1993.* (JMSDF)

claimed by that country, was widely reported, but the talks failed; work on *Varyag*, on the fourth *Slava* class cruiser *Admiral Lobov*, also at Nikolayev, and on the nuclear-powered cruiser *Pyotr Veliki* at St Petersburg continued very slowly. Construction of *Sovremenniy* class destroyers has been slowed, and of the *Neutrashimy* class frigates almost halted; the *Udaloy II* class (NATO 'BALCOM 12') destroyer programme is believed to have been terminated at one ship. The eighth 'Krivak III' frigate is for sale. The new light frigate *Gepard* is running trials.

D(ii). Ships Entering Service During the Year

These are listed in Table 5 (the figures for Russia and China are approximate).

D(iii). Reconstructions

(a) United States. *Inchon* (LPH-12) is to be converted as the USN's first dedicated MCMV command, control and support vessel, with provision for

Table 5. *New Ships Entering Service, 1 April 1992 to 31 March 1993 (Russia, China in 1992)*

Type	USA	Russia	UK	France	China	India	Japan	Italy
CV (large)	CVN-73	–	–	–	–	–	–	–
CV (medium)	–	–	–	–	–	–	–	–
CV (small)	–	–	–	–	–	–	–	–
CAH	–	–	–	–	–	–	–	–
CG	CG-67	–	–	–	–	–	–	–
	CG-68							
	CG-69							
DD	DDG-52	1 *Sovremenniy*	–	–	–	–	1	1
FF (fleet)	–	1 'Krivak III'	1 Type 23	–	2	–	2	–
(escort)	–	3 'Grisha V'	–	3	–	–	–	–
SSBN	SSBN-738	–	–	–	–	–	–	–
SSGN	SSN-760	1 'Oscar II'	–	–	–	–	–	–
	SSN-761							
SSN	SSN-764	1 'Akula'	–	–	–	–	–	–
	SSN-765	1 'Sierra II'						
SS (all)	–	3 'Kilo'	*Ursula*	–	1	1	1	–

Note: *One of above 'Kilos' for export.*

THE NAVAL YEAR IN REVIEW

The Swedish GRP minehunter Landsort. *Four of this type are being built for Singapore, for delivery by 1995.* (Karlskronavarvet)

The Greek destroyer Themistocles, *ex-*USS Berkeley *(DDG-15), commissioned 1 October 1992; the previous* Themistocles, *an ex-US* Gearing *class destroyer, was discarded in 1992.* (Hellenic Navy)

up to eight MCM helicopters. AAW upgrade proposals were announced for twenty-two FFG-7 frigates.

D(iv). Fleet Depletions (decommissionings, transfers, etc)

(a) United States. USS *Ranger* (CV-61) paid off in 1993, the first supercarrier to be de-activated. The last *Coontz* class DDG will go in 1993, and six *Leahy* class CG in 1994; only eight *Knox* class remain, all in the NRF. Four *Charles F Adams* DDG, three *Knox* class frigates and two re-engined *Asheville* class PG were transferred to Greece between June and December 1992. Taiwan received three *Knox* class frigates, and four will go to Turkey, one as a replacement for *Muavenet*; a Turkish preference for the FFG-7 class was turned down. USS *Goldborough* (DDG-20) was sold to Australia for spares.

(b) United Kingdom. The last two

The USN's test craft Sea Shadow *at its first daylight testing off southern California, 11 April 1993. Designed to test the application of advanced modern techologies to surface ship design,* Sea Shadow *was built by Lockheed in the mid 1980s; stealth technology results have been used in the DDG-51* Arleigh Burke *class destroyers, and SWATH technology results in the TAGS-19.*
(USN)

Batch 2 TA *Leander*s were retired in early 1993; the first three Type 21 decommissionings will take place later this year, the remainder in 1994; RFA *Blue Rover* was sold to Portugal, 31 March.

(c) Germany. Thirty-seven warships and two support ships of the former East German *Volksmarine* sold to Indonesia included sixteen 'Parchim I' light frigates, twelve 'Frosch I' class landing ships and nine 'Kondor II' class minesweepers. Four 'Kondor I' class were sold to Tunisia, two to Malta and one to Guinea-Bissau.

(d) Netherlands. Three *Kortenaer* class frigates were sold to Greece, for transfer between 1993 and 1995.

(e) Russia. A major scrapping programme is in hand, removing obsolescent and non-operational types. The first 'Delta I' has been withdrawn, and the 'Echo II' and 'Juliett' cruise missile submarine classes are being paid off. Of major surface ships, *Minsk* and *Leningrad* were scrapped in 1992 (the former in India), in addition to the incomplete *Ulyanovsk*; three 'Kresta II' class cruisers were deleted in 1992, and only one 'Kresta I', two 'Kyndas' and eight 'Kashins' remain in service. The apparent aim is to have 65 per cent of operational warships under twenty years old.

A model of Le Triomphante, *first of France's new generation ballistic missile submarines (SNLE-NG). The first two boats will be completed with the M4 missiles; unit 3 will be the first with the new M5, which will be retrofitted in the earlier two.*
(DCN)

Russia supplied Lithuania with two 'Grisha III' class light frigates and two fast attack craft in exhange for housing to be constructed in the Kaliningrad enclave. Eight Caspian Flotilla warships, including two 'Petya II' class light frigates, were transferred to the new Azerbaijan Coastguard in August.

(f) Singapore. The former RFA *Sir Lancelot*, sold privately in 1989, was acquired by the Singapore navy and renamed *Perseverance*.

E. NAVAL WEAPON SYSTEMS

Salient developments in naval weapon systems are listed below.

E(i). Missiles, including Ballistic Missiles

(a) United States. The naval requirement for the TSSAM missile was reduced from 8650 to 7450; plans to fit it on naval aircraft may be dropped. A 'smarter' Tomahawk cruise missile is being developed and a new tactical missile for surface fire support, designated NATACMS, is being evaluated.

(b) United Kingdom. A possible tactical role for the Trident missile is being investigated; 44 missiles had been purchased by the end of 1992, out of an eventual 100 at a total cost of $1.0b. Future developments in anti-ballistic missile systems may act against proposed reductions in warhead totals.

(c) France. All ballistic missile submarines are now fitted with the M4 missile. A replacement for the cancelled Franco-German ANS supersonic anti-ship missile is being investigated.

E(ii). Maritime Aircraft

Multinational. The NH 90 helicopter design and development contract was signed in September; the first prototype should fly in 1995.

(a) United States. A $3.9b contract was awarded for the design, manufacture and testing of the F/A-18E/F: five E, two F; the new types will have 25 per cent greater wing area than the F/A-18C/D and are intended for 28 per cent greater range. The A-X became the A/F-X; this and the USAF's MRF may be replaced by a single new joint attack fighter, cheaper than the former and more capable than the latter. The first production AV-8B Harrier II Plus was delivered in April 1993. The first Advanced STOVL strike fighter should fly around 2000, with production models in service in 2010–2015; contracts were signed for two separate technology demonstration programmes. The UK agreed to assist in development. The P-3C Update IV was cancelled, because of the disappearance of the Soviet surface and submarine threat. Reversing previous policy, the DoD will from FY93 support V-22 tiltrotor development, although new performance requirements may permit the substitution of a helicopter as the H-46 replacement.

The Thai ASW corvette Khamronsin, first of three completed in 1992. The hull is based on a 'stretched' Vosper Thornycroft 'Province' design. (RThaiN)

(b) United Kingdom. The first FRS2 Sea Harrier was delivered, 2 April 1993.

(c) Italy. Thirteen Harrier II Plus were ordered at a cost of $393m, for delivery from 1995, taking the Italian Harrier II total to eighteen; there is an option on eight more.

(d) France. The prototype Rafale M landed on *Foch*, 19 April 1993, and took off 20 April. Delivery of the first production aircraft is expected in 1996. Four E-2C Hawkeye AEW aircraft will be purchased for the new carrier.

(e) Spain. Eight Harrier II Plus were authorised at a cost of $237m for delivery from 1996. The eleven Harrier II in service will be converted to Plus standard in the future.

(f) Russia. The air wing for the *Admiral Kuznetsov* is likely to comprise Su-33 fighters, mainly for air defence, and Su-25UTG trainers. About twenty Su-33, a variant of the Su-27K, will be acquired, their armament to include the Kh-41 Mosquito ASM, an air-lauched version of the SS-N-22. A Western partner is sought to complete development of the Yak-41 'Freestyle'; no order has been placed, and one of the two flying prototypes crashed on *Admiral Gorshkov*, 5 October 1991.

E(iii). Anti-Aircraft and Anti-Missile Warfare (AAW)

Multinational The development of air defence missile systems for future European AAW frigates continues under a multiplicity of acronyms. The suggested merger of the British LAMS (Local Air Missile System, part of the FAMS programme) and Franco-Italian SAMP/N (medium range naval air defence system, part of the overlapping FSAF programme) into a single programme before full-scale development starts in early 1994 would require design compromises; the systems have a degree of commonality, including the Aster 30 missile, but LAMS is intended to be more capable. Spain pulled out of LAMS in 1991, but may re-enter. The go-ahead was given for the third development phase of the SAAM (Franco-Italian point defence missile system, using the Aster 15 missile) and SAMP/T (medium range land-based air defence) systems. The contract definition phase of the Evolved Sea Sparrow Missile (ESSM) is to go ahead. Britain and Italy are to collaborate on a feasibility study for a Very Short Range Air Defence (VSRAD) missile for surface ship defence.

(a) United States. There is interest in developing an anti-tactical ballistic missile capability in modified AEGIS platforms. General Dynamics is test-

The Russian carrier Admiral Kuznetsov, *approaching operational status in the Northern Fleet.* (USN)

The Thai frigate Chaopraya, *first of four built in China. Official estimates state that this type will be 85–90 per cent as efficient as the Meko 200 type at 25 per cent of the cost, but quality and damage control standards are low.* (RThaiN)

The Thai frigate Kraburi, *third of four built in China: two have a flight deck for a helicopter aft, and two have a second gun mounting.* (RThaiN)

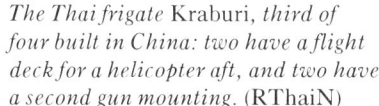

ing an integrated combined RAM and Phalanx CIWS. The RAM would intercept at 3.5nm–4.0nm ranges, whence 'leakers' or non-radiative threats would be passed to the Phalanx 20mm CIWS, range 0.5nm–0.8nm.

(b) United Kingdom. An improved fuse for the Sea Dart missile is being developed, the last in a series of modifications intended to improve performance against anti-ship missiles.

(c) France approved a contract for equipping the carrier *Charles de Gaulle* with the new SAAM point-defence missile system, making it the first of the FSAF family of missiles to enter service.

E(iv). Anti-Submarine Warfare (ASW)

A new Franco-Italian torpedo delivery system being offered to Saudi Arabia and UAE consists of an Otomat 2 surface-to-surface missile delivering a light ASW torpedo.

E(v). Other Weapon Systems

In the United States, Loral is to study a laser-based system to counter electro-guided anti-ship missiles. The Marine Corps may take up a modified 'Magic Lantern' ML-90 laser-based airborne minehunting system.

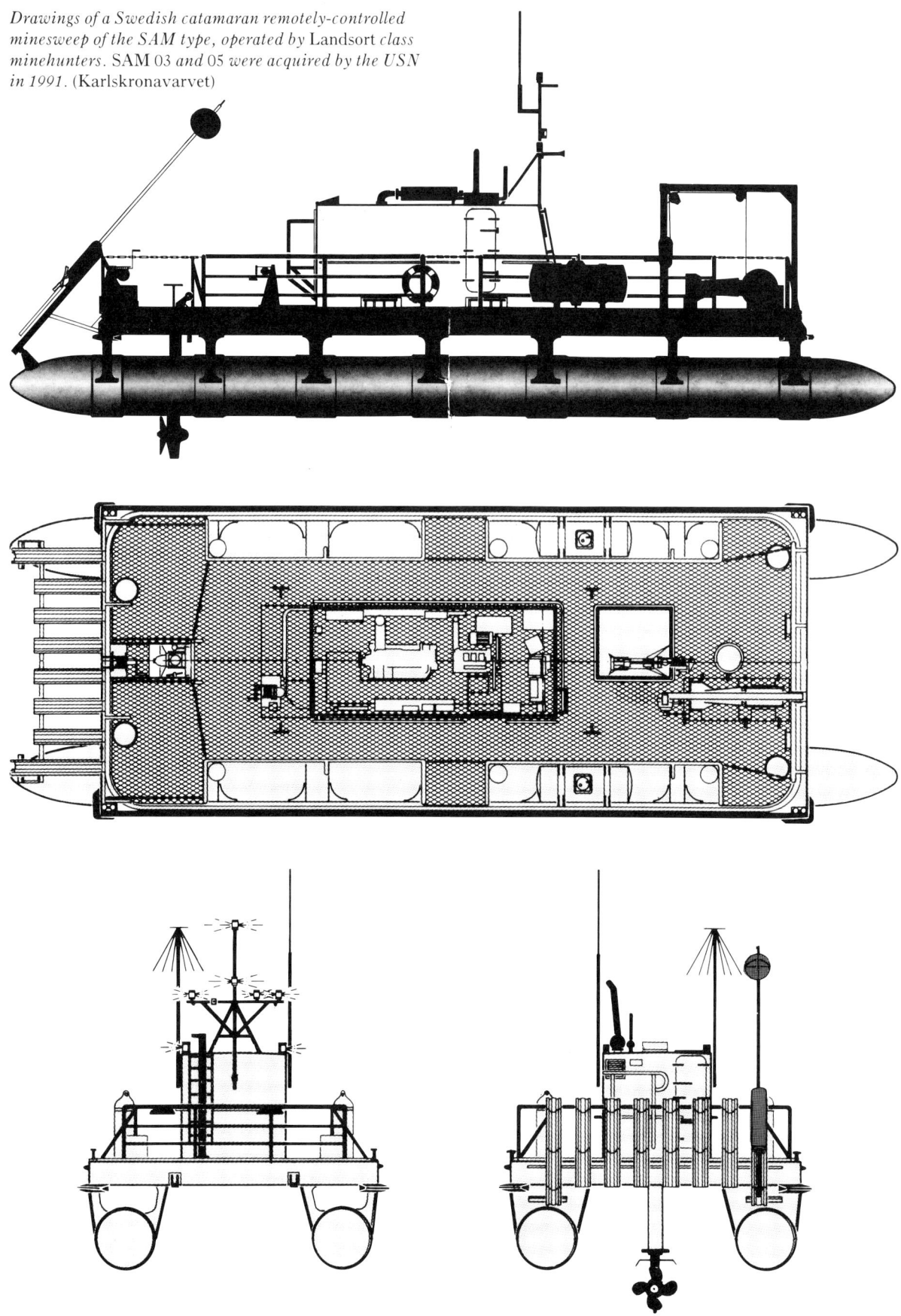

Drawings of a Swedish catamaran remotely-controlled minesweep of the SAM type, operated by Landsort *class minehunters. SAM 03 and 05 were acquired by the USN in 1991.* (Karlskronavarvet)

HMS Roebuck, *the RN's most recent survey ship, completed in 1986. At present, some survey work is done by chartered vessels with Naval Parties embarked, and new ships are projected.* (Brooke Marine)

F. NAVAL EVENTS

F(i) Areas of Conflict and Naval Actions

(a) The Gulf. International sanctions against Iraq under Security Council resolution 687 continued; in September, the Russian destroyer *Admiral Vinogradov* and tanker *Boris Butoma* were added to the Gulf patrol (two Soviet warships had patrolled the Gulf before the conflict, but had not assisted in enforcing the arms embargo). On 22 July, renewed military action was threatened after a UN arms inspection team was forced to abandon a search attempt in Baghdad. An amphibious warfare battle group arrived in the Gulf, and the *John F Kennedy* CVBG was ordered to the Mediterranean on 28 July (aiming at three carrier battle groups on station), but a deal was reached before military action was taken. To protect the Shi'ite Muslims in the south, a 'no-fly' zone was established south of the 32nd parallel (27 August), enforced by Allied planes from Saudi Arabia and naval planes from USS *Independence* in the Gulf. There were more than 500 sorties by US, British and French aircraft in the first five days, but flights were rapidly scaled down to 50–60 missions daily because of Iraqi inactivity.

Tension increased again after the US landings in Somalia; Iraqi planes entered the 'no-fly' zone and weapons were seized from storage sites on the Iraq-Kuwait border. On 13 January 1993, 118 American, British and French aircraft, including 38 strike aircraft, bombed Iraqi targets; the 34 naval aircraft that participated were launched from USS *Kitty Hawk*, heading a twelve-ship CVBG in the Gulf. No aircraft was lost. On 17 January thirty-two Tomahawk cruise missiles were launched from the cruiser *Cowpens* and destroyers *Hewitt* and *Stump* in the Gulf, and the destroyer *Caron* in the Red Sea; seventy-five aircraft launched attacks in the south on the 18th, hitting targets not completely destroyed earlier.

In September, Iran, perhaps pre-empting American base facility requests from UAE, took effective control of the strategic island of Abu Musa, near the Strait of Hormuz; the arrival of Iran's first submarine at the end of October focused attention on the changing balance of power and the new arms race in the region. USS *Topeka* was due to arrive in the Gulf, 3 November, shortly after the arrival of the Iranian boat; five months later, HMS *Triumph* became the first British nuclear submarine to enter the Gulf. Egypt threatened to attack any Iranian warships sent to the Red Sea if Iran tried to establish a naval base at Port Sudan.

(b) The Mediterranean. The year started with the UN peacekeeping force, UNPROFOR 1, deployed in Croatia and Sarajevo. Sanctions were imposed on the rump Yugoslavia (Serbia and Montenegro) by the Security Council, 30 May; on 9 July European defence ministers and on 10 July the WEU agreed to send

HMS Ark Royal, *heading the British Task Force in the Adriatic. In mid 1993, the Task Force included the frigates* Brazen *and* Abraham Crijnssen *(RNethN) and the* RFA Argus, Olwen *and* Fort Grange. *The* RFA Sir Percivale *and* Resource *at Split and the Sea Kings of 845 Squadron ashore provided support for the UN humanitarian programme, while* HMS Cardiff, *in STANAVFORMED, was enforcing Resolution 820 in the Strait of Otranto.*

warships to the Adriatic to enforce them. On 14 July, the US cruiser *Biddle*, supporting US relief flights to Sarajevo, locked its fire control radar on to five suspect aircraft descending towards it and USS *Iwo Jima*. All five were challenged by radar; one, a British relief flight, identified itself and the other four, believed to be Yugoslav military aircraft, withdrew. By late July, NATO and WEU forces were on station in the Adriatic for monitoring merchant ship movements, but without the power to stop, board or search. The naval forces monitoring the trade embargo in late October were four warships from the WEU (Italy, France, Spain, Belgium) and three or four on rotation from NATO's STANAVFORMED.

As the refugee problem increased, the Security Council approved the use of whatever means, including force, necessary to deliver humanitarian aid, 13 August; individual nations and regional alliances were empowered to take the necessary military action. There was to be a ban on military flights, but no punitive measures against violators were approved. As part of the humanitarian relief force UNPROFOR 2 to protect aid convoys, British troops arrived at Split in RFA *Sir Bedivere*, 27 October.

To increase the effectiveness of sanctions, the Security Council approved a naval blockade of Yugoslavia in the Adriatic Sea and the Danube River, 16 November, and on 20 November NATO and the WEU agreed to enforce it. Under the same rules of engagement as in the Gulf conflict, warships were able to halt and inspect all ships entering or leaving Yugoslav waters, firing warning shots, but were not allowed to destroy ships. Blockade patrols started at 5pm on 22 November, with seven NATO and five WEU warships operating in adjacent areas, the former in the southern Adriatic under Operation 'Maritime Guard' and the latter in the Strait of Otranto under Operation 'Sharp Fence'.

As the situation worsened, 14

January, HMS *Ark Royal* left Portsmouth to head the British Adriatic Task Force; the RN force was held on 12-hour alert to cover withdrawing land forces, or to reinforce them; the French carrier *Clemenceau* left Toulon for the Adriatic later in the month. USS *Guam* also arrived at the end of January, with the carrier *John F Kennedy* from the eastern Mediterranean, but agreement could not be reached on possible military action to halt fighting. The blockade on the Danube was ineffective, with extensive sanctions-breaking reported – tug crews pushing barges laden with oil threatened to dump or fire cargoes in the river – until the US lent three fast patrol boats to Romania and three to Bulgaria, and sent customs officers to assist in checking documentation; the last major sanctions-breaking on the river was in February.

On 28 April, NATO warships were empowered to fire non-explosive 'inert charges' and to enter former Yugoslav territorial waters as last resorts in enforcing sanctions. More than 500 ships had already been stopped and searched by that date.

(c) Northern Europe

The Channel and Channel Approaches. Anglo-French fisheries disputes intensified during the year: on 28 March off Guernsey, three members of a boarding party from HMS *Brocklesby* were abducted by the trawler *Calypso* and landed at Cherbourg. Later that day, the RN training ship *Blazer*, on a courtesy visit to Cherbourg, was seized by fishermen, the crew secured below and the White Ensign burnt. As a deterrent, the crews of the fishery protection ships were then each reinforced by six Royal Marines; the *Calypso* was arrested and taken to St Peter Port, 2 April, and the skipper fined.

Three fishing boats containing daytrippers came under live 4.5in shellfire from HMS *Southampton* in the English Channel off Weymouth on 10 June, shells landing within 200m of one boat. An enquiry established that the destroyer's surveillance radar could not spot the boats at almost 13km distance. In mid November, anglers came under pistol and rifle fire from NATO minesweepers off Folkestone.

The first trip of Japan's plutonium carrier, *Akatsuki Maru*, to France caused numerous protests. In spite of comprehensive safety measures, including the specially built Maritime Safety Agency escort *Shikishima*, several countries banned the ship from their waters. Some 2000 French police, naval commandos and frogmen protected the loading of 1.7t of plutonium at Cherbourg, on 7 November, and the Greenpeace ship *Moby Dick* was boarded by commandos. Another Greenpeace ship, *Solo*, followed the Japanese ships and collided with *Shikishima* off Brittany.

Sweden. In September, a suspected foreign submarine was attacked with depth charges and grenades. This was 1992's first attack, although an intruder had been reported in May. Russia admitted previous intrusions

RFA Fort Victoria *fitting out at Harland and Wolff's Belfast shipyard. The House of Commons Select Committee on Defence castigated the lamentable progress of* Fort Victoria, *32 months behind schedule and more than £60m over budget; in July 1992 she entered Cammell Laird's Birkenhead shipyard for final fitting out.*
(Harland and Wolff)

The Japanese frigate Abukuma, *first of a class of six named after Second World War cruisers.* (JMSDF)

The Japanese ocean minehunter/sweeper Yaeyama, *first of a new wooden-hulled class, was completed in March 1993.* (JMSDF)

and offered to assist future Swedish submarine hunts.

Arctic. The Greenpeace ship *Solo* sailed to investigate and publicise nuclear dumping in the Kara Sea, 8 October; arrested by the Border Guard ship *Ladoga* and taken to Murmansk, *Solo* had to tow the Russian ship when its engine failed, 14 October.

(d) 'PACRIM' and Pacific Ocean

South China Sea. Merchant ships leaving Hong Kong were intercepted, fired on and boarded by Chinese marine police from the nearby Lema islands; one was seized and taken to the port of Zhuhai. The islands contain a naval base, but ships had previously been able to pass through the area unimpeded. Steps were taken by Royal Navy patrols and the Hong Kong police to deal with the large-scale smuggling of stolen luxury cars from Hong Kong to the Chinese mainland in specially built speedboats.

Malacca Strait. At least eighty piratical attacks on ships were reported in the first eight months of 1992, mostly around SE Asia. British ships were advised to keep vigilant, to have ready an anti-attack plan, and to repel boarders with water hoses, but the carrying of firearms was not sanctioned. An anti-piracy coordination centre was set up in Kuala Lumpur, Malaysia, on 1 October.

South Pacific. Since declaring independence unilaterally in 1990, the island of Bougainville has been cut off from the outside world by Papua New Guinea's naval blockade.

North-East Pacific. USS *Cowpens* (CG-63) inadvertently 'threatened' to shoot down a Qantas airliner which strayed over a multi-national naval exercise east of Hawaii, 14 July; the ship operator unintentionally sent out the message on the international dis-

An artist's cut-away impression of the Australian Type A471 submarine Collins, *the first of six, begun at Adelaide in February 1990 and scheduled for launching in August 1993. Kockums will deliver bow and stern units for early units.* (Kockums)

The new Angolan coastal patrol craft Mandume, *first of four built by Bazan at San Fernando.* (Bazan)

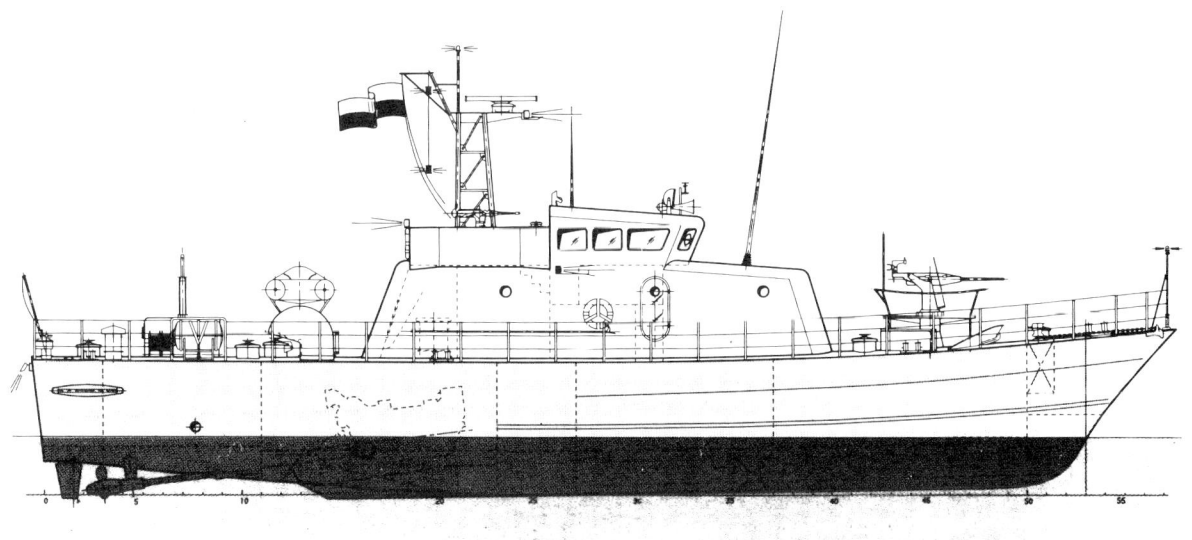

The French patrol frigate Floreal *on trials; Syracuse 2 SATCOM remains to be fitted between the funnels.* (Alsthom)

tress network instead of the exercise network.

(e) Africa

Somalia. The vicious clan warfare and disastrous famine in Somalia produced military intervention under UN co-ordination. The USS *Ranger* CVBG, with the assault ship *Tripoli*, LSD *Juneau* and LST *Rushmore*, assembled, and a vanguard of 1800 US Marines spearheaded the bloodless amphibious occupation of Mogadishu on 9 December, securing port and airport as the first step of Operation 'Restore Hope'. Kismayu was similarly occupied without opposition, 20 December. The total of US troops was nearly 30,000, with 8000 from other nations. UN forces took over on 1 May 1993.

Liberia. In Liberia, three years of ECOMOG peacemaking has had little success, the peacemakers having become associated with one warring faction. ECOMOG forces, supported by Liberian units and two Nigerian warships, attempted to lift the siege of Monrovia without success in October. In November, the Greek freighter *Konkar Pioneer*, apparently assisting the rebels by exporting their goods, was seized and impounded by a Nigerian gunboat off Buchanan, and on 15 November Nigerian warships again shelled rebel positions around Monrovia. US warships off Liberia may have helped the Nigerians by indicating targets.

(f) Caribbean. At least five US warships, twelve coastguard cutters and twelve aircraft and helicopters were used to turn back the flow of refugees from Haiti that followed the 1991 military coup. In May 1992, harsh US restrictions classified most refugees as economic, not political, and returned the former to Haiti. Clinton's election pledge to grant political refugee status to Haitian refugees started a greatly increased exodus, and had to be rescinded.

F(ii). Major Casualties at Sea, 1 April 1992 to 31 March 1993

(a) The Indonesian landing ship *Amurang* sank 900km south-east of Jakarta, with the loss of one crew member and 13 missing, June 1992.
(b) On the night of 16–17 September, USS *Sturgeon* (SSN-637) snagged the nets of the fishing boat *Lupina C* in the North Channel, between Scotland and Northern Ireland, snapping hawsers and towing the boat backwards. The submarine's captain and officer of the watch were disciplined for breaking safety rules established in the Clyde area after the 1990 sinking of the trawler *Antares*, and the USN paid $100,000 (£64,000) in damages.
(c) The bridge of the Turkish des-

troyer *Muavenet* was hit by two NATO Sea Sparrow missiles from USS *Saratoga* (CV-60) during night exercises, Aegean Sea, 1 October, killing five crew members including the captain. The captain and seven crew members of the US ship faced a disciplinary hearing; the crew members are believed to have mistaken a practice drill for a real attack.

(d) USS *Grayling* (SSN-646) collided with a Russian 'Delta III' or 'Delta IV' SSBN in international waters 195km north of the Kola peninsula on 20 March. Neither boat was seriously damaged, although the Russian was hit within 20m of its reactors.

F(iii). Footnotes.

(a) Work continues at Hartlepool on the restoration of the 1817 frigate *Trincomalee*, until 1987 in Portsmouth harbour as the training ship *Foudroyant*.

(b) An expedition to seek the wreck of George Garrett's submarine *Resurgam*, sunk on tow off North Wales in September 1880, is planned by his great-grandson.

(c) The remains of HMAS *Canberra*, sunk off Savo Island in August 1942, were located just before the 50th anniversary of the battle.

(d) Antonio Marceglia, who disabled the battleship *Queen Elizabeth* with explosive charges at Alexandria, 18–19 December 1941, died on 14 July, aged 76; Luigi Durand de la Penne, who disabled the *Valiant* on the same occasion, died in January 1992.

(e) The 50th anniversary of the Battle of the Atlantic was commemorated in celebrations at Liverpool, including a Fleet Review, 26–31 May 1993.

(f) In November 1992, the destroyer *Gloucester* paid the first British naval visit to an Albanian port since 1938.

(g) Moscow revealed details of an accident to the early 'Hotel' class SSBN *K-19*, also called *Polarii Krug*. On 18 June 1961, while submerged in mid Atlantic, a seal failed, leaking reactor coolant and starting a fire. A fourteen-member volunteer damage control party prevented a possible reactor meltdown, but all died from radiation overdoses. Russian reports listed many other serious accidents to Soviet nuclear submarines, and collisions between Soviet and NATO submarines. In 1985, there was a reactor explosion during the refuelling of an 'Echo II' class SSGN at Chazhma Bay, Soviet Far East; reactor meltdowns took place in an 'Echo I' class SSN in 1979 and a 'Charlie' SSGN, both again in the Far East. Seven non-fatal collisions were listed between 1968 and 1987, and the report speculated that such collisions might have helped cause the loss of the 'Golf' class SSB *PL-722* in the Pacific in February 1968 and of the 'Yankee' class SSBN *K-219* off Bermuda in October 1986.

(h) Israel is to renew the search for the submarine *Dakar*, lost with all hands on its delivery voyage, January 1968. A 1987 search near the Egyptian coast failed to find any remains.

(j) A search of Russian documents failed to find any record of the Hull trawler *Gaul*, lost without trace in 1974. Relatives of crew members have suggested that she had been seized and held as a suspected 'spy ship'.

(k) A fire aboard HMS *Turbulent* at Devonport on 30 April 1992 hospitalised 24 crew members and dockyard workers with smoke inhalation and is reported to have taken three hours to bring under control; two awards were later made for brave conduct in 'potentially lethal' conditions.

G. MISCELLANEOUS

(a) In April 1993, 117 USN and USMC officers were charged with sexual misconduct offences against 83 woman pilots during a September 1991 Reunion of Gulf War aviators at

HMCS Toronto, *the fourth Canadian Patrol Frigate (CPF) on sea trials, September 1992.*
(Saint John Shipbuilding Ltd)

The British SSN Turbulent, seen here in May 1993, suffered a serious fire in April of the previous year. (L & L van Ginderen)

Las Vegas. This 'Tailhook' scandal caused the resignation of Navy Secretary Lawrence Garrett, June 1992.

(b) Obsolete Royal Australian Navy Mk 23 torpedoes are being fitted with special sonar and transducers to measure the thickness of the Antarctic ice shelf. Launched from the research ship *Aurora Australis*, each torpedo will travel under the ice at a depth of 15m for about 10km.

(c) Rosshelf, a new Russian consortium, has proposed a major project using ex-naval nuclear submarines for Arctic oil prospecting and drilling, and for under-ice oil transportation. This would be an enormous engineering task; another formidable difficulty would be persuading foreign investors to part with the necessary billions of dollars.

(d) Six young sperm whales, trapped in the former naval base of Scapa Flow for almost five weeks, were herded to freedom through Hoxa Sound by a flotilla of small boats, March 1993.

INDEX

Page numbers in italics refer to illustrations; those in bold refer to tables.

Abdiel class (GB) 210
Abukuma (Japan) *246*
Abyssinia 146
Abyssinia (Ind) 15–16
Achilles (NZ) 105
Active, HMS 204–5
added weight 194–6
Admiral Graf Spee (Ger) *105*, 105–12, 107, *108*, *110–111*, 130
Admiral Hipper (Ger) 133–4, 138–40
Admiral Kuznetsov (Russia) 240, *240*
Admiral Lobov (Russia) 236
Admiral Scheer (Ger) 130–42, *131–5*, *137*, *139–141*
Admiral Sénès (Fr) 99
Admiralty *see* Great Britain
Admiral Vinogradov (Russia) 243
Adventure, HMS 86, *88*, 210
Africa 14, 16, 109, 217, 233, 248
Agassiz, HMS 209–10, *210*
Agincourt, HMS 62
Agosta class (Fr) 233
aircraft 39–40, 79–83, 113, 130, 137–9, 140, 142–3, 145–6, 155, 168, 173, 186, 191–2, 207, 218, 239–40, 243, 248
  anti-aircraft armament 60–1, 79–83, *83*, 85, 96, 100, 104, 140, 143, 146–7, 152, 156–8, 173, 234, 240–1
  anti-aircraft ships 146–7
  civil 247–8
  launching systems *33*, 33–45, *34–5*, *37–38*, *41–42*, 80
  VTOL 45, 171, 174, 240
  see also airships, flying-boats, radar, RAF
aircraft carriers 80, *80*, 81, 146–7, 171, 207–8, 219, 222, 226
  carrier escorts *see Akizuki* class
airships 80–1
Ajax, HMS 49, 52, **60**, 62–3, 105
Akagi (Japan) 147
Akatsuki Maru (Japan) 245
Akizuki class (Japan) **154**, *143–64*, *146*, *155*
  armament 148–9, 151–8, **154**, *156–7*
  building programme **148**, 149–51, **152**
  machinery **163**, 159–64, *160–1*, *164*
    engines **163**, 160–1, *162*
    ventilation *163*, 163–4, *165*
  staff requirements *143*, *147*, 147–8
  trials *159*, 164
Akizuki (Japan) **151**–2, *143–4*, 160
Albacore (US) 73
Aldersdale (GB) 138
Aleksandr Sibiryakov (Russia) **142**, *138*, 139
Alsterufer (Ger) 135
Altmarck (Ger) 105
Amagumo (Japan) **151**
Amsterdam (Holland) 235

Amurang (Indonesia) 248
Andalusian (GB) 133
anti-submarine warfare *see under* submarines
Aogumo (Japan) **151**
Argentina 94, 174–5, 192, 233
Argus, HMS 36, 79–80
Ark Royal, HMS 245
Arleigh Burke (US) 223
Armstrong Whitworth 49, 68, **68**, 74
Asagochi (Japan) **151**
Asashio (Japan) 119
Asdic 81, 84, 88
  see also sonar
Asheville class (US) 237
Asquith, H H 47
Assaye (Bombay Marine) 12
Atlanta class (US) 146, 148
Atlantic, battle of *202–5*, 249
Atlantic Conveyor (GB) 174
Atlantis (Ger) 134, 192
Audacious, HMS **50**, **60**, 61–2
Aurora Australis (Australia) 250
Australia 11, 226, 237, 250
Austria–Hungary 47, 96
Avenger, HMS 174
Azerbaijan 230

Baden (Ger) 77, *78*, 82
Bainbridge-Bell, Flt/Lt L H 108–12 *109*
Bainbridge (US) 178
Balao class s/m (US) 172
Baretta (US) *118–19*
Barneveld (Holland) 142, 134
Barracuda see 'K' class submarines
Barra Head 74
Barrow 74
battlecruisers 86
  designs, 1920–1 *79*, *79*
battleships 77, 79–80, 82, 89–90, 94, 146, 217, **220**
  pocket battleships *see Panzerschiffe*
  1st Battle Squadron, Home Fleet 61
  2nd Battle Squadron, Home Fleet 61–3
  4th Battle Squadron, Home Fleet 61
  Battle Squadron 1 (Germany) 53
Beach, Captain E L 168, 174, 177–82, *182*, 187
Beardmores 49, 212
Beatty, Admiral 65, 80, 87
Beaverford (GB) 132, **142**
Belgium 225
Benbow, HMS 49, **60**, 61–2
Benigumo (Japan) **151**
Beresford, Admiral Sir Charles 198
Berwick, HMS 137–8

Biddle (US) 244
Birmingham, USS *33*, 34
Bismarck (Ger) 130
Bisson class (Fr) 90
Bixio class (It) 94
Black Prince, HMS 49, *198–9*
Blériot, Louis 38–9
Blucke, Sqn/Ldr R S 105
Blue Rover, RFA 238
Blyth Flotilla 73
Boadicea, HMS 198, *199*
Bogert, John L 42
Bombay Dockyard 9–14, *10*, *13*, 14, 16–17
Bombay Marine 9–15
Bon, Admiral de 94
Boothby, Capt FLM 35
Boris Butoma (Russia) 243
Bouclier class (Fr) 90–1
Bouclier (Fr) 101
Bougainville 247
Bourrasque class (Fr) 100–1, 103
Bourrasque (Fr) 90, *92–3*, 97–8, **98**, *102*
Bowen, E G 108
Brandenburg (Ger) 224
Brazil 62, 233
Bretagne class (Fr) 89, 101
Bretagne (Fr) **54–5**, 63
Bridgeman, Sir Francis 51, 60
Bringenwald, A 192
Britannia, HMS 204–5
British Advocate (GB) **142**, 134
British Aerospace 45
British Loyalty (GB) 126
Brocklesby, HMS 245
Brodie, Lt J H 39–40, 44
Brown, Lt-Cdr Eric M 43
Buckley class (US) 171
Bulla, *Kapitänleutnant* 192
Bulldog, HMS *203–4*
Burke, Admiral A 177
Bushnell & Co 20

'C' class cruisers (GB) 85
Calypso (Fr) 245
Cammell Laird 68, 74, 87
Campania, HMS 35, *35*, 36
Campbeltown Flotilla 73
Canada 43, 188, 191, 209, 218, 223–4, 232
Canada, HMS **54–5**, 61–2, 62
Canadian Cruiser (GB) 134, **142**
Canberra, HMAS 87, 135, 249
Canterbury (NZ) *231*
Carney, Admiral 172, 176
Caroline, HMS 63
Caron (US) 243
Castex, Vice-Admiral 104
Castilian (GB) 15

251

'Castle' class (GB) 191, 210
Catclaw (US) *118–19*
Centurion, HMS *50*, 51–3, **60**, 63
Centurion (US) 234
Cerberus (GB) *14*
Chacal (Fr) 98, **99**
Chambers, Captain WF 36–7
Chantiers Normand 91
Chaoprava (Thailand) *241*
Chaplin, Cdr FL 34
Charlemagne (Fr) 39
Charles de Gaulle (France) 241
Charles F Adams class (US) 237
Charleston 26, 30
Charlestown 207
Chatfield, Lord 81
Chatham Dockyard 13, 68, **68**, 71, 73, 83, 198–9
Che Ju (Korea) 230
Chesapeake (GB) 12
Chile 233
Chiles, Admiral H 183–4, 187
China 233; 236, 9, 16, 198, 219, 232–3, **234**, 235, 247
Chitose s/m (Japan) 118–19
Chiyoda (Japan) 113, 118–20, 125, 127
Churchill, Winston 35, 51, 61
City of Dalhart (USA) 39
Clemenceau (Fr) 245
Clerk, Sir George 14
CMBs 86
Cochrane, HMS 198–9
Cole, AP 61
Collins (Aus) *247*
Colossus, HMS 49
Colville, Admiral S 198
Confounder, HMS 215
Conneau, Enseigne J 42
Connecticut class 34
Connecticut (US) 234
Conqueror, HMS 46, 49, **60**, *62*, 62
Conqueror (US) 186
contre-torpilleurs 94–6, *96*
Coontz class (US) 237
Cornwall, HMS 202
Cornwallis, HMS 49
corvette engine 189–91
corvettes 9, 188–91, *188–91*, 230
Coventry, HMS 146
Cowpens (US) 243, 247
Cramp, William 21, 25
Crimean War 11–15
cruisers 47, 80, 85–6, 89, 94, 105, 143, 146
 nuclear 236
Cuckoo torpedo bombers 79, *80*
Curlew, HMS 146
Curtatone (It) **96**
Curtiss, Glen 36–7, 43
C–V cruisers (GB) 47
Cyclone (Fr) **98**

'D' class cruisers (GB) 85
'D' class s/m (GB) 74
Dahl, Captain 105
Dahlgren, John 27, *28*
Dakar (Israel) 249
Dalhousie (Bombay Marine) *9*, 9–18, *11*, 12–14
Dalmuir yard *212–13*
Danaide, HMS 202
Daphne class s/m (Spain) 235
Dartmouth, HMS 198
David, CSS 30, *31*
DDG–51 class (USA) **233**, 234
Defensora (Brazil) *202*

Delaware , USS 58
Delhi class (India) **233**
Denmark 225
depth charges 84, 100–1
4th Destroyer Flotilla 53
destroyers 53, 63, 85, 94, 143, 147, 174, 194–6, 211, **220**, 226, **233**
 see also 'Z' class
Deutschland (Ger) 130, *133*
Devonport Dockyard 59, 68, 210, 249
Dido, HMS 133
Dido class (GB) 146, 148
Dithmarschen (Ger) 138
Dönitz, Admiral 192
Doran, Flt/Lt 130
Dorsetshire, HMS 133
Dragon, HMS 81, 133
Drake, HMS 200
Drakensburg (SA) *203*
Dreadnought, HMS 49–1
dreadnoughts 46–64
 'super–dreadnoughts' **54–5**, *58–9*, 89
Dudrick, Walter 177
Duncan class (GB) 49
Dunkerque (Fr) 130
Duquesa (GB) **142**, 133–4

'E' class cruisers (GB) 85
'E' class s/m (GB) 74
Eagle, HMS 80
Eastep, Cdr D 172–4, 185
East India Company 9, 11, 13–14, 16–17
Echo II class (Russia) 171
éclaireurs d'escardre 94
Edsall class (US) 171
Egypt 232, 243, 249
Ellyson, Lt T G 36, *36*, 38
Elphinstone, Lord 12–14
Elswick yard 62, 79
Ely, Eugene *33*, 34, 36, 42
Emden (Ger) 136
Emerald, HMS 135
Emperor of India, HMS **60**, 61–2, *64*
Endeavour (NZ) *231*
Endymion (GB) *12*
Enseigne Gabolde (Fr) *89*, 91, 101
Enseigne Roux class (Fr) 90
Enterprise, HMS 135
Erebus, HMS 77
Erich Steinbrinck (Ger) 139
Ericsson, John 20
Erin, HMS **54–5**, 61–3
Estonia 230
Ethan Allen class (US) 167
Eurofeld (Ger) 133–4
Euryalus, HMS 12, **12**, 13
Exeter, HMS 105–6, 111, 174
Exocet missiles 174

Falck, Captain 192
Falklands, battle of (1914) 112
Falklands War (1982) 174–5
Fegan, Captain E S F 130–1
Fehler, *Kapitän-zur-See* J H 192
Fiji, HMS 135
Finland 226
fire control 51–2, 77, 95–6
Fisher, Admiral Lord 49, 88
Fiske, R/A Bradley A 42
Fleet Air Arm 81, 207
Floréal (Fr) *248*
Flower class (GB) 189–91, 209–10
Flyfisken class (Den) *226*
flying boats 80–1

Foch (France) 240
Formidable, HMS 133
Forrestal (US) 187
Fort Victoria, RFA *245*
Foudre (Fr) 40, 42–3
France 13, 17, 19, 84, 89–99, 168, 189, 206–7, 226, 232–3, **234**, 235, **236**, 240–1, 243, 245
 aircraft launching systems 39–41, 43
 as threat to Britain 12–14, 17
Fresno City (GB) 132, **142**
Friedman, N 171–2, 174
Friedrich Eckholdt (Ger) 139
Friedrich Ihn (Ger) 137
Frigate 2000 *223*
frigates **220**, 9, 17, 189, 206–7, 211, 229, 231, 233, **234**, 237–8, 249
 paddle frigates 11–12, 15
Froude, R E 87
Fullagar (GB) 87
Fushimi, Admiral 116–17
Fuyugumo (Japan) **151**
Fuyukaze (Japan) **151**
Fuyutsuki (Japan) 145, 150, **151–2**, *155*, 156, 163, *165*

'G3' battlecruiser design *79*
Galena (US) 20–1
Gallantin (GB) *203*
Gambier Bay (US) 209
Garnier, Cdr HK 196–9
Garrett, George 249
Garrett, L 250
Gato class s/m (US) 172
Gaudalcanal, battle of 113
Gaul (GB) 249
Gearing class (US) 171
General Electric 176–8
George Ireland (GB) **84**
George Washington s/m (US) 167, 177–8
Gepard (Russia) *235*, 236
Germany 47, 61–3, 81, 94, 96, 119, 188, 191–4, 209, 213–14, 216, 224, 233, 238
 Air Force 81, 105
 radar 105–12
 rivalry with Britain 47, 49
Gill, Major 84
Gladiator, HMS 15
Gladstone, WE 16
Glasgow, HMS 135
Glory, HMS 207–8
Gloucester, HMS 249
Gneisenau (Ger) 135
Goldborough (US) 237
Gorgon, HMS 82
Graham, Sir James 12
Grahame-White, Claude 42
Grappler (Bengal Marine) 11
Grayling (US) 249
Great Britain 9, 14–15, 17, 19, 46–88, 95–6, 107–12, 146, 211, 214–15, 218, 223, **236**, 237–8, 239–41, 243, 245, 247, 249
 and *Admiral Scheer* 134–40, 142
 construction 10–17, 143, 233–5, **234**
 inter-war technical developments 77–88
 nuclear submarines 174–5
 radar 105–9, 111–12
 see also Royal Naval Air Service, RAF
Greece 225, 237–8
Greenfish (GB) 194
Gremyaschiyiy (Russia) *204*
Grigorios (Greece) **142**, 134
Grosser Kurfurst (Ger) 62
Gruber, *Fregattenkapitän* 140
Guam (US) 245

INDEX

Guidoni, Capitano Alessandro 42
Gulf War 186, 243–4, 249–50

H–145 (Ger) **96**
Hae (Japan) *151*
Hagerty, J 182
Haiti 248
Halibut s/m (US) 176–7
Halifax class (Canada) 232
Hall, Cdre S S 69
Halmahera 114
Hanazuki (Japan) **151**–**2**, *148*, *152*
Harada, Captain 117, 120
Harland & Wolff 74, 189
Harriss, James B 41
Harugumo (Japan) **151**
Harushio s/m (Japan) 227
Harutsuki (Japan) **151**–**2**, *158*
Hastings (Bombay Marine) 11
Hatsuaki (Japan) table151
Hatsuharu class (Japan) 156–7
Hatsunatsu (Japan) **151**
Hatsuzuki (Japan) **151**–**2**, *153*
Hawkins, HMS 135
Hawkins class (GB) 85
Hayaharu (Japan) **151**
Hayakaze (Japan) **151**
Hazuki (Japan) **151**–**2**
Helena (USA) 109
Hellendorn, *Kapitänleutnat* 192
Hemond, H 168
Henry Jennings (GB) **84**, 84–5
Hercules, HMS 11
Herd, J F 108
Hermann Schoemann (Ger) 137
Hermes, HMS 80, 133, 135
Hesperides (Spain) *221*
Hewitt (US) 243
Hibernia, HMS 34
Himmler, H 193
Hinoki (Japan) *165*
Hipper, Admiral 63
Hitler, A 140, 192–3
H J Bull (GB) 188
Holland 207, 224, 235, 238
Holt, W J 209
Hong Kong 247
Hood, HMS 79, 83, *84*, 130, 133
Hori, S/Lt 119
Hosho (Japan) 147
10th Hussars 11
hydrophones *see* sonar

I–400 class s/m (Japan) 167, 175
Iltis (Ger) 136
Imperieuse, HMS 13
Imperieuse class (GB) 12
India 9, 11–12, 15–17, **233**, 233, **236**, 238
  Mutiny 9, 13–14, 16, 18
  naval shipbuilding in 9–13
Indian Ocean 9, 11, 17
Indomitable, HMS 198–9, 200
Indonesia 233, 238, 248
Inflexible, HMS 200
Invincible, HMS 46, 62, *171*, 200
Invincible class (GB) 49
Iran 230, 235, 243
  1856–7 expedition 13
Iraq 243
  see also Gulf War
Ireland 61, 194
Iron Duke, HMS 52, **54–5**, **60**, 61, *62*, *63*, *223*
Iron Duke class (GB) 53, 53–63, *64*
Israel 231, 249

Italy **236**, 53, 94, 96, 146, 224, 233, **234**, 240–1
  dreadnoughts 47
Iwasa, S/Lt 125
Iwo Jima 40
Iwo Jima (US) 244

J–1 s/m (GB) 67
Jaguar class (Fr) 97–8, **99**, 100–1, 101–4
Jaguar (Fr) 98, **99**
Japan 143–65, 168, 192–4, 207, 214, 226, 245
  as enemy of Great Britain 75, 85
  Imperial Japanese Navy 143–64
    naval air force 143
    midget submarines 113–27
    naval strategy 114, 116
Jauréguiberry (Fr) 39
Jehuboy, Jamsetsee 17
Jellicoe, Admiral Sir John 49, 51, 58, 61, 65, 80, 87
Jervis Bay (GB) 130, 132–3, **142**
jet propulsion 84, 84–5
John F Kennedy (US) 243, 245
Jones, RV 112
Juneau (US) 248
Juno, HMS 15, *16*
Jupiter (USA) *see Langley*
Jutland, battle of 49, 53, 60, *62*, *62*, 77

'K' class s/m (GB) 168, 172, *179–80*, 180
'K' class s/m (Russia) 76
Kaba class (Japan) 94
Kagero class 160
Kagero (Japan) 158
Kaiser (Ger) 64
Kaiserin (Ger) *64*
kamikaze 168
Karel Doorman (Holland) 225
Katayama, Cdr 117, 121
Kato, Cdr 117, 120
Kenbane Head (GB) **142**, 132
Kent, HMS 87–8
Kent class (GB) 85–6
Keppel, HMS 98
Kerama Retto Islands 40
Kerr, R/A Mark 53
Kessler, General U 192, *193*
Ketty Brövig (GB) 134
Khamronsin (Thailand) 239
Kilo class (Russia) 229
Kilroy, Lt CP 109–10
King Edward VII, HMS 196, *196–8*, 198
King George V, HMS 51–3, **60**, 62–3, 136, 138
King George V class (GB) 50–2, 50–3, *52*, 61, 63
Kirov (Russia) 142
Kishimoto, Captain 116–17
Kitakaze (Japan) **151**
Kitty Hawk (US) 243
Kiyotsuki (Japan) **151–2**
Knox class (US) 237
Kochi (Japan) **151**
kohyoteki 113–27, *116*, *121*, *124–127*
  technical data 120–4, *123*, *122*, 125–7, *128*
  transport ships 113, 118–19
  Type A 113, *114*, *117*, 120
  Type B 113–14, 127
  Type C 113–14, *114*, 125, 127
Köln (Ger) 140
Kongo (Japan) 228
König (Ger) **54–5**, *62*
Konkar Pioneer (Greece) 248

Konovalov, Captain 142
Korean War 207–8, 218
Kormoran (Ger) 135
Kortenauer class (Holland) 238
Koyama, Captain 120
Kraburi (Thailand) *241*
Krancke, *Kapitän zur See* 130, 132, 135
Krassin (Russia) 139
Kure Maru (Japan) 125
Kure Navy Yard 117–18, 125

'L' designs *see under Orion* class
L–3 s/m (GB) 67
L–3 s/m (Russia) 142
L'Adroit class (Fr) 98
La Fayette class (Fr) 232
La Fayette (Fr) *221*
Lafon, Enseigne 40–2
La Gloire (Fr) 19
Lamotte-Piquet (Fr) 94
Landsort class (Sw) *242*
Landsort (Sw) *237*
Langley, USS 42
Langsdorff, Captain 107
Latvia 230
Leahy class (US) 237
Leander class (GB) 238
Leander (NZ) 135
Leatham, V/A 135
Leeke, Sir Henry 11
Leighton, Cdr D *174*, 174–8
Leipzig (Ger) *107*, 136, 140
Leningrad (Russia) 238
Lenin (Russia) 139, 168, 221
Leone (It) **94**
Léopard (Fr) **99**, 104
Le Triomphante (France) 238
Lexington class (US) 216, *216*
Leygues, Georges 94
Leyte Gulf 40
Liberal (Brazil) 232
Liberia 248
Liberty (US) 186
Lightning, HMS 197, 198
Linotype Corporation 52
Lion class (GB) 49
Lithuania 230
Littorio class (Italy) 53
Liverpool, HMS 61
Lockhart, Lt 74–5
London Treaties 114, 118–19, 143
Long Beach (US) 177–8
Lorraine (Fr) 41, *41*
Los Angeles class (US) 167, 177
Los Angeles s/m (US) 175, 186
Low, Cdr 9
LSTs 39–40, *40*
LST(3) 190
Luhu class (China) **233**
Lupina (GB) 248
Lützen see Deutschland 130
Lützow (Ger) 60, 62, 130, 136, *136*, 138, 140, 142
Lynx (Fr) **99**, *103*
Lyon class (Fr) 89, 91

'M' designs *see under Orion* class
Macon airship 39
Macrow, George 49
Magdala (Ind) 15–16
Maidan (GB) **142**, 132
Makedonia (Greece) 225
Maksim Gorky (Russia) 142

Malaysia 233, 247
Mallory, Stephen R 19
Malta 63
Mandelblatt, James 171–2
Mandume (Angola) *247*
Manila Bay 40
Marceglia, A 249
Mar del Sur (Spain) *235*
Markgraf (Ger) 62
Mark VII torpedo 85
Marlborough, HMS 52, **60**, *61*, 61–3
'Masker' 84
Matsu class (Japan) 159
Maudsley, Sons & Field 13
Mauritius 17
Maux, M 43
McCall, Captain 107
MCDV class (Canada) 223–4, *224*
McKenna, Reginald 49, 53
Meanee (BM) *10*, 12
Meendsen-Bohlken, *Kapitän-zur-See* 136
Meko 200 class (Turkey) 235
Menzel, Lt-Col 192
Merrimack, USS 19
Merttonia (Canada) 209
M89 (Fr) 91
M90 (Fr) 91
Michitsuki (Japan) *tables*151–2
Middleton , HMS 204
Midway 127
'Migraine' 172–3
Miller & Ravenhill company 12
mines 85–6, 142, 210
minesweeping 208–9, 218, *218*, 223, 238, *242*
Minotaur, HMS 198, *200*
Minsk (Russia) 238
missiles 81, *82*, 174, 219–0, 229, 231–2, 238–9, 239–1, 241
 anti-missile systems 240–1
Mistral (Fr) **98**
Mizuho s/m (Japan) 118
ML-497 (GB) 84, *85*
Moby Dick (GB) 245
Moltke (Ger) 63
Monadnock , USS *31*
Monarch , HMS 46, 49, **60**, 62, 82–3
Monitor, USS 20, 27
Monmouth, HMS 198
Montevideo 106–7, 109
Mooney, E 47, 51, 60
Moore, A G H W 60–1
Mopan (GB) 130, **142**
Morin, Captain G 179, 183–5
Mountbatten, Lord *174*
Moy, HMS 198
MTBs *see* torpedo boats, 'T' class
Muavenet (Turkey) 237, 248–9
Muscat, Imaum of 11

Nacken (Sw) 228
Naiad, HMS 133
Robert Napier & Sons 13
NATO 219
Natsukase (Japan) 151
Natsuzuki (Japan) 151–2, *150*, *153*
Nautilus s/m (US) 175
Nautilus (US) 172
Nelson , HMS 79, 82–3, 85, 88, 133
Neptune , HMS 133
Neptune class (GB) 47
Neustrashimy (Russia) *229*
Neutrashimy class (Russia) 236
Nevada (US) 53
New Ironsides, USS 19, 19–32, *20*, **21**, 23–4, *24*, 24–5, *26–27*, *30–31*

armament 25, 27, *28–29*, 32
 combat 29–32
New Zealand 233
Nieschling, Lt-Col 192
Nigeria 248
Nigeria, HMS 135
Niizuki (Japan) **151–2**
Nishikaze (Japan) **151**
Nisshin (Japan) 125
Nordmark (Ger) 133–5, **142**
Norfolk, HMS 133
Norfolk (US) 173
Normandie class (Fr) 89–90
Normandie (Fr) 105
North British Diesel Co 71
Norway 136–137, 192, 225, 234
Novorossiysk (Russia) 229
Nowrojee, Jehangir 15
nuclear disarmament 220
nuclear material 174–8, 193–4, 221–2, 245
 see also Triton and *under* submarines

'O' class s/m (GB) 85
Ocean, HMS 234, 207, 235
Ogasawara Islands 113, 116
Ohio class (US) 177, 183
Okaze (Japan) **151**
Okinawa 40, 114, 168
Okitsukaze (Japan) **151**
Oktyabrskaya Revolutsiya (Russia) 142
Olmeda, RFA 202
Oman 231
Orage (Fr) **98**, *100*
Orion, HMS 46, 47–52, **54–5**, 60, 62
Orion class (GB) 47–52, **54–5**, *62*, 62–3
 'L' designs 47, 49–50, **54–5**, *58*
 'M' designs 46–7, 50–1, **54–5**, *56–7*, 59, 61
Orsa (It) 220
Osprey , HMS 84
Ostfriesland (Ger) 82
Otsuki (Japan) **151–2**
Ouragan (Fr) **98**, *101*

P-59 (GB) 84
Pakistan 233
Panthère (Fr) **99**, *102*
Panzerschiffe (Ger) 105–6, 130
Papua New Guinea 247
Paterson, H G 35
Paul Jacobi (Ger) 137, 142
PC-43 (GB) 84
Pearl Harbour 40, 113, 120–1, 124–5
Pears, R/A Sir Edward 16
Pégoud, C A 38, *38*
Pembroke Dockyard 68, **68**
Penne, L G de la 249
John Penn & Sons 13
Pennsylvania, USS 36, 42
Permit class (US) 186
Persia *see* Gulf War, Iran
Petya II class frigates (Russia) 229
Philadelphia 32
Philippines 233
 battle of 114
Phoebe, HMS 133
Pietsch, *Oberleutnant* 132, 135
Pinguin (Ger) 134–5
Platypus, HMS *73*, 74
Polarii Krug (Russia) 249
Pollen, A H 52
Pond, R/A Charles F 36
Porter, Admiral David 32
Port Hobart (GB) **142**, 133
Portland (Ger) 135

Portsmouth Dockyard 49, 59
PQ 17 convoy 138–9
Pretoria Castle (GB) 133
Prince of Wales, HMS 77, *78*
Principe de Asturias (Spain) 171
Prinz Eugen (Ger) 130, 137, 140, 142
Pueblo (US) 186
Punjaub (Bombay Marine) 12
Purvis, MK 109–10
Pyotr Veliki (Russia) 236

Q-ships 65
Qatar 231–2
Quarto class (It) 94
Queen Elizabeth, HMS 77, 249
Queen Elizabeth class (GB) 49, 60, 91
Queen Mary, HMS 50–1

'R' class submarines (GB) 65, 65–76, *66–7*, **68–9**, *70–73*, **74**, *75*, 85
Rabaul 114
radar 105–12, *106–7*, 173–4, 244
 radar pickets *168*, 168–9, 171–4, 179, 183
 see also under submarines
radio 72, 85, 96, 105, 136
 RDF 105
Raeder, Admiral 138, 140
RAE Farnborough 43, 105
RAF 39, 80–1, 87, 105, 108, 130, 136, 142, 207
Raikes, Admiral 133
Ramadan (Egypt) *230*
Ramillies, HMS 126
Ranger (US) 237, 248
Rangitiki (GB) 132
Rantaupandjang (Holland) 142, 135
Rasenack, F W 111
Rawlins, Captain R 183–4
Ray (US) 183
Reed, Sir Edward 15
Reed, William 191
Renown, HMS 130, 133
Repulse, HMS 130, 133
Repulse, HMS (1866) 16
Requin s/m (US) 172–3
Reshadieh (Turkey) *see Erin*
Resurgam (GB) 249
Revolutsioner (Russia) 139
Ribbentrop, J von 193
Riblet, Byron C 42
Richard Beitzen (Ger) 137, 139
Rickover, Admiral 168–9, *174*, 174–5, 176–7, 179, 183, 186
River class minesweepers (GB) 223
River Plate, battle of 105, 111
Robinson, Richard 189
Rodney, HMS 112, 133
Roebuck, HMS 243
Rokoryu Incident 119
Roxburgh, HMS 198, *200*
Royal Flying Corps 108
Royalist, HMS 63
Royal Naval Air Service 79–80, 83, 87
Royal Navy *see* Great Britain
Royal Sovereign, HMS 46, *62*, 77, 79
Ruf, F 192
Rurik (Russia) 200
Rushmore (US) 248
Russia 14, 43–4, 76, 114, 137, 139–40, 142, 171, 175, 182, 186, 209, 212, 219, 221–2, 228–30, **233–4**, 235–6, **236**, 238–9, 240, 243, 245–6, 249
Russo-Japanese War 94

INDEX

S–113 destroyer (Ger)
Sackville (Canada) 191
sail 13, 16, 206–7, 215
Sailfish s/m (US) 172, 176
Sailfish (US) *171*
Saint Louis (Fr) 39
Saipan 40
Salisbury class (GB) 171
Salmon s/m (US) 172–4, 176, 186
Samson, Lt Charles 34
Sandefjord (Norway) 134, 142
San Demetrio (GB) 132
Sandown class (GB) 232
Sandrart, Colonel von 192
Saratoga (US) 249
Saudi Arabia 232, 241
Savo Island, battle of 218
Scapa Flow 63
Scharnhorst (Ger) 130, 135
Schlesien (Ger) 140
Schnellboote 217
Schniewind, Admiral 138
Schulz, Kapitänleutnant 135
Schwann, Wing Captain Oliver 35
Scorpene s/m (Spain) 235
Scorpion s/m (US) 177
Scott, HMS **94**
Scott class (GB) 91
Scott (US) 193
screw propulsion 11–15
  see also jet propulsion
Seaawolf s/m (US) 222
Seagull, HMS 87
seaplanes 35, 80
Sea Shadow (US) 238
Seawolf s/m (US) 174–6, 234
Sekido, Lt 119–20
Semyon Dezhnev (Russia) 139
Seydlitz (Ger) 62–3
Shakespeare class (GB) 91
Shanghai 114, 119
Shankush s/m (India) *229*
Sheffield, HMS 174
Shikishima (Japan) 245
Shimakaze (Japan) 147, 150
Shimotsuki (Japan) **151–2**
Shoji, Genzo 192–3
Shokaku (Japan) 145
Shomitsuki (Japan) 160
Shropshire, HMS 135
Simoun (Fr) 98, **98**
Singapore 14, 233, 239
Sir Bedevere (GB) 244
Sir Lancelot (GB) 239
Siroco (Fr) **98**
Sjaelland (Den) *202*, *204*
Skate class s/m (US) 172, 176, 179
Skipjack class s/m (US) 176
Skipjack s/m (US) *175*, 177–8, 180
SkyHook 45
Skyraiders 173
Slavutych (Ukraine) *229*
Smith, F/O D C 108
Société Anonyme pour l'Aviation et ses Derives 39
Solo (GB) 247
Solomon Islands 114, 155
Somalia 243, 248
Somerset, Duke of 14, 16
Somerville, Admiral Sir James 111, 133
sonar 71–2, 75, 81, 84–5, 88, 250
Soryu (Japan) 145
Souchon, *Kapitän-zur-See* 192
Southern Pride 191
South Africa 233
Southampton, HMS 245

Southern Gem (GB) 188
Southern Pride (GB) 188
South Korea 233
Sovremenniy class (Russia) 236
Spain 226, 235, 240
Spee, Admiral Maximilian von 112
Spencer Robinson, Captain Sir Robert 15
Speybank (GB) 134
Spinax (US) *168*
Spruance class (US) 171
Stanpark (GB) 134, **142**
Strasbourg (Fr) 130
Stronghold, HMS 36, *36*
Stump (US) 243
Sturgeon class (US) 186
Sturgeon (US) 248
St Vincent, Earl 9–10
14th Submarine Flotilla 73
submarines 65–76, 81, 84–5, 89, 113, 142–64, 167–87, 191–4, 226, 230, 232–3, 235, 238, 243, 245–6, 249
  aircraft carrying 167, 171, 175
  anti-submarine warfare 66, 80–1, 84–5, 87, 100–1, 143, 147, 234, 241
  midget *see* Kohyoteki
  nuclear 167–87, 221, 223, 233–5, 249–50
  radar pickets *see under* radar
  see also 'D' class, 'E' class, 'K' classes, 'R' class, torpedoes, *Triton*
Sueter, Admiral 35, 79–80, 87
Sukomo Bay 118–19
Superb, HMS 62, 82
Sutton (US) 193
Suzutsuki (Japan) **151–2**, *153*
Swan, HMAS 227
Swatton, PO 72
Sweden 228, 245–6

'T' class torpedo boats (Ger) 135–7, 138, 140, 142
Taiho (Japan) 147
Taiwan 233, 237
Tang class s/m (Russia) 175, 182, 229
Tannenfels (Ger) 134
Taymir (Russia) 139
TB–98 (GB) 85
Tempête (Fr) ß98ß
Tench class s/m (US) 172
Tennyson d'Eyncourt, Eustace 59, 83, 85–6, 87
Terror, HMS 77
Teruzuki (Japan) **151–2**
Teste, Lt 41
Teviot (GB) 84
Texas, USS **54–5**
Thalia, HMS **15**, 15–16, *17*, 18
Thanet, HMS 36
Themistocles (Greece) *237*
Theseus, HMS 208
Thienemann, *Kapitän zur See* 140
Thomas Ansell (GB) 84
Thomas Fielden (GB) 13
Thor (Ger) 134
Thorp, James M 38
Thresher (US) 184
Thunderer, HMS *46–48*, 49–50, **60**, *62*, 62
Thurston, Sir George 61
Tiger, HMS 59
Tigre (Fr) *92–3*, *96*, **99**, *99*, 101, *103*
Tigris, HMS s/m 140
Tirpitz (Ger) 130, 136–9
Tizard, Sir Henry 111
Tomonaga, Captain Hideo 123, 192–3
Tomozuru (Japan) 119, 147, 158
1500 Tonnes class (Fr) 91–104
  armament 98–101, 104

Topaz (Fr) *206*
Topeka (US) 243
Tornado (Fr) **98**
Toronto (Canada) 249
torpedo boats 86, 89–91, 94, 136–7, 138, 140, 214, *214*
  see also 'T' class torpedo boats
torpedoes 49, 60, 62–3, 85, 91, 94–5, 100, 116, 118, 143, 148, 151, 179, 250
  air–dropped 42, 79, 87
  anti–submarine 67, 70
  spar 30, *31*, 32
  torpedo protection 53, 60, 83, *84*
torpilleurs d'escadre 90–1, 94, 96
Toshihide, Cdr Asama 116
Tovey, Admiral 136–7
Toyoda, V/A 120
6th Training Flotilla 73
Tramontane (Fr) **98**
Trewellard (GB) **142**, 132
Tribesman (GB) **142**, 133
Trident class s/m (GB) 223, 234–5
Trident s/m (GB) 137
Trigger (US) 182
Trincomalee, HMS 249
Tripoli (US) 248
Triton s/m (US) **178**, *167*, 167–87, *169–170*, *172–3*, *177–8*, *182–3*, *184–5*, *186–7*
  function 185–6
  operational career 181–6
  radar 173–4, 179
  reactor *174*, 174–8, 187
  transworld voyage 168–9, 182–3
Triumph, HMS 243
Trombe (Fr) **98**
Tromp class (Holland) 235
Turbulent, HMS 249
Turbulent s/m, HMS 249, *250*
Turkey 225, 235, 237, 248
Turner, V/A Richmond Kelly 40
Type 42 destroyers (GB) 174
Type 21 frigates (GB) 233, 238
Type XXI submarines (Ger) 72
Typhon (Fr) **98**, *103*

UAE 231, 241, 243
U–boats (Ger) 65–6, 74–5, 84, 91, 98, 135, 191–4, *192*, *193*, 193, *194*, 218
  sunk **66**, **76**, 80
Udaloy II class (Russia) **233**, 236
Ulyanovsk (Russia) 238
Unicorn, HMS 207–8
Unitas I (GB) 188
Urazuki (Japan) **151**
Uruguay 106–7, 109–11
USA 17, 19–32, 39, 43, 81–4, 86, 105, 109, 149, **222**, 233, **234**, **236**, 236–7, 239–41, 243–50
  Civil War 19, 32, 211
  and Japan 113–14, 116, 120, 125, 143, 214
  and *U–234* 193–4
  US Army 39–40
  US Marine Corps 40, 241, 248–50
USSR *see* Russia

Valerian Kuibishev (Russia) 139, **142**
Valiant, HMS 249
Vanguard class s/m (GB) 234
Varyag (Russia) 235–6
Vega Helguera, Julio 109
Vengeance, HMS 77
Victorious, HMS 137–8
Vietnam 279

Virginia, CSS 19
VTOL *see under* aircraft
Vulcan, HMS 73
V & W class destroyers (GB) 85, 91, 96, **96**

Wadsworth, Captain F 179, 183–4, 187
Wakatsuki (Japan) *151–2*, 156
Walker, Sir Baldwin 12, 14–15
Warrior, HMS 19, *43*, 43, *44*
Washington Treaty 63, 84–6, 114, 118, 143
Watkins, R/A F 176
Watson-Watt, Sir Robert 108
Watts, Philip 49, 51, 53
Wellesley, Commodore 15
Westinghouse 175–6, 178
Westminster, HMS 235

Wilhelmshaven 105, 130, 140
Wilkins, A F 108
Willem van der Zaan (Holland) *219*
Willis, John 15
Wood, Sir Charles 12, 14
wooden hulls 12–13, 15
Woolwich Dockyard 15–16
World War I 34, 65–75, 91, 94
World War II 105–13, 125–42, 150, 191–4, 202–5
Worthington, A W 47
Wurmbach, *Kapitän zur See* 130

X–1 s/m (GB) 85, *86*

Yaegumo (Japan) **151**

Yaeyama (Japan) *246*
Yamato class (Japan) 147
Yamazuki (Japan) **151**
Yanagi (Japan) *165*
Yildiz class (Turkey) 235
Yoizuki class (Japan) *149*
Yoizuki (Japan) **151–2**
Yokoo, Captain 116–17
Yubari (Japan) 158
Yugiri (Japan) *236*
Yugoslavia 219–20, 243–4
Yukigumo (Japan) **151**

'Z' class destroyers (Ger) 136–7, 140, 142, 194–5, *195*
Zaafaran (GB) 138